Professionalism in the Information and Communication Technology Industry

John Weckert

Centre for Applied Philosophy and Public Ethics (CAPPE)

Charles Sturt University

Richard Lucas

Information Systems

University of Canberra

Professionalism in the Information and Communication Technology Industry

John Weckert

Centre for Applied Philosophy and Public Ethics (CAPPE)

Charles Sturt University

Richard Lucas

Information Systems

University of Canberra

E PRESS

Practical Ethics and Public Policy Monograph 3

Series Editor: Michael J. Selgelid

Published by ANU E Press
The Australian National University
Canberra ACT 0200, Australia
Email: anuepress@anu.edu.au
This title is also available online at http://epress.anu.edu.au

National Library of Australia Cataloguing-in-Publication entry

Author: Weckert, John.

Title: Professionalism in the information and communication technology industry /
John Weckert, Richard Lucas.

ISBN: 9781922144430 (pbk.) 9781922144447 (ebook)

Notes: Includes bibliographical references.

Subjects: Information technology--Employees--Professional ethics.
Telecommunication--Employees--Professional ethics.
Employees--Attitudes.

Other Authors/Contributors: Lucas, Richard.

Dewey Number: 174.93034833

Cover design and layout by ANU E Press

Contents

Section IV. ICT governance

Section V. Ethics education

Section VI. Codes of ethics

Section VII. ICT and society

Acknowledgements

Many of the papers in this volume were presented at a workshop that was part of an Australian Research Council (ARC) Linkage Project (LP0560659), and the rest were solicited from other leading researchers in the field. We wish to thank the ARC and the Australian Computer Society (ACS) for their support for the initial project. We would particularly like to thank the participants in the project, John Ridge, past president of the ACS and Chairman of the ACS Foundation, Professor Jeroen van den Hoven of Delft Technical University and Dr Jeremy Moss of the University of Melbourne. We especially thank Dr Yeslam Al-Saggaf of Charles Sturt University, who played a major role in interviews and analysis in the early part of the project.

Introduction

Professionalism: It's NOT the job you DO, it's HOW you DO the job (anonymous)

Professionalism, in the sense described in the quotation above, is arguably more important in some occupations than others. It is vital in some because of the life and death decisions that must be made, for example, in medicine. In others, the rapidly changing nature of the occupation makes efficient regulation difficult and so the professional behaviour of the practitioners is central to the functioning of that occupation. The central idea behind this book is that this process of rapid change is relevant to information and communications technology (ICT). The technology changes so quickly that regulation will always lag behind.

The contributors to this volume come from a variety of backgrounds, and this range of contributions is intentional. It is not just academics who have important things to say about professionalism in ICT. Some contributors are or were ICT practitioners, one is a retired judge from the High Court of Australia, some are ICT academics and others are philosophers. Underlying all chapters, except the first, however, is the concept of ICT professionalism. The first chapter, by the Hon Michael Kirby, focuses on regulation. What it does, by showing the difficulties in regulating new technologies — and ICT in particular — is highlight the importance of professionalism in ICT. If regulation is difficult, the behaviour of professionals is paramount. The rest of the sections all contain chapters that approach aspects of professionalism from different perspectives.

The second section contains reflections on professionalism and ethics by experienced ICT practitioners and the third focuses on professionalism itself. The fourth considers ICT governance and its relation to professionalism. Section five contains two discussions of ethics education, something considered by the Australian Computer Society (ACS) as a necessary component of education for ICT professionals. Codes of ethics are a common component of a professional's armoury and these are discussed in Section Six. Finally, in Section Seven, ethical concerns about ICT in society more generally are considered, together with the role that professionals have in this context.

These sections, considered as a whole, present a richer picture of what an ICT professional is or ought to be. In fact, the overall argument of the book is that, given the problems of regulation, if the ICT industry is to make its proper contribution to society, those in the ICT industry must be professionals in the sense of behaving professionally.

Each section of the book contains an introductory piece by the editors outlining not only the chapters, but more importantly, setting the scene for those chapters and making the links between the various sections. Each section's introduction also contains information on the chapters' authors to assist the reader in understanding the perspective from which the authors are approaching their topic.

Section I

Regulating technology

On the need for professionalism in the ICT industry

If information and communications technology (ICT) is to fulfil its potential in improving the lives of all, then the importance of the professionalism of its practitioners cannot be overemphasised. This is, of course, true of all occupations; but, there is an additional reason to highlight this in the case of ICT and other new technologies. In his paper, the Hon Michael Kirby says that Justice Windeyer, one of his predecessors in the High Court of Australia, 'once declared of the relationship between law and medical technology, that the law generally marches in the rear and limping a little'. Assuming that the situation is the same for ICT, and we have good evidence for this, we have strong reasons for emphasising the importance of professionalism in ICT. Kirby raises a number of problems in regulating technologies, particularly new and rapidly changing technologies, a central one of which is clearly ICT, and these suggest that the problem is even worse than that stated by Windeyer. Kirby writes:

> From the perspective of the law, they [biotechnology and ICT] present a common difficulty that, no sooner is a conventional law made to address some of their features, and to regulate those deemed necessary for regulation by reference to community standards, but the technology itself has changed. The law in the books is then in great danger of being irrelevant, in whole or part. Language written down at one time may have little, or no, relevance to events that happen soon thereafter (see Chapter One, this volume).

Kirby is convincing in showing that serious problems exist in the attempts to regulate new technologies, including ICT. The one already mentioned is its rapidly changing nature and the difficulty of the law keeping pace, but there are others too, for example, the global nature of ICT. Apart from laws regarding international trade or conflict, most laws do not extend beyond national boundaries and many are even narrower, applying to only a specific region of a nation. National borders are largely irrelevant to ICT; local regulations have only limited power. For example, regulating pornographic material on the Internet, if that is deemed necessary, is difficult, because the sites hosting the material could be outside the jurisdiction of the legal system that formulated the laws. This leads to another of Kirby's worries about regulation. To effectively block relevant material, regulation for the control of pornography and other material must be draconian. A major problem with this approach is that it almost always lessens the value of the Internet by blocking non-offending material as well. Additionally, it raises concerns about freedom of speech and expression.

The regulatory issues that are raised by Kirby do not show that regulation should not be attempted. There is a need to curb excesses and limit abuses of this technology, but there are no easy answers to the question of how best to achieve this. The interesting question is framed around how we can best ensure that ICT serves the interests of society in general, rather than merely those of the few, vested interests, given the problems with regulation? The answer proposed in this book is professionalism. There is a strong argument that professionalism is particularly important in new and rapidly changing technologies. This may sound idealistic, and based on the false assumption that professionals will always behave well, even in the absence of regulation, but it seems to be the best approach given that regulation has such difficulties.

Later in the book, the concepts of professionalism, a professional, and a profession will be considered more closely, but, informally, a professional is someone who takes his work seriously; is an expert, at least to some extent relative to the population at large; and, can take a 'big picture' view of his work and see it in the context of society and life more generally. One aspect of professionalism would be seeing the relative importance of one's work and its ethical and social aspects. Being a professional even in this informal sense, is different from being, in the words of well-known computer ethicist Don Gotterbarn, a 'gun for hire' (Gotterbarn, nd) — someone in the ICT industry who sees work as purely a source of income.

This discussion is important not only because of Kirby's arguments about the problems of regulation of ICT, but also because there is some evidence that many working in ICT do not see themselves as professionals, and nor do they see professionalism as significant. This will be developed further in Section Three and is mentioned by John Ridge in Chapter 2.

The two main facets of professionalism to be emphasised here are moral responsibility and trust, both of which are central to the notion of a professional and to maximising an industry's, or occupation's, social value.

First, responsibility. Taking moral responsibility for one's own actions, or what amounts to more or less the same thing in this context, being morally accountable for what one does, places certain restraints on behaviour. Legal responsibility or accountability obviously places restraints on what one does, but here we are more interested in a situation where few regulations exist.

There is a strong link between professionalism and moral responsibility (something also developed in Section Three). A professional has special, and socially useful, knowledge or skills. This places that individual in a position of power relative to those who lack that knowledge, but who have a need for it. They are dependent on the professional. This makes them vulnerable to some extent with respect to the professional and, as individuals, we have greater moral responsibilities towards those who are vulnerable (see Goodin, 1985, for a detailed discussion). So, with respect to those with whom they relate professionally, professionals have moral responsibilities over and above their normal moral responsibilities as human beings. An ICT industry, then, comprising people who see themselves as professionals and acknowledge the responsibility that such a role entails, should require less regulation.

Another reason why professionalism is important is that professional behaviour engenders trust and a trusting environment has a number of advantages over one that relies on regulation, enforcement and compliance. Living and working successfully in groups requires a degree of trust. I must trust that others will act in ways that are conducive to harmonious living and working, and they must trust that I will do the same. Groups function better the more trust that exists between individuals and, without any trust, they could not function at all (Putnam, 1994). The necessity for mutual trust applies also to economic efficiency. Where there is lack of trust, there must be monitoring and surveillance, filling out of documents and keeping of records, and someone must oversee at least some of these. This is largely unproductive work. Robert Putnam and James Coleman (1990) talk of trust as social capital, the ability of people to work together for common purposes:

> Like other forms of capital, social capital is productive, making possible the achievement of certain ends that would not be attainable in its absence. … For example, a group whose members manifest trustworthiness and place extensive trust in one another will be able to accomplish much more than a comparable group lacking that trustworthiness and trust (Coleman, 1990: 302, 304).

Without trust, people cooperate only under a system of formal rules, which are often called 'transactions costs'. These costs are, in effect, a tax on the lack of trust. Francis Fukuyama, talking about trust in society in general, says that costs incurred by police, lawyers, prisons, and so on are 'a direct tax imposed by the breakdown of trust in society.' He continues:

> People who do not trust one another will end up cooperating only under a system of formal rules and regulations … This legal apparatus, serving as a substitute for trust, entails what economists call 'transaction costs.' Widespread distrust in a society, in other words, imposes a kind of tax on all forms of economic activity, a tax that high-trust societies do not have to pay (Fukuyama, 1995: 11, 27–28).

Our concern is not with society in general, of course, but those working in the ICT industry, an industry that, if Kirby is right, is difficult to regulate. Being in an industry that is difficult to regulate has some advantages for professionalism. Less regulation means more autonomy (at least in a generally law-abiding society) and this gives more scope for 'doing the right thing' and being trustworthy. In this environment, the individual is better able to display professionalism.

We keep our word and do the right thing, for a variety of reasons. Often, hopefully, we do the right thing simply because it is the right thing; we are basically moral people. Sometimes, of course, rules and regulations, and the threat of punishment if we disobey them, play an important role in determining our actions. Another motivating factor is self-interest. Good behaviour can be rewarded in many ways and, as a result, reputation or esteem are influential incentives. Most of us want people to think well of us. The thought that esteem or something very like it, plays an important role in modifying our behaviour, has a long history and is expressed clearly by David Hume:

> Our reputation, our character, our name are considerations of vast weight and importance; and even the other causes of pride; virtue, beauty and riches; have little influence, when not seconded by the opinions and sentiments of others (Hume, 1975: 316).

Andrew Alexander and Seumas Miller (Alexander and Miller, 2010: 100) talk of the 'virtuous triangle' — reputation — self-interest — ethics. It is in our self-interest to be highly regarded, to have a good reputation, which is enhanced by behaving well. This point is discussed extensively in the literature on trust. In general, people like to be trusted and are perturbed if they are not trusted. This desire to be seen as trustworthy is, in turn, an incentive to behave in a trustworthy manner. The desire to have a reputation of trustworthiness is a matter of self-interest that can be achieved through ethical behaviour. While this motivation is not always paramount and can be overridden by other considerations, for example, financial gain when 'we can get away with it', it is an important consideration, one that must not be overlooked.

While the desire for esteem is important it can also, however, be a double-edged sword. In some contexts, this desire can motivate bad behaviour. I might want to impress my friends with my toughness by displaying aggressive behaviour, or my colleagues with my financial prowess by engaging in clever but dishonest business deals. In the right contexts, however, it can be a useful motivation for good.

What follows from all this? Because of the difficulties of regulation in the ICT industry, as discussed by Kirby, professionalism has special importance. The issue, then, is how best to promote professional conduct. One motivating factor is esteem, or reputation, but this, while important, is not enough. What is required is an environment that is conducive to, and encourages, professional and good behaviour. Such an environment is of course supported by some level of regulation together with a disciplinary mechanism to punish breaches of the regulation. A more positive element is a structure that rewards good behaviour, perhaps with promotion, bonuses or increased salary. Codes of ethics or conduct and education are also important in guiding behaviour and sections five and six address this. Such mechanisms, taken together, are often referred to as integrity systems. An integrity system is 'an assemblage of institutional entities, roles, mechanisms and procedures, the purpose of which is to ensure compliance with minimum ethical standards and promote the pursuit of ethical goals' (Miller, 2007: 354). This is discussed further in the introduction to governance (Section Four).

The difficulties of regulating ICT highlight the need for further study of professionalism, governance, and integrity systems in the industry. The purpose of this book is to contribute to that study.

References

Alexander, A & Miller, S, 2010, *Integrity systems for occupations*, Ashgate.

Coleman, JS, 1990, *Foundations of social theory*, Harvard University Press, Cambridge, Mass.

Fukuyama, F, 1995, *Trust: the social virtues and the creation of prosperity*, Penguin Books, London.

Goodin, R, 1985, *Protecting the vulnerable: a reanalysis of our social responsibilities*, University of Chicago Press.

Gotterbarn, D, nd, 'Computer practitioners: professionals or hired guns', <http://csciwww.etsu.edu/gotterbarn/>

Hume, David, 1975, *A treatise of human nature*, LA Selby-Bigge (ed), Clarendon Press, Oxford.

Miller, S, 2007, 'Institutions, integrity systems and market actors', in J O'Brien (ed), *Private equity, corporate governance and the dynamics of capital market regulation*, Imperial College Press, pp 339–70.

Putnam, RD, 1994, *Making democracy work: civic traditions in modern Italy*, Princeton University Press.

Biography

The Hon Michael Kirby. When he retired as a Justice of the High Court of Australia in 2009, he was Australia's longest serving judge, having first been appointed to judicial office in 1975. His work (1975–84) as inaugural chairman of the Australian Law Reform Commission brought him to public notice, including for his engagement with the interface of law, science and technology.

He has served on many international and United Nations bodies, chairing two expert groups of the OECD on issues of informatics and participating in the UNESCO international bioethics committee, which addresses the law and ethics of the human genome project. In 1991 he was awarded the Australian human rights medal and in 2010 he was co-winner of the Gruber Justice Prize.

1. The fundamental problem of regulating technology[1]

The Honourable Michael Kirby AC CMG[2]

[T]he continued rapid advance in science is going to make life difficult for judges. We live in an age of breakneck technological change that will thrust many difficult technical and scientific issues on judges, for which very few of them (of us, I should say) are prepared because of the excessive rhetorical emphasis of legal education and the weak scientific background of most law students.

RA Posner, 'The role of the judge in the twenty-first century', *Boston University Law Review*, vol 86, 2006, p 1049.

Present at the creation

Preposterous claims: Dean Acheson, one-time Secretary of State of the United States of America, called his memoirs *Present at the Creation* (1969). It was a clever title, laying claim to having been at the important meetings during and after the Second World War in which the new world order was established.

The claim was faintly preposterous, given that the Second World War grew out of the first, and bore remarkable parallels to other conflicts dating back to the Peloponnesian Wars in ancient times. All history, and all technology, grow out of the giant strides that preceded their current manifestations. We forgive Acheson because (unlike some of his predecessors and successors) he was an elegant and sophisticated man, significantly concerned about improving the condition of the world and the welfare of its inhabitants.

I make an equally preposterous claim that I was present at the creation of the central problem that occasioned the TELOS[3] conference to discuss the challenge

---

1 This chapter is based on the author's report on the TELOS conference, 8 April, 2007, London. The report was delivered orally at the close of the conference. Originally published in R Brownsword & K Yeung (eds), *Regulating technologies: legal futures, regulatory frames and technological fixes*. This text has been revised and updated.
2 Justice of the High Court of Australia (1996–2009). One-time chairman of the Expert Group of the OECD on Transborder Data Flows and the Protection of Privacy (1978–1980). Formerly a member of the World Health Organisation Global Commission on AIDS and of the UNESCO International Bioethics Committee.
3 TELOS — Centre for the Study of Technology, Ethics and Law in Society, King's College School of Law, London.

presented to legal regulation by the advent of modern biotechnology and information technology. The claim is absurd because such technologies have advanced by reason of the genius of technologists and scientists, who stand on the shoulders of their predecessors, also dating back to ancient times.[4]

In one of the closing talks at the conference, Professor Mireille Hildebrandt described the advances that had occurred in the communication of ideas in medieval times following the perfection of spectacle glasses and the invention of the printing press. The former allowed the monks, who spent their years inscribing religious texts, to extend their working lives beyond presbyopia. Yet it was the printing press that released words (and hence the ideas represented by words) from the calligraphy of the monks. For holy men, the words were written to be said or sung. But after Caxton, printed words took on a life of their own. Their meaning could be gathered without mouthing the sounds they conjured up. In a forerunner to the urgencies of the present day email and social networks, words could be read four times more quickly than they could be said. A revolution in communications had begun. It continues into our own times.

Acknowledging the lineage of contemporary technologies, the changes upon which the TELOS conference concentrated were information technology and biotechnology. They are major features of the contemporary world. From the perspective of the law, they present a common difficulty that, no sooner is a conventional law made to address some of their features, and to regulate those deemed necessary for regulation by reference to community standards, but the technology itself has changed. The law in the books is then danger of being irrelevant, in whole or part. Language written down at one time may have little, or no, relevance to events that happen soon thereafter.

Regulating biotechnology: This is the sense in which I claim to have been present at the creation of the two nominated technologies. It came about in this way.

In 1975, soon after I was first appointed to federal judicial office in Australia, I was seconded to chair the Australian Law Reform Commission (ALRC). The commission, a federal statutory body, was created after the model of Lord Scarman's Law Commissions in the United Kingdom (Kirby, 2006: 449; Murphy, 2009). Our task was to advise the Australian Parliament on the reform, modernisation and simplification of Australian federal law.

One of the first inquiries assigned to the ALRC concerned an issue of biotechnology. The Attorney-General required us to prepare a law for the Australian Capital Territory (ACT) (a federal responsibility) to deal with the

---

4   Sir Isaac Newton in a letter to Robert Hooke, 5 February 1675/6 wrote: 'If I have seen further it is by standing on the shoulders of giants'.

issues presented to the law by human tissue transplantation (ALRC, 1977). The project was initiated in July 1976. The commission was obliged to report no later than 30 June 1977. The timetable was very tight.

In the event, the commission fulfilled its mandate. It produced its report on time. Within Australia, the report proved highly successful. Not only did it result in the adoption of a law on this aspect of biotechnology for the ACT;[5] the draft legislation attached to the ALRC's report was soon copied in all parts of Australia.[6] Such was the universality of the issues that we addressed that the report was also quickly translated into languages other than English and used overseas in the development of the laws of other countries.

The report described the then rapid advances that had occurred in transplantation surgery. The earliest attempts in this technology were dated back 2000 years. Instances of the transplantation of teeth in England at the close of the eighteenth century, of successful bone transplantation at the close of the nineteenth century and of transplantation of organs such as the kidney dating from the early 1950s (Woodruff, 1968: 380, 521–25), indicated that this was an area of human activity that probably required fresh legal thinking. One of the events that had propelled the Federal Attorney-General in Australia into action on this subject was the worldwide controversy that had surrounded the first transplantation of a human heart in South Africa in December 1967 by Dr Christiaan Barnard. The recipient died 18 days later from pneumonia. But successful long-lasting operations quickly followed.

The ALRC was quite pleased with itself for getting its report completed on time. After all, there were many difficult and controversial legal topics of regulation to be addressed. These included:

- whether a system of 'opting in' or 'opting out' should be accepted to permit the
- removal of human tissue from the source
- whether legal minors should be permitted to give consent, as for a sibling recipient
- and, if so, under what conditions
- whether payments for human organs should be forbidden in all circumstances
- whether organs might be taken from prisoners and other dependent persons for transplantation
- whether tissue might be removed from coroner's cadavers

---

5   *Transplantation and Anatomy Act 1978* (ACT).
6   *Human Tissue Transplant Act 1979* (NT); *Transplantation and Anatomy Act 1979* (Qld); *Human Tissue Act 1982* (Vic); *Human Tissue and Transplant Act 1982* (WA); *Human Tissue Act 1983* (NSW); *Transplantation and Anatomy Act 1983* (SA); *Human Tissue Act 1985* (Tas).

- whether blood was to be treated separately or as just another human tissue
- how 'death' should be defined for legal purposes, as a precondition to the removal of vital organs for transplantation.

As the ALRC was producing its report, it became aware of a 'major medical development ... expected within the near future — possibly the next two or three years'. This was described as 'the fertilisation of human egg cells outside the human body'. The process of *in vitro* fertilisation (IVF) and embryo transplantation was therefore mentioned in the report. However, the ALRC recognised that the fertilisation of the ovum of a woman by the use of donor semen, whether *in utero* or *in vitro*, raised issues different in kind from those presented by the transplantation of particular organs and tissues. Whether or not embryo transplantation literally fell within its terms of reference, the ALRC felt bound to exclude the subject from its report and proposed legislation. If there were to be an inquiry into IVF, it would require a separate reference (ALRC, 1977: 18–19 [41]–[42]).

Similarly, the ALRC had become aware, even at that time 30 years ago, of the potential of transplantation of foetal tissue. It noted that work on foetal tissue transplants 'may have already begun in Australia' (ALRC, 1977: 20 [45]–[46]). Already 'right to life' organisations and others had made submissions calling for legal prohibitions. Reports in Britain (Peel, 1972), the United States (National Commission for the Protection of Human Subjects of Biomedical and Behavioural Research, 1975), and New Zealand (Royal Commission on Contraception, Sterilisation and Abortion, 1977), were mentioned. Once again the subject was sidestepped.

The ALRC inquiry afforded a vivid illustration of how, in the regulation of technology, events rarely, if ever, stand still. Even between the time that the ALRC initiated its project on human tissue transplantation law and the time it reported, the technology had marched on. Draft legislation prepared to address other topics was unsuitable, and plainly so, for the more sensitive and complicated issues emerging from IVF and foetal tissue transplants. Before long, Louise Brown was born in England. Eventually, special laws on IVF were adopted in Australia, as elsewhere.[7] As I was to learn in my judicial capacity, such laws and the issues involving the availability of IVF for unmarried or same-sex recipients, invoke strong feelings, conflicting demands and different regulatory responses in different places.[8]

Regulating information technology: Soon after the completion of the law reform project on human tissue transplants, the ALRC was asked to prepare

---

7   See, for example, *Infertility Treatment Act 1995* (Vic); *Reproductive Technology (Clinical Practices) Act 1988* (SA); *Human Reproductive Technology Act 1991* (WA).

8   Re *McCain*; Ex parte Australian Catholic Bishops Conference (2002) 209 CLR 372.

recommendations on reform of the Australian law governing the protection of privacy. This too led to a major inquiry although, in this case, the object was the preparation of proposals for federal legislation, suitable for enactment by the Australian Parliament. In the result, a number of reports were delivered on the topic.[9] The major report, delivered in 1983, dealt with many aspects of privacy protection under federal law.

As befitted its delivery on the brink of 1984, a major focus of the 1983 report was the new information technology. Even at that time, that technology had significantly changed the way in which information was collected and distributed and the amount of personal information that could be communicated.

Because of the currency of the Australian inquiry, I was sent as the Australian representative to a group of experts convened by the Organisation for Economic Cooperation and Development (OECD) in Paris. That expert group was formed to make recommendations to member countries of the OECD on guidelines for the protection of privacy in the context of trans-border data flows. In the event, I was elected to chair the OECD expert group. It conducted its inquiry between 1978 and 1980, drawing upon principles already developed in relation to automated and non-automated data systems by the Nordic Council, the Council of Europe and the then European Economic Community. In the result, guidelines were agreed to by the OECD (1980). They were to prove influential in the development of the national laws of member states, influencing the design and contents of such laws in countries with legal systems as diverse as Australia, Canada, Japan and the Netherlands and corporate practice in the United States. The Australian *Privacy Act*, based on the ALRC report, was enacted by the Australian Parliament in 1988.[10]

Annexed to the Australian *Privacy Act*, in Schedule 3, were 'national privacy principles'. As the Act declared in its Preamble, its purpose included compliance by Australia, as a member of the OECD, with the recommendation of the Council 'that member countries take into account in their domestic legislation the principles concerning the protection of privacy and individual liberties set forth in Guidelines annexed to the recommendations'. The Act recited that Australia had 'informed that organisation that it will participate in the recommendation concerning those Guidelines'.[11] Hence, the national privacy principles adopted by the new federal law.

A difficulty soon became apparent. It did not arise out of any defect in the understanding of the OECD expert group, or of the ALRC in its recommendations to the Australian Government and Parliament, concerning the technology

---

9   See, ALRC 1979(a), 1979(b), 1983.

10   *Privacy Act* 1988 (Cth).

11   *Privacy Act* 1988 (Cth), Preambles 4 and 5.

then deployed. That technology, however, quickly changed in its potential. Moreover, it did so in a way that rendered an assumption, expressed in the OECD guidelines and the Australian national privacy principles, out of date (at best) and irrelevant (at worst).

Illustrating the issue by reference to the 'use and disclosure' principle, the second in the Australian national privacy principles, this principle stated:

> 2.1 An organisation must not use or disclose personal information about an individual for a purpose (the secondary purpose) other than the primary purpose of collection unless:

> (a) Both of the following apply:

>> (ii) The secondary purpose is related to the primary purpose of collection and, if the personal information is sensitive information, directly related to the primary purpose of collection;

>> (ii) The individual would reasonably expect the organisation to use or disclose the information for the secondary purpose; or

> (b) The individual has consented to the use or disclosure; or

> (c) If the information is not sensitive information and the use of the information is for the secondary purpose of direct marketing [certain provisions follow]; or

> (e) The organisation reasonably believes that the use or disclosure is necessary to lessen or prevent:

>> (i) A serious or imminent threat to an individual's life, health or safety; or

>> (ii) A serious threat to public health or public safety; or

> (f) The organisation has reason to suspect that unlawful activity has been, is being or may be engaged in ...; or

> (g) The use or disclosure is required or authorised by or under law; or

> (h) The organisation reasonably believes that the use or disclosure is reasonably or necessary for one or more of the following by or on behalf of an enforcement body.

> [Provisions on law enforcement follow].

The basic hypothesis of the OECD guidelines (and therefore of the ALRC recommendations) and the *Privacy Act* was that personal information that was collected should ordinarily be used for the purpose for which it was collected and that such purpose should be made known to the individual at the time of the collection.[12] Then, along come search engines, including Google and Yahoo. The specification of purposes of collection and the limitation of use and disclosure by reference to such purposes went out the window.[13]

This is the sense in which I assert that I was present at the creation of the problems addressed in the TELOS conference on the regulation of new technologies. Accepting as paradigm instances the cases of biotechnology and information technology that I have described, the difficulty (in some cases near impossibility) was soon apparent in drafting any law of the conventional kind that would not quickly be overtaken by events. In part, legal texts might be overtaken by advances in technology of the kind that I have described. But, in part too, changes in social attitudes, themselves stimulated by advances in technology and a perception of the utility of the advances, made it more difficult than in other fields of law to draw a clear line in the sand.

The caravan of controversy: Take for example, *in vitro* fertilisation. In 1976, when the ALRC report *Human tissue transplants* was written, many earnest debates were conducted over the suggested ethical quandary of transplantation of ova fertilised by a husband's sperm. These debates were quickly replaced by new ones concerned with the use of non-husband (donor) sperm. Such debates are now rarely raised, even in esoteric legal circles. Today the ethical (and legal) debates in Australia and elsewhere are more likely to be concerned with the availability of IVF to single parents and to same-sex couples. Thus, the caravan of controversy moves on. A law drafted too early may freeze in time the resolution of earlier controversies, which may later be regarded as immaterial or insignificant.

Napoleon reportedly observed a principle of never responding to letters for at least a year. He adopted this principle on the footing that, if the problem still existed a year later, it would be time enough for it to receive the Emperor's attention. Whether by default, or by design, many issues presented to the law by contemporary technology appear to receive the same treatment. One suspects that, in many instances, it is because of the complexity and sensitivity of the issues rather than a strategic policy of lawmakers to postpone lawmaking or clarification of regulation until the contours of the necessary law have become clear.

---

12 *Privacy Act* 1988 (Cth), Schedule 3: 'Privacy Principle 1 (Collection:)'.
13 Another illustration arises out of the enactment of provisions requiring that confessions and admissions to police, by suspects in custody, should be recorded on 'videotape'. See, for example, *Criminal Code* (WA), s 570D(2)(a). The change to digital technology necessitated amendment of such laws to substitute a requirement for 'audio-visual recording'. See *Criminal Investigation Act* 2006 (WA), s 118(1).

Five paradoxes

1. *Doing the best without experts*: Having laid the ground for my competence to provide a summation of the London TELOS conference, I will start by identifying a number of paradoxes, or at least curiosities, that emerged during the debates. The first of these curiosities is a reflection not only on my own limited competence but also on the limited competence of everyone else.

There are no real experts on the subject of regulating technologies. They do not exist in the United Kingdom, the United States, Australia or elsewhere. It is much easier to find an expert on the intellectual property implications of biotechnology and information technology than it is to find someone skilled in considering what new law, if any, should be adopted to deal with a particular issue presented by technology and how it should be devised. Easier by far to find an expert on income tax or unjust enrichment or international human rights law than to find scholars, judges or even legislative drafters who can claim to be experts in the interface of law and technology.

It is true that we had the privilege at TELOS of an opening address by Professor Lawrence Lessig, then professor of law at Stanford Law School in the United States. He was founder of that school's Center for Internet and Society. Lessig's book *Code and other laws of cyberspace* (now updated by *Code V2*) blazed a trail. He launched the host organisation, TELOS. On the interface of cyberspace and the law, he is something of a guru. His launching speech, like his books, challenged us all to think afresh. His novel thesis is that 'Code', or the architecture of technological systems, will sometimes incorporate regulatory imperatives into information technology, obviating any real choice on the part of the user as to whether or not to conform to the law.

In the High Court of Australia we came face to face with this reality in *Stevens v Kabushiki Kaisha Sony Computer Entertainment*.[14] The case concerned a claim by Sony Corporation of breach of a 'technological protection measure' installed by it in the program of its computer games. Sony asserted that the measure was protected under the Australian *Copyright Act* 1968. Sony argued that Stevens had unlawfully sought to circumvent the device incorporated in computer games that it produced and sold on CD-ROM for use in its PlayStation consoles.

---

14   (2005) 224 CLR 193; [2005] HCA 58.

Applying a strict interpretation to the expression 'technological protection measure', the court held that Sony's device did not fall within the statute. I agreed in this analysis.[15] The case was a vivid illustration of the way in which, for copyright, contractual and other legal purposes, attempts are now often made to incorporate regulatory provisions in the relevant technological codes. It is a new development, although I suppose one might see primitive attempts directed at the same object in the safety provisions incorporated in the design of houses, bridges and aeroplanes. The computer PlayStations simply take this development to a higher level of sophistication and technological capability. Lessig identified this new development. Inevitably, his expertise did not include all of the current major technologies, still less the way in which law can regulate them.

I, too, am no expert in the design of laws. True, in the High Court of Australia, I participated in a final national court that sometimes declares new laws. I worked for a decade in national law reform, as I have described. True, also, I have participated in the drafting of international guidelines, such as those of the OECD.[16] This, however, is hardly an intensive preparation for the complex and highly technical task of drafting laws for, or under, a legislature. I have become rusty since, in my law reform days, I worked with former parliamentary counsel on the draft legislation annexed to the ALRC's reports.

Nor can it be said that the academics present at the conference had any special skills (at least skills that any of them revealed) in drafting statutes and subordinate regulations. Professor Brownsword confessed to beginning his academic career by teaching contract law, with later experience in consumer and environmental law. Whilst the latter fields are overburdened with a mass of regulation, it is different to use and interpret such laws, on the one hand, and on the other to design and draft them. Many participants in the conference were, to use the words of Judy Illes, trained as 'bench scientists'. Although the experience of authentic scientists and technologists was essential to an understanding of the problem, it did not necessarily provide the best guidance for legal solutions.

VI Lenin once declared that the person who writes the minutes of an organisation usually ends up controlling it. His work as general secretary of the Soviet Communist Party obliges us to take this advice seriously.

---

15   (2005) 224 CLR 193 at 246 [186].

16   Also as chair of the UNESCO International Bioethics Committee drafting group for the *Universal declaration on bioethics and human rights*, adopted by the General Conference of UNESCO, Paris, October 2005. See Andorno, 2007: 150.

We may complain about the absence of law concerned with new and cutting-edge technology. We may acknowledge our own imperfections for addressing the gap. We may recognise, with Lessig, that regulation in the future may not necessarily come in the form of written instruments made by or under the legislature and published in the *Government Gazette*.

Nevertheless, the issue tackled in the TELOS conference was undoubtedly one of the greatest importance for the future of the rule of law in every society. Despite the manifold interpretations of those whom it invited to its conference, TELOS may, in the long run, have a paradoxically disproportionate impact on perceptions of how technologies may be regulated and used in regulation, simply because it is one of the first organisations to tackle this issue generically. It surveys what is substantially a blank page. Increasingly the content of law, like the content of life, will be concerned with technology and with its many consequences for society. The importance of the chosen topic, therefore, belies the comparatively little that is written, said and thought about it. Paradoxically, then, those who first lay claim to expertise may participate in a self-fulfilling prophesy.

2. *Too much/too little law*: The second paradox is that most of us recognise that the failure to provide law to deal with the impact of particular technologies is not socially neutral. Effectively, to do nothing is often to make a decision.

Thus, for the law to say nothing about reproductive cloning of human beings, for example, (assuming that outcome to be technically possible) is to give a green light to experiments in that technology. In so far as the law expresses prohibitions supported by sanctions that uphold the command of a sovereign power, silence may, for once, imply consent or at least non-prohibition. Thus, if there is no law to prohibit or regulate reproductive cloning or hybridisation or xeno-transplants, scientists and technologists at their benches may decide to experiment. Nothing then exists to restrain them except their own ethical principles, any institutional ethics requirements, the availability of funding and the prospects of a market. A scientist or technologist may proceed out of sheer curiosity, as when David Baltimore so beneficially investigated a simian retrovirus a decade before the discovery of the immuno-deficiency virus in human beings.

The scientist or technologist may, of course, do this in the hope of cashing in on a potentially lucrative therapeutic market. One such market certainly exists in respect of therapies to overcome human infertility. Reproductive human cloning might, potentially, be one such therapy. Some of its supporters treat with contempt the supposed moral objections to this form of scientific advance (Robertson, 2001: 35; Shikai, 2002: 259). They point to earlier resistance to other reproductive technologies such as artificial insemination donor (AID), artificial insemination husband (AIH), IVF and surrogacy arrangements.[17] Most of these objections have faded away as society becomes more used to 'non-natural' ways of securing a desired pregnancy in a particular patient.

The recognition that inaction in the face of significant technologies may amount to making a decision coexists with our appreciation, as observers of the law, that premature, over-reaching or excessive lawmaking may, in some cases, be an option worse than doing nothing. It may place a needless restriction upon local scientists and technologists, obliging them to take their laboratories and experiments offshore.

In a big world with diverse cultures, religions and moral beliefs, it is never difficult to find a place offering a regulation-free zone in exchange for investment dollars. Just as bad is the possibility that laws are solemnly made and then ignored or found to be ineffective, as was temporarily the case with the 'technological protection measure' considered in the Australian Sony litigation. Following the decision of the High Court of Australia in that case, and under pressure from the US government under the *United States–Australia Free Trade Agreement*, Australian law was changed. The new law represented an attempt to overcome the High Court's decision in the *Sony Case*, although in a somewhat different way.[18]

Many participants in the TELOS conference, whether expert in matters of biotechnology or information technology, revealed themselves as legal libertarians. They were so mainly because of their recognition of the common potential of premature, over-reaching and ill-targeted laws to diminish experimentation, burden innovation and cause economic and other inefficiencies. Thus, Han Somsen presented a number of compelling arguments about the dangers of the 'precautionary principle' (Andorno, 2004). Whilst this principle appears to be gaining increasing

---

17   The New South Wales Law Reform Commission in 1988 recommended a prohibition on surrogacy arrangements that was not implemented. Surrogacy arrangements are, however, regulated in some Australian jurisdictions: *Parentage Act* 2004 (ACT); *Surrogate Parenthood Act* 1988 (Qld); *Family Relationships Act* 1975 (SA); *Surrogacy Contracts Act* 1993 (Tas); and *Infertility Treatment Act* 1995 (Vic).

18   For the story of the change of law following the decision in the *Sony* case, see de Zwart (2007: 7); in contrast, see D Brennan, 'What can it mean "to prevent or inhibit the infringement of copyright"?: — a critique of *Stevens v Sony*', *Australian Intellectual Property Journal*, vol 17, 2006, p 86. See also *Copyright Amendment Act* 2006 (Cth) implementing the new scheme said to be required by art 17.4.7 of the *Australia–United States Free Trade Agreement*.

acceptance in the international community, particularly in respect of protection of the global environment, it carries risks of its own. If taken too far, it could instil a negative attitude towards science and technology and encourage excessive regulation in the attempt to avoid *any* risks. Life is risky. Most technological innovations carry some risk. An undue emphasis on precaution, for fear of *any* risks, would not be good for science or technology or for the global economy or for innovation in thought as well as action.

The second paradox is thus more of a contradiction or tension, and difficult to resolve. At the one time we must accept that doing nothing to regulate technologies involves making a decision. Yet we must also recognise that sometimes doing nothing will be a better option than making laws that impede innovation and burden efficiency.

3. *First Amendment and copyright law*: An early illustration of the second paradox arose in Lessig's opening address, which was concerned with the potential of 'Code' (or information technology architecture) to play a part in regulating technology in ways more universal and immediately effective than most laws are.

An instance, frequently mentioned, is the installation of filters designed to prohibit access to materials considered 'harmful to minors'. Many countries now have legal regulations forbidding access to, or possession of, child pornography. Available software may prevent access to sites providing such images. But sometimes they may do so at a cost of over-reaching prohibitions. The burden on free communication may outstrip the legitimate place of legal regulation, forbidding access not only to child pornography but to lawful erotic materials or discussion about censorship itself or to websites concerned with subjects of legitimate interest, such as aspects of human sexuality, women's rights and even children's rights.

Whereas the law will commonly afford avenues of appeal and review of decisions that purport to apply legal norms, an over-reaching 'protective' software program may afford no such rights of challenge. Those concerned with the human right of free expression are naturally anxious about the potential of 'Code' to re-institute excessive censorship in society, just when we thought we had grown out of that habit.

Like most American lawyers, Lessig approached these issues from the standpoint of the First Amendment to the US Constitution.[19] This

---

19 Relevantly, the First Amendment states: 'Congress shall make no law … abridging the freedom of speech, or of the press'.

upholds a very high level of unrestricted and unregulated freedom of communication. The rest of the world tends to be less absolutist in this respect.[20] It recognises that, whilst 'free' expression and access to a 'free' media constitute important human rights, they are not unlimited. They have to be harmonised with other fundamental human rights. These include the right to individual honour and reputation and to protection of privacy and family relationships.[21] They also include protection of the legitimate rights of inventors.[22]

Lessig expressed concern about the balance that has been struck in the United States between rights to free expression and the right to copyright protection that impinges on free expression.[23]

In an international meeting such as the TELOS conference, we were not, as such, concerned with the particularities of US law, including the way the constitutional law of that country reconciles free expression and lawful copyright protection. On the other hand, because of the dominance of the US media and its hegemony in entertainment and popular culture, what is done in that country to regulate information technology obviously has consequences worldwide. Just as, in earlier decades, the hard copy issues of *Playboy*, circulating in huge numbers around the world, broke down the prevailing culture of censorship, carrying First Amendment values virtually everywhere, so today the inbuilt 'Code' or architecture of information systems may carry US legal protections for US copyright holders far beyond the protections that the laws of other countries afford them.[24]

This consequence can present legal and practical problems for the regulation of technology in jurisdictions enjoying different capacities to contest the balances struck by the Constitution and laws of the United States. In smaller economies, there may be no real choice. Upholding the local constitution and its values may, as a matter of practicalities, be impossible. Consumers may be presented with no real option. If they buy the software that drives the PlayStation, they may find that it reflects US

---

20   For example, *ABC v Lenah Game Meats Ltd* (2001) 208 CLR 199 at 283 [202] ([2001] HCA 63); *Dow Jones and Co Inc v Gutnick* (2002) 210 CLR 575 at 626 [115] ([2002] HCA 56).
21   *International Covenant on Civil and Political Rights* (1976) arts 17.1, 17.2 and 19.3.
22   cf *Universal Declaration of Human Rights* (1948) art 27.1; *International Covenant on Economic, Social and Cultural Rights* (1976), art 15.1(b) and (c).
23   Cf *Nintendo Co Ltd v Sentronics Systems Pty Ltd* (1994) 181 CLR 134 at 160; *Grain Pool of WA v The Commonwealth* (2000) 202 CLR 479 at 531 [133], fn 266 ([2000] HCA 14) referring to *Graham v John Deere & Co* 383 US 1 at 6 (1966); *Feist Publications Inc v Rural Telephone Service Co* 499 US 340 at 348 (1991) and L Lessig, *Code and other laws of cyberspace* (1999), pp 131, 133–34.
24   *Stevens v Kabushiki Kaisha Sony Computer Entertainment* (2005) 224 CLR 193 at 256 [216] citing L Lessig, *Code and other laws of cyberspace* (1999); see Fitzgerald, 2005: 96. See also *Metro-Goldwyn-Mayer Studios Inc v Grokster Ltd* 73 USLW 4675 (2005).

constitutional and copyright laws. Indeed, such software may exceed even the protections afforded by those laws. It is in this sense that 'Code' and architecture may challenge the previous assumption that, within its own borders, each nation state is entitled, and able, to enforce its own laws, reflecting its own values. In Australia, we gained a glimpse of things to come in the *Sony* litigation. But it was only the beginning.

The debate that Lessig recounted between First Amendment values and the current state of US copyright law presents a microcosm of similar conflicts in every society. There is an element of the paradoxical about it in the United States. This is because, as Lessig put it, intellectual property law in that country has been able, to some extent, to slip under the radar of First Amendment values. To a large extent, intellectual property law has developed separately and, in part, inconsistently. This point was noted by me in my reasons in *Sony*. Eventually, in the United States, Britain, Australia and elsewhere, it will be necessary to face directly the tension between enlarging copyright protection (including through the use of the technological architecture of information technology) and adhering to high levels of free communication, unimpeded by undue governmental regulation (such as by copyright law[25]).

The conflict recounted by Lessig presents a paradox, visible to non-Americans and to American lawyers themselves.[26] The country that has been foremost in promoting values of free expression and the free press has also lately been foremost in promoting, extending and enforcing the intellectual property rights of its own creators, 'inventors' and designers. This is not only true in the context of information technology. It is also true in the case of biotechnology, as the closely divided decision of the Supreme Court of the United States in *Diamond v Chakrabarti*,[27] and its progeny, demonstrate. Lessig appreciated, and highlighted, this paradox. It appears in an acute form in the United States. But it has its counterparts everywhere.

4. *Technology's democratic deficit*: A fourth paradox derives from the way in which contemporary technology at once enhances, and diminishes, democratic governance. No one at the TELOS conference questioned the importance of science and technology in the current age. Similarly, no one questioned the desirability of rendering laws, and regulation more generally, available and accountable to the people from whom authority

---

25   *Grain Pool* (2000) 202 CLR 479 at 531 [133]; *Sony* (2005) 79 ALJR 1850 at 1886 [216].
26   *Graham v John Deere Co* 383 US 1 at 6 (1966).
27   477 US 303 (1980); compare with Kirby (2001: 64).

to govern society is ultimately derived. On balance, however, does technology enhance or reduce democratic accountability for the state of the resulting regulations?

In some respects, there can be no doubt that technology has improved communication that is essential to converting the formalities of electoral democracy into the realities of genuine accountability of the governors to the governed. Radio, television, worldwide satellite communications, the internet, podcasts, blogs, Facebook, YouTube, and Twitter have revolutionised the distribution of information about those persons and institutions whose decisions affect the regulation of our daily lives. In this sense, democratic governance has moved from small, town hall assemblies of earlier times into huge, national and international forums both public and private.

Paradoxically the very quantity of information has resulted in its manipulation and presentation in a way that is often antithetical to real democratic accountability. The technology stimulates a demand for the simplification and visualisation of messages, the personalisation of issues, the trivialisation of conflict, the confusion between fact and opinion and the centralisation and 'management' of news. So-called 'spin' and 'infotainment' are characteristics of media in the present age. They tend to concentrate power in a way that even George Orwell could not have imagined.

Several speakers at the TELOS conference referred to yet another feature of contemporary technology that can be inimical to democracy. This is the incorporation of regulation in the technology itself that goes beyond what is strictly required by local law, yet without effective opportunities for those affected to challenge the regulation so imposed. Who can, or would, challenge the over-inclusive software designed to bar access to Internet sites selected as 'harmful to minors' but sometimes operating in an over-inclusive way?

When serving on the High Court of Australia, I found that the website of the Archbishop of Canterbury was barred to my use. My staff were unable to access one of the Archbishop's addresses. This was presumably because a filter, instituted to deny access to websites deemed undesirable, had erected a bar. Ostensibly, this was because, in the manner of these times, one or more of his Grace's addresses dealt with issues of sex, specifically homosexuality. In fact, that was exactly why I wanted the speech. I was surprised to find that, at the same time, the Vatican website was accessible without any restriction. This may say something either about the prudence of the then Pope's choice of language, the power of the Roman Catholic Church in such matters, or the religion of the filter programmer. I gave directions that led to the filter being over-ridden. I secured a copy of the desired speech. But many might not be so lucky.

Given the importance of technology to the current age, how do we render those who design, install and enforce such programs accountable to the democratic values of our society? As 'Code' enlarges and replaces the old-style legal regulation of technology, how do we render its architects answerable to the majority views of the people? How, if at all, are transnational corporations, like Sony for instance, rendered responsible to the democratic values of the nations in which their products are used?

These are legitimate questions because the fourth paradox is the coincidence, at the one time of history, of technologies that vastly enhance access to information that jumped the Berlin Wall, bringing messages of freedom, at the same time as they sometimes diminish genuine debate, enlarge unreviewable 'technological' corporate decisions and expand the capacity to 'manage' news in a way inimical to real transparency and accountability of decision-makers to the people.

5. *Vital but neglected topics*: I reach my fifth, and final, paradox. The TELOS conference addressed one of the most important issues for the future of the rule of law in every country. Because of the elusiveness of much contemporary technology to effective regulation, large and increasing areas of activity in society find themselves beyond the traditional reach of law as we have hitherto known it. When regulation is attempted, as I have shown, it will often be quickly rendered ineffective because the target has already shifted. Typically, in the past, the drawing up of laws has been a slow and painstaking process. Consulting governments and those primarily affected, not to say the people more generally, takes time. In that time, the technology may itself change, as I have demonstrated from my experience with human tissue transplantation and privacy laws. Now, new regulation are being developed in the form of what Lessig calls 'Code'. Yet, this form of regulation is not so readily susceptible, if susceptible at all, as conventional laws in the past have been, to democratic values and to the participation (or even appreciation) of most of those affected in the moral choices that determine the point at which the regulation is pitched.

If, on the same weekend in London, King's College School of Law had convened a conference on revenue law, it would have filled a convention hall. A month earlier, in Hobart, Tasmania, I addressed more than 600 lawyers and accountants at such a conference. Similarly, a conference on the law of unjust enrichment would attract hundreds of contributors, with their differing opinions. Even a meeting on the rule against perpetuities would probably have attracted more participants than the inaugural conference of TELOS. Yet, in truth, the issues addressed by TELOS were more important for our societies and their governance than virtually any of the other topics that the legal discipline could offer.

It sometimes falls to small groups, particularly in professions, to lead the way and to bring enlightenment to the many. This, then, is the fifth paradox — at least it is an oddity. Such an important topic as the regulation of burgeoning technologies in modern society should engage the interest and attention of all who claim to be lawyers, sociologists and philosophers and express an interest in the health of the rule of law. Yet, for the moment, and for most such observers, this is *terra incognita*. The contributions at the TELOS conference suggest that it will, and should, not be so for long.

Seven lessons

1. *Recognise a basic dilemma*: Certain general lessons therefore stand out from the presentations at the TELOS conference. Some of them have already been touched on.

   The first is that, the regulation of technology faces a fundamental dilemma hitherto relatively uncommon in the law. This is that, of its character, technology is normally global. Law, being the command of an organised community, is traditionally tied to a particular geographical jurisdiction. Whereas in recent years the need for extraterritorial operation of municipal law has been recognised, and upheld,[28] the fact remains that the focus of most national law is the territory of the nation. By way of contrast, the focus of regulating technology must be the technology itself.[29] Sometimes, that feature of the technology will make effective regulation by national law difficult, or even impossible.

   It is into this context that direct enforcement by 'Code', written into software programs or otherwise imposed, adds a new dimension to global technology. The values and objectives of transnational corporations may be even more unresponsive to national regulation than the rules of a municipal legal system are. Moreover, 'Code' of this kind may opt for caution and over-inclusion so as to avoid dangers to markets in the

---

28   *Re Aird; Ex parte Alpert* (2004) 220 CLR 308 at 344–350 [114]–[133]; ([2004] HCA 44) referring to the case of the *SS Lotus* (1927) Permanent Court of International Justice, Series A, No 10, Judgment No 9, pp 18–19 and Martinez (2003: 429).
29   *Dow Jones* (2002) 210 CLR 575 at 615–619 [78]–[92].

least right-respecting countries. The contractual arrangements entered between the government of the People's Republic of China and the corporations selling access to Yahoo and Google in China, which were described during the conference, illustrate the willingness of the latter to succumb to the demands of the former so as to avoid endangering a lucrative economic market for their products. In this way the provider, but also the users, are subjected to forms of censorship that might not be tolerated in other societies. A smaller country, with a smaller market, is unlikely to exert the same clout. Considerations of economics rather than of legal principle, ethical rules or democratic values may come to predominate in such cases.

2. *Recognise that inaction is a decision*: In the past, proponents of technological innovation have often favoured containment of law and a 'libertarian' approach to developments of technology. Yet, most lawyers recognise that there are limits. Unless such limits are clearly expressed, and upheld in an effective way, the absence of regulation will mean, effectively, that the society in question has made a decision to permit the technological advances to occur, without impediment.

Those who are cautious about adopting any form of the precautionary principle may yet recognise the need for some restraints. Thus, unlimited access to child pornography will probably offend most people and sustain a call for regulation of the Internet to prohibit or restrict access to such sites. However, that will still leave room for debate about the detailed content of such regulation: the age of the subjects depicted; any permissible (computer graphic or cartoon format rather than human) images; the means of enforcing the law; and the provision of effective sanctions.[30] Cases on these issues, and on any constitutional questions that they present, are now quite common.[31]

Likewise with biotechnology. Views may differ over whether regulation is necessary, or even desirable, to prohibit therapeutic cloning, reproductive cloning or the use of human embryonic stem cells. Yet, non-binding prohibitory resolutions and declarations have been adopted in the organs of the United Nations on this subject.[32] Even those nations, like the United Kingdom, that have not favoured prohibitions or moratoriums on experiments with human cloning for therapeutic

---

30   *Bounds v The Queen* (2006) 228 ALR 190 at 197 [26], 211 [94]; ([2006] HCA 39).

31   *The Queen v Fellows and Arnold* [1997] 2 All ER 548; *The Queen v Oliver* [2003] 1 Cr App R 28 at 466–467 [10]; cf *Lawrence v Texas* 539 US 558 at 590 (2003).

32   See Macintosh (2005: 135–36), describing the resolution of the General Assembly of the United Nations of 8 March 2005. This approved a Declaration, proposed by the Sixth Committee, to 'prohibit all forms of human cloning inasmuch as they are incompatible with human dignity and the protection of human life'. The General Assembly vote was 84 to 34 in favour with 37 abstentions.

purposes might accept the need to prohibit, or restrict, some bio-technological experiments. Hybridisation and xeno-transplantation of tissue across species clearly require, at the very least, restrictions and safeguards so as to prevent cross-species transmission of endogenous viruses. To do nothing is effectively to decide that nothing should be done. It does not necessarily amount to a decision to 'wait and see'.

This is why the regulation of technology is such an important subject. It is not one that can be ignored, simply because the subject matter, and the available regulatory techniques, are difficult and controversial.

3. *Recognise the limited power to regulate*: A third lesson, derived from the first two, is that the normal institutions of legal regulation often appear powerless in the face of new technology. This is clear in the case of attempts to regulate new information technology. So far as the Internet is concerned, the regulatory values of the United States inevitably exert the greatest influence on the way the Internet operates and what it may include. This means that both First Amendment and copyright protection values, established by the laws of the United States, profoundly influence the Internet's present design and operation. An attempt by another nation's laws (such as those of France) to prohibit transnational publication that is offensive to that country's values (such as advertising Nazi memorabilia) may face difficulties of acceptance and enforcement in the Internet.[33]

The same is true of biotechnology. The Australian Parliament initially enacted the *Prohibition of Human Cloning Act* 2002 (Cth) and the *Research Involving Human Embryos Act* 2002 (Cth). These were part of a package of laws aimed at the consistent prohibition in Australia of human cloning and other practices deemed unacceptable at the time. Both Acts were adopted on the basis of the promise of an independent review two years after the enactment. Such a review was duly established. It was chaired by a retired federal judge, the Hon John Lockhart. The review presented its report on December 2005. It recommended an end to the strict prohibitions of the 2002 legislation; the redefinition for legal purposes of the 'human embryo'; and the introduction of a system of licensing for the creation of embryos for use for therapeutic purposes (Australian Government, 2005).

Initially, the Australian Government rejected the recommendations of the Lockhart review. However, following strong political, scientific and media reaction, a conscience vote on an amending Act, introduced by a

---

33   *League Against Racism and Anti-Semitism (LICRA), French Union of Jewish Students, v Yahoo! Inc.* (USA), Yahoo France [2--1] *Electronic Business Law Reports*, 1(3) 110–20 (The County Court of Paris).

previous health minister, was allowed. In the outcome, the amendments were enacted. But, they were passed by the Australian Senate with only a tiny majority.[34]

The main argument that secured this outcome in Australia was the recognition of the pluralistic nature of the society; widespread reports on the potential utility of the research and experimentation; and the expressed conviction that experimentation would proceed in overseas countries with results that, if they proved successful, would necessarily be adopted and utilised in Australia.[35] Interestingly, both the then prime minister and the leader of the federal opposition (soon to be his successor) voted against the amending Act.[36]

The global debates on the regulation of experiments using embryonic stem cells have often been driven by countries that, to put it politely, are not at the cutting edge of the applicable technology.[37] On the other hand, the United States, certainly during the administration of President George W Bush, also adopted a conservative position on these topics in United Nations forums. As happened in Australia, this was to change on the election and re-election of President Barack Obama.

4. *Recognise differentiating technologies*: So far as regulation of technologies is concerned, the TELOS conference established the need to differentiate technologies for the purpose of regulation. It is not a case of one response fits all. Self-evidently, some forms of technology are highly sensitive and urgently in need of regulation. Unless the proliferation of nuclear weapons is effectively regulated, the massive destructive power that such technology presents has the potential to render all other topics theoretical. Similarly, some aspects of the regulation of biotechnology are sensitive, including the use of embryonic stem cells and germline modification. For some, the sensitivity derives from deep religious or other beliefs concerning the starting point of human existence. For others, it arises out of fears of irreversible experiments that may go wrong.

Somewhat less sensitive is the regulation of information technology. Yet this technology too presents questions about values concerning which

---

34   In the Australian House of Representatives, the vote was 82:62; see Australia, House of Representatives (2006: 127). In the Senate the vote was 34:31; see Australia, Senate (2006: 48).
35   See, for example, Australian, (2006: 15); Finkel & Cannold (2006: 9); Sydney Morning Herald (SMH) (2006: 11); Carr (2006: 13).
36   Mr Howard spoke at the House of Representatives on 6 December 2006, (Australia, 2006: p 117); Mr Rudd also spoke to the House (Australia, 2006: 119).
37   Thus, Honduras was the national sponsor of the United Nations ban on human cloning, reproductive and therapeutic. See Macintosh (2005: 134).

people may have strong differences of opinion. To outsiders, Americans seem to imbibe First Amendment values with their mother's milk. Lawyers from the United States sometimes have to be reminded that their balance between free speech and other human rights is viewed by most of the world as extreme and disproportionate.

5. *Recognise different cultures*: Most of the participants in the TELOS conference came from the developed world. They therefore reflected general attitudes of optimism and confidence about the outcome of rational dialogue and the capacity of human beings ultimately to arrive at reasonable responses to regulating technologies, on the basis of calm debate.

This is not, however, universally true. The conference in London coincided with a declaration by the Roman Catholic Bishop of Birmingham, the Most Rev Vincent Nichols, that Britain was facing a period of secular revulsion. This response was attributed to impatience with the instances of violence attributed to those with religious beliefs and the apparent obsession of some Christian churches with issues of sexuality and gender.

There is no doubt that the current age bears witness to many instances of religious fundamentalism. Modern secular democracies can usually prepare their regulations of technology without undue attention to such extremist considerations. But when the considerations come before international institutions, they may have to run the gauntlet of fundamental beliefs. Such religious beliefs are by no means confined to Islam. They also exist in Christianity, Judaism, Hinduism and other world religions. Because, in such instances, religious instruction is attributed to God and derived from human understandings of inerrant religious texts, it may brook no compromise and even no debate.

Recognising the coincidence of galloping technology and the force of religious fundamentalism is necessary to an understanding of what can be done in different countries to respond effectively to aspects of technology that challenge orthodox religious beliefs. In the Australian parliamentary debates on the amendment of the 2002 moratorium on human cloning and use of embryonic tissue, many of the legislators addressed the extent to which it was legitimate, in a pluralistic society, to allow beliefs, even of a majority, to control the design of national legal regulation. Yet, if such beliefs are treated as irrelevant, what other foundations can be provided for a coherent system of moral principle? In some societies such issues simply do not arise. The Taliban in Afghanistan would not entertain an open debate on topics that are treated as concluded by a holy text. The diversity of regulatory responses to new technology, therefore, grows out of the different starting points in each society.

6. *Basing regulation on good science*: In the early days of the HIV pandemic, I served on the Global Commission on AIDS of the World Health Organisation. One of the members, June Osborn, then a professor of public health in the University of Michigan, taught the importance of basing all regulatory responses to the epidemic upon good science. The danger of responses based on assumptions, religious dogmas, intuitive beliefs, or popular opinion were that they would not address the target of regulation effectively.

    The intervening decades have suggested that the countries that have been most successful in responding to HIV/AIDS have been those that have observed Osborn's dictum (Plummer & Irwin, 2006: 1). The same is true of the subjects of biotechnology, information technology and neuroscience examined in the TELOS conference. All too often, science and technology shatter earlier assumptions and intuitions.

    For example, the long-held judicial assumption that jurors, and judges themselves, may safely rest conclusions concerning the truth of witness testimony on the basis of the appearance of witnesses and courtroom demeanour has gradually evaporated because scientific experiments shatter this illusion.[38] One day, by subjecting witnesses to brain scans, it may be possible to demonstrate objectively the truthfulness or falsity of their evidence. One lesson of the Illes paper for the TELOS conference is that we have not yet reached that position. If, and when, it arrives, other issues will doubtless be presented for regulators. We are not there yet. But any regulation must recognise the need to remain abreast of scientific knowledge and technological advances.

7. *Addressing the democratic deficit*: This brings me to the last, and most pervasive, of the lessons of the TELOS conference. Technology races ahead. Often its innovations quickly become out of date. Laws addressed to a particular technology are overtaken and rendered irrelevant or even obstructive. Nowadays, scientific knowledge, technological inventions and community values change radically in a very short space of time.

    Within less than two years of the initial laws, demands were made for reversal to the Australian federal prohibition on therapeutic cloning. Within five years, the prohibition was repealed. In such an environment, there is an obvious danger for the rule of law. It is impossible to expect

---

38   See, for example, *Fox v Percy* (2003) 214 CLR 118 at 129 [31]; ([2003] HCA 22).

of legislatures, with their many responsibilities, that they will address all of the technological developments necessary or useful for regulatory purposes. The average legislator often finds such issues complex and impenetrable. They are rarely political vote-winners. They struggle to find a place in the entertainment and personality politics of the present age as well as with the many other competing questions awaiting political decision-making. This leaves a gap in democratic decision-making in this sphere of regulation. It is a gap that is being filled, in part, by 'Code', which incorporates regulations designed by inventors of information systems themselves in the structure of such systems but without a democratic input or the necessity of human moral judgment.

The democratic deficit presented by contemporary technology is thus the largest potential lesson from the TELOS conference. In an age when technology is so important to society, yet so complex and fast moving that it often defies lay understanding, how do we adapt our accountable lawmaking institutions to keep pace with such changes? One means, undertaken in Australia, is by the use of consultative mechanisms such as the ALRC (Chalmers, 2005: 374)[39] or independent inquiries, such as the Lockhart committee (Cooper, 2006: 27; Stobbs, 2006, 247; Karpin, 2006: 599). In such cases, the very process of consultation and public debate promotes a broad community understanding of the issues, an appreciation of different viewpoints and an acceptance of any regulations adopted, even when they may give effect to conclusions different from one's own.

Adapting the legislative timetable and machinery to the challenges of modern governance is a subject that has engaged law reform bodies and executive government for decades. In Australia, proposals for some form of delegated legislation have been made to increase the implementation of such reports. Often they lie unconsidered for years, or indefinitely, not because of any real objections to their proposals but because of the legislative logjam (Mason, 1971: 197). In the United Kingdom, suggestions for a fast-track system for implementing reports of the law commissions have been under review for some time.[40]

In the face of radically changing technologies and the danger of a growing democratic deficit, it will obviously be necessary to adapt and supplement the lawmaking processes we have hitherto followed in most countries. Various forms of delegated legislation may need to

---

39   Important recent reports of the ALRC in the field have included *Essentially yours: the regulation of human genetic information in Australia* (2003).
40   See Kirby (2006: 466). Such a fast track mechanism was adopted in the United Kingdom in 2009.

be considered. So may the enactment of over-arching laws, expressed in general terms, which will not be quickly reduced to irrelevancy by further technological change.[41] Addressing the weaknesses in democratic accountability of large and complex modern government is an important challenge to legal and political theory.[42] The TELOS conference demonstrated once again the ingredients and urgency of the problem. It will take more conferences, and more books like this, to provide the solutions appropriate to the differing systems of government operating in different countries.

The future

Consideration of these remarks will demonstrate, even to the most sceptical, the variety of the issues raised at the TELOS conference, the importance of the topics considered and the danger of doing nothing to envisage, and carry forward, the efficient regulation of technology where that course is judged beneficial and necessary.

TELOS will doubtless go on to establish a network amongst those lawyers and others who are interested in the developments of technology of special relevance to the law and concerned about the potential democratic deficit identified during the deliberations.

Future conferences will need to broaden the scope of the technologies addressed, so that they include participants with expertise in nuclear technology, the technologies of energy and global climate change and of explorations of the biosphere and outer space. They will need to widen the participation from other parts of the world, including Russia and India, both countries of significance because of their technological capacity. Participants from poorer countries will be essential so as to reflect the diversity of humanity.

There will also be a need to deepen the examination of the law so as to include case studies of effective, as well as ineffective attempts to regulate technology by municipal law in addition to those attempts that are now emerging from international agencies designed to address global technology on a trans-border basis. Finally, it will be necessary to extend the fields of expertise of participants. The involvement of political philosophers, of persons who sometimes advocate more vigorous regulation, of civil society organisations, law reformers, politicians and legislative drafters would enlarge the pool of expertise in essential fields.

---

41 Issues considered in R v Quintaralle (on behalf of Reproductive Ethics) v Human Fertilisation and Embryology Authority [2005] UKHL 28 at [25]; compare with Brownsword, (nd: 20).
42 ibid.

The TELOS conference demonstrated that regulating technologies is not a matter appropriate to purely verbal analysis of the traditional legal kind. We cannot find the way ahead by simply reading the judicial reasoning of our predecessors, however learned they may have been. In default of more effective solutions, the common law system authorises judges to fill the gaps left by lawmakers.[43] Sometimes this is necessary. But a more coherent solution is desirable. TELOS has opened a dialogue as to how that solution may be offered.

A great judge, and one of my predecessors in the High Court of Australia, Justice Windeyer, once declared of the relationship between law and medical technology, that the law generally marches in the rear and limping a little.[44] Windeyer was a soldier as well as a judge. He knew what he was talking about when he used this metaphor. In the years since he offered his description the gap that he discerned has widened. The institutional problem has deepened. That is why TELOS is so important. It is why the subject matters here examined concern nothing less than the future of law itself.

References

Acheson, D, 1969, *Present at the creation: my years at the state department*, WW Norton, Inc, 1969.

Andorno, R, 2004, 'The precautionary principle: a new legal standard for a technological age', *Journal of International Biotechnology Law*, vol 1, pp 11–19.

——, 2007, 'Global bioethics at UNESCO: in defence of the Universal declaration on bioethics and human rights', *Journal of Medical Ethics*, vol 33, pp 150–54.

Australia, House of Representatives, 2006, *Debates*, 6 December.

Australia, Senate, 2006, *Debates*, 7 November.

Australian, 2006, 'Let the debate begin: Australia should lead, not lag, in regenerative medicine', *Australian*, 7 August, p 15.

Australian Government, 2005, *Legislation review: Prohibition of Human Cloning Act 2002 and the Research Involving Human Embryos Act 2002*. Reports, Canberra.

---

43   Recent illustrations include judicial decisions in cases of 'wrongful birth' and 'wrongful life'. See eg *Cattanach v Melchoir* (2003) 215 CLR 1; ([2003] HCA 38) and *Harriton v Stevens* (2006) 226 CLR 52; [(2006)] HCA 15); cf *McKay v Essex Area Health Authority* [1983] QB 1166; *Gleitman v Gosgrove* 227 Atl Rep (2d) 689 (1967) and *Curlender v Bio-Science Laboratories* 165 Cal Rptr 477 (1960).
44   *Mount Isa Mines Ltd v Pusey* (1970) 125 CLR 383 at 395.

Australian Law Reform Commission (ALRC), 1977, *Human tissue transplants*, report no 7, 1977.

——, 1979 (a), *Unfair publication: defamation and privacy*, report no 11, Australian Government Publishing Service, Canberra.

——, 2003, *Essentially yours: the regulation of human genetic information in Australia*, report no 96.

——, 1979 (b), *Privacy and the Census*, report no 12, Australian Government Publishing Service, Canberra.

——, 1983, *Privacy*, report no 22, Australian Government Publishing Service, Canberra.

Brennan, D, 2006, 'What can it mean "to prevent or inhibit the infringement of copyright"?: — a critique of Stevens v Sony', *Australian Intellectual Property Journal*, vol 17, pp 81–98.

Brownsword, R, nd, 'Interpretive re-connection, the reproductive revolution and the rule of law', unpublished, 20 pp

Carr, B, 2006, 'Age-old objections must not be allowed to delay this revolution', *Sydney Morning Herald*, 25 July, p 13.

Chalmers, D, 2005, 'Science, medicine and health in the work of the Australian Law Reform Commission', in D Weisbrot & B Opeskin (eds), *The promise of law reform*, Federation Press, pp 374–87.

Cooper, D, 2006, 'The Lockhart review: where now for Australia?', *Journal of Law and Medicine*, vol 14, pp 27–44.

de Zwart, M, 2007, 'Technological enclosure of copyright: the end of fair dealing?', *Australian Intellectual Property Journal*, vol 18, pp 7–38.

Finkel, B & Cannold, L, 2006, 'Day for stem cells and the hope of finding cures', *Sydney Morning Herald*, 7 August, p 9.

Fitzgerald, B, 2005, 'The PlayStation mod chip: a technological guarantee of the digital consumer's liberty or copyright menace/circumvention device?', *Media and Arts Law Review*, vol 10, pp 85–98.

Karpin, I, 2006, 'The uncanny embryos: legal limits to human reproduction without women', *Sydney Law Review*, vol 28, pp 599–623.

Kirby, MD, 2001, 'Intellectual property and the human genome', *Australian Intellectual Property Journal*, vol 12, pp 61–81.

——, 2006, 'Law reform and human rights — Scarman's great legacy', *Legal Studies*, vol 26, 2006, pp 449–74.

Macintosh, KL, 2005, 'Human clones and international human rights', University of Technology, *Sydney Review*, vol 7, pp 134–56.

Martinez, J, 2003, 'Towards an international judicial system', *Stanford Law Review*, vol 56, pp 429–529.

Mason, AF, 1971, 'Law reform in Australia', *Federal Law Review*, vol 4, p 197.

Murphy, T (ed), 2009, *New technologies and human rights*, OUP.

National Commission for the Protection of Human Subjects of Biomedical and Behavioural Research, 1975, *Report and recommendations: research on the fetus*, US Department of Health, Education and Welfare, Bethesda, Maryland.

New Zealand. *Royal Commission on Contraception, Sterilisation and Abortion*, 1977, Contraception, sterilisation and abortion in New Zealand, Government Printer, Wellington.

Organisation for Economic Cooperation and Development, 1980, *Guidelines on the protection of privacy and transborder data flows*, Paris.

Peel, J, 1972, *Code of practice on the use of foetuses and foetal material for research*, HMSO, London.

Plummer, D & Irwin, L, 2006, 'Grassroots activities, national initiatives and HIV prevention: clues to explain Australia's dramatic early success in controlling the HIV epidemic', *International Journal of STD and AIDS*, vol 17, pp 787–93.

Robertson, JA, 2001, 'Why human reproductive cloning should not in all cases be prohibited', *Legislation and Public Policy*, vol 4, pp 35–43.

Shikai, YM, 2002, 'Don't be swept away by mass hysteria: the benefits of human reproductive cloning and its future', *Southwestern University Law Review*, vol 33, pp 259–84.

Stobbs, N, 2006, 'Lockhart review into human cloning and research involving human embryo — closing the gap', *Queensland Lawyer*, vol 26, pp 247–51.

Sydney Morning Herald (SMH), 2006, 'A greater moralilty at stake on the decision of stem-cells research', *SMH*, 14 August, p 11.

Woodruff, MFA, 1968, *The transplantation of tissues and organs*, Chas Thomas, Illinois.

Section II

Practitioners' perspectives

An initiation into ICT[1] professionalism

The previous section emphasised problems in adequately regulating new technologies, particularly in the information and communications technology (ICT) industry, and the demands that this places on ICT practitioners to behave professionally. The three papers in this section are all based on the industry experiences of ICT practitioners and contain important reflections on the industry. As such, they are not typical academic papers. Rather, they give insights into how a number of thoughtful practitioners view their work and professionalism in the industry. In many occupations a common way of learning, or being initiated into that occupation, is through apprenticeship or an extended period of work experience often undertaken during a course of study. The initiates learn from the experts. This section can be taken as proposing a kind of apprenticeship approach to learning, not so much the content of ICT but the attitudes and concerns that experienced practitioners have. It is primarily a matter of seeing what these practitioners think, particularly about professionalism in the industry, and using these insights to discover more about the industry.

The first two chapters focus on professionalism and the third provides a different perspective to ethical concerns in ICT. In the first paper, John Ridge, a former president of the Australian Computer Society (ACS), discusses the fact that, generally, ICT is not seen as a profession by society at large nor, more surprisingly, by ICT practitioners themselves. This he sees as a problem because ICT is so important to almost all industries and for the prosperity of Australia, and 'professionalism is absolutely fundamental to the effect practice of ICT'. The core of the problem, he believes, is that the vast majority of ICT practitioners do not see ICT as a profession. That they do not is borne out by low memberships of relevant professional bodies such as the ACS (relative to the total numbers working the industry), by the lack of any certification or licensing of ICT practitioners, and by the view of many that professional organisations are elitist and exclusive. This last point is true, but a profession — by its very nature — is elitist and exclusive. Most of us are happy for medical professional bodies to be elitist. Ridge believes that the move to ICT being accepted as a profession by society must begin from within the industry. The practitioners must start to see themselves as part of a profession, something he, as ACS president at the time, hoped would happen when the ACS was admitted to the Australian Council of Professions (ACP), now Professions Australia, in 2002.

---

1   While we use the more common ICT, some authors prefer the older IT and we have respected their wishes in their contributions.

Neville Holmes, in the second paper, while an experienced ICT academic, writes more from a practitioner's perspective. He does not so much discuss whether or not ICT is a profession but focuses on what he believes are special responsibilities of the profession. He makes two main points. First, he thinks that programming should be seen as a technical skill and not part of the profession of software engineering and, second, he believes that professionalism needs to be taken beyond software engineering to computing in society in general. ICT professionals should focus on people, not on the technology. He sees three important ethical responsibilities for computing professionals. The first is to convince people that technology is not responsible for problems, people are. Blaming computers has become common but is clearly just a simple, and pernicious, way of avoiding responsibility. Second, and related to this, the profession needs to redefine itself in the light of developments that have taken place over the last 10–20 years, where computers have become ubiquitous in society. And, finally, the professions must take more responsibility for the way that ICT is used in society. The technology should be used for more than just making things cheaper and easier.

Mark Haughey, a long time practitioner in the public sector, in the final paper of the section, suggests a different way of considering ethical problems in ICT, informed by his professional experience rather than the academic literature. Instead of concentrating on general social issues he focuses on the development of information technology (IT) systems, taking as the starting point *specification*, *time* and *resources*. There is a lack of certainty in the relationships between these 'three potential moving surfaces.' For example, a lack of understanding of the specification can lead to problems with the project time frame and with cost; a time frame that is too short can lead to cutting corners and it is at this point that unethical practices arise. Haughey gives a variety of suggestions to minimise the potential problems including better project management, breaking the project into manageable pieces, the management of expectations and more engagement between the players.

Biographies

Mark Haughey has held a number of senior IT executive roles with Australian Government agencies, large and small.

Since graduating from the University of Adelaide in the late 1970s, Mark has worked exclusively in IT. He has extensive experience in technical support, quality assurance, procurement, service management, and governance.

In 2007 he became the chief information officer for the Workplace Authority and later with the Fair Work Ombudsman. A major achievement in this time was the creation of an in house application development and maintenance capability.

Mark has recently taken on the position of executive director IT projects with the Fair Work Ombudsman.

Neville Holmes worked as a patent examiner for two years, as a systems engineer with IBM Australia for 30 years, then as a lecturer at the University of Tasmania for 12 years, before shifting back to the mainland. He was a founding member of the Victorian Computer Society and has been a member of the editorial board and a columnist for *Computer*, the house magazine of the IEEE Computer Society, since 2000.

John Ridge AM is executive director of the ACS Foundation, a position he has held since 2005. From its launch in August 2001 until late 2004 he was the chairman of the foundation.

He was previously president of the Australian Computer Society, the professional body for ICT in Australia, in 2000 and 2001. Since first becoming involved in the ACS in 1992, he has held a number of offices at both the branch and national levels. He was awarded the Order of Australia in 2007 for his contribution to the ICT industry and education.

An initiative of the ACS, the ACS Foundation was established to encourage both private and public sponsorship of ICT education through scholarships and research projects. It has raised in excess of $35 million and awarded over 3500 scholarships.

2. The maturing of a profession

John Ridge AM
Australian Computer Society (ACS) Foundation

There is a significant difference between a person being considered to be 'professional' in their approach to conducting business, whatever that business may be, and a person being part of a profession and therefore considered to be a professional. This difference has created enormous misunderstanding and confusion within the information and communications technology (ICT) sector, and hampered its progress towards being recognised as a profession. Being part of a profession is a vocation requiring knowledge of some department of learning or science, for example, medicine, law, engineering, architecture, accountancy and, more recently, ICT.

Technology, and particularly ICT, more than anything else in recent times, has forever changed our lives and will continue to do so at an ever-increasing rate. Today, anyone with an Internet connection can access overwhelming amounts of information and enjoy instant, global communications. Technology is also the primary enabler and driver of growth in just about every other industry. Without a vibrant, dynamic and innovative ICT sector, Australia will be unable to achieve world-leading innovation in these other industries. ICT is driving the significant innovation and advances in medicine, science and mining, amongst other fields. Without ICT the mapping of the human genome would not have been possible. ICT has an important and positive influence on the economy and on productivity, where it is responsible for around 17 per cent of productivity gains. Growth in ICT has an economic multiplier factor of between five and 10 times in other associated areas. This means that in terms of both dollars invested and employment in ICT, there is a factor of five to 10 times the benefit flowing into other parts of the economy.

Yet, despite the fundamental importance of ICT to so many other industries and to the economy, very few know or recognise that ICT has the status of a profession and that ICT professionals belong to one of the learned or skilled professions. In January 2000, the Australian Computer Society (ACS) was admitted to the Australian Council of Professions — since renamed Professions Australia — as the membership body formally representing the ICT profession. Professions Australia is the peak body for all professions in Australia and includes representatives of medicine, law, accounting, engineering and architecture, amongst other professions. The ACS's membership was the culmination of more than five years work by a committed team, and it meant that ICT had finally come of age! The final decision was made by the 30 or so other professional associations represented by Professions Australia. This recognised ICT as a profession, making the ACS the first computer society in the world to achieve this status in its own right. As president of the ACS at that time, I was privileged to share the excitement at this achievement.

The ACS is one of a number of professional bodies for ICT that offers membership to ICT practitioners. Obtaining the status of a profession requires professional bodies to set suitable standards of knowledge and codes of conduct. Professionalism is absolutely fundamental to the effective practice of ICT.

The ACS is involved in the development and monitoring of educational programs and accredits ICT courses that meet the standard of the ACS Professional Standards Board (Core) Body of Knowledge (CBOK). It has set standards of knowledge and experience for professionals and has codes of ethics, conduct and professional practice.

Any organisation would want its ICT staff to be professional in their practice, and to be recognised and accepted as being professionals. The benefits to organisations and society include practitioners having an adequate level of knowledge, providing a higher level of products and services, and abiding by professional standards and codes of ethics. It would be plausible to think that organisations, as a way of guaranteeing professionalism, would seek to employ ICT practitioners who are part of the profession and members of a professional body.

One problem is that membership of a professional body is not a prerequisite for ICT practitioners to practice, unlike other professions such as law or accounting. In most of these professions, membership of a professional body is inextricably linked to the ability to practice, either in reality or by common perception and, in some cases, the loss of that professional membership results in the loss of the right to practice. The absence of this professional obligation in ICT can result in practitioners employing haphazard levels of knowledge, with inadequate understanding of codes of ethics and professional practice.

To be clear, as with many other disciplines there is a 'full spectrum' of ICT practitioners, ranging from people who have relevant university degrees and a minimum level of experience, through TAFE and industry certifications, to those who have gained their skills from on-the-job training and experience. The role of the professional body is to establish the standard criteria to differentiate a practitioner from a professional.

The ACS currently has more than 22,000 members, which, on a per capita basis, places it amongst the top two or three computer societies in the world. Approximately 20 per cent of that number would be practitioners, and the remainder would be professionals. Depending on the data used to calculate the number of ICT professionals in Australia, however, that membership represents approximately 6.5 per cent of the total. So, it is reasonable to conclude that only a small proportion of ICT practitioners understand or recognise the value of belonging to a professional body. Compare this to other professions in which almost 100 per cent of practitioners belong to their appropriate professional body. Changing this attitude will require the combined approach and resources of the whole ICT industry — employers, practitioners, academics, government and industry associations.

Another problem, which may be due to the limited membership of the relevant professional body, is that ICT practitioners and society in general do not regard ICT as a profession. ICT's recognition as a profession in 2000 did not generate significant publicity and ICT's public profile remains slight.

The ICT industry needs to promote its status as a profession. It needs to promote the value of ICT and how technology has changed and will continue to change the practice of other industries. And it needs to promote professionalism in the ICT industry in general. Changing attitudes and public perception will have beneficial effects on enrolments of students in ICT and the quality of those students.

I was excited when the ACS was admitted to Professions Australia. It meant that ICT had come of age as a profession and was recognised as such. Parents who want a professional career for their children could now consider ICT as a legitimate choice. More importantly, ICT practitioners would be proud to be recognised as being part of a profession, joining their professional body, and adhering to a recognised code of conduct and ethics.

Jo Coldwell in 'Professional ethics and responsibilities' considers this issue in some depth. She concludes that ICT is, 'an immature profession which will develop into a fully recognised one as governments and employers recognise the benefits of a registration process that becomes an expected prerequisite to employment within the industry' (Coldwell, 2008). Using the example of an architect, Coldwell discusses three components that are required for a person to become registered, which include: demonstrated meeting of the academic requirements; completion of two years practical experience; and, completion of a professional practice component, which focuses on legal, contractual and business matters. She also talks about the eight characteristics of the infrastructure surrounding and supporting a profession such as architecture. Six of those eight components currently exist for ICT and are met by the ACS. The remaining two components pertain to licensing or registration. There are currently no requirements for ICT practitioners to be licensed or registered in Australia (or even to be members of a professional body) in order for them to practice.

The ICT profession or industry may not yet be ready to implement a licensing or registration process. It should start, however, with professional membership as a requirement to practice as a first step. As membership becomes more pervasive, widespread and accepted, registration may well be the next logical step.

Recent surveys show that a majority of ICT practitioners do not truly understand, recognise or value being a part of a profession or holding professional membership. This attitude stems from a lack of understanding of what being a part of a profession represents. Most practitioners would consider themselves as being professional and seek to act professionally in their day-to-day activities, but they may not necessarily consider themselves as being a part of a profession, or recognise ICT as a profession.

Organisations, like the ACS, need to do more to make the public, current ICT practitioners and students aware that ICT is a recognised profession. This can be achieved through extensive media coverage. As practitioners and professionals, we need to promote the view that ICT is a profession and, as such, it requires members of the profession to belong to a professional body.

The standard for the Core Body of Knowledge, which underpins professional accreditation, must be driven by our industry in collaboration with academia. We need to speak with a single voice and be championed by our ICT executives who are able to bring local and global perspectives to bear on the issue. Imagine how much more effective this would be than the current system, where there are separate advisory groups (advising, for example, on issues related to curricula and skills relevant to industry) at each of the 38 universities performing a similar function independently. Consider if we all worked together as one how efficiently we could effect change and improve productivity. Course curricula must be meaningful to business and industry requirements, and training and professional development needs to be nationally consistent.

The longer the industry takes to implement a campaign to make people aware that ICT is a recognised profession, the longer it will take to attract high-calibre students to the industry — an industry on which all other industries rely. The maturing of the ICT profession now needs to be widely communicated.

Reference

Coldwell, J, 2008, 'Professional ethics and responsibilities', in D McDermid (ed), *Ethics in ICT: an Australian perspective*, Pearson Education Australia, Frenchs Forest, NSW, pp 277–302.

3. Some ethical imperatives for the computing profession[1]

Neville Holmes

Introduction

This chapter focuses on the ethical responsibilities of the computing profession. Governance should be based on ethics and be imposed by authority. The ethical aspects of digital technology should, therefore, be understood before its governance can be expected to be effective. Also, to be respected, governance of the industry should be administered by a knowledgeable authority; that is, by a professional body encompassing the entirety of digital technology. This technology has been developed and exploited in a quagmire of commercialism and hyperbole in which intellectual property law and market dominance determine the direction of development much more than the nature of the technology available and capability of the technologists involved. For its ethics to be understood and its governance to be properly designed, a very broad viewpoint must be taken of digital technology. This essay is intended to take such a viewpoint.

Technology and ethics

It is nowadays almost trite to say that our world is the way it is because of technology. Everything we do — watching, eating, listening, travelling — is embedded in technology, very often digital technology. Ethically, such a claim is grossly misleading and, in principle, completely false. In a fundamental way, technology can carry no direct responsibility for anything.

Technology

Technology is something people use. Consider the following extract from the entry for *technology* in an authoritative dictionary:

1   This essay expands on my presentation to the EtGovICT2008 symposium, Canberra, 1–2 May 2008. It draws on essays that I have written since 2000 for a monthly column, 'The profession', which appeared for more than 10 years in Computer, the house journal of the IEEE Computer Society (<http://www.computer.org>). For educational or research purposes only, these essays are archived at <http://tinyurl.com/wnhrkyv>

a. A discourse or treatise on an art or arts; the scientific study of the practical or industrial arts

b. Practical arts collectively

c. A particular practical or industrial art

Oxford English Dictionary (*OED*) (2nd edition)

In a technological context, *art* has a personal meaning:

I. Skill; its display or application

OED

Thus, technology is, by definition, personal because skill is personal. Technology is practiced by technologists and applied industrially by engineers, though the distinction between the two is usually blurred. Products of technology are used by other people who are, thus, the indirect users of technology. People decide what technology is used for and must, therefore, take responsibility for the effects of such use.

Ethics

Responsibility might bring blame or praise, or something in between. The nature of this spectrum is the subject matter of *ethics*, that is, of

II. 2. The science of morals; the department of study concerned with the principles of human duty

4. In a wider sense: The whole field of moral science, including besides Ethics properly so called, the science of law whether civil, political, or international

OED

The adjective *moral* is defined as

a. Of or pertaining to character or disposition considered as good or bad, virtuous or vicious, of or pertaining to the distinction between right and wrong, or good and evil, in relation to the actions, volitions, or character of responsible beings; ethical

OED

The nub here is that morality, the focus of ethics, has to do with the entirely personal ideas of right and wrong, of good and evil. These ideas cannot be applied to technology, or to the machines or processes proceeding from the use of technology. Only users of technology, be they direct or indirect users, can be held responsible for any good or evil resulting from its use.

Anthropopathism

The problem with being responsible for evil or wrongdoing is the human desire to avoid the responsibility, at least in people with a normal social sensitivity. The complexity of technology seems to provide an excuse for a bad outcome arising out of its use. The driver of a car causing an accident might seek to blame the car itself for its imputed malfunction, or the other driver, or the road conditions. Malfunction might well be the fault of the manufacturer or designer or maintainer. The other driver might well be responsible. For all that, there is usually some level of responsibility that must be borne by the driver at fault.

With computers, it's a bit different. There isn't another driver, and the software and hardware manufacturers use complexity to avoid blame for any but the most blatant fault. So, the computer, its software, or both, get the blame for malfunction. When a business or government department using a complex computing system blunders, blaming the system seems to be standard practice. There is a distinction to be remembered between cause and responsibility.

The greatest social role of the computer is as a scapegoat. This possibility was foreseen by computing pioneers many decades ago. In developing an international standard vocabulary for the profession, they put the most important terms first, data and information,[2] and deliberately designed their definitions to bring out the contrast between people and technology.

> **Data.** A representation of facts or ideas in a formalized manner capable of being communicated or manipulated by some process.

> **Information.** In *automatic data processing* the meaning that a human assigns to data by means of the known conventions used in its representation.

> *IFIP–ICC Vocabulary of Information Processing 1966*

The distinction can be more simply made by saying that 'only people can process information; machines can only process data'.

The computing profession has a moral responsibility to discourage the use of anthropopathic language when discussing computers. One place to start would be with the adoption of the standard vocabulary, in particular with

---

2   These were originally first and third, with *signal* second, but are lately first and second.

respect to the most important terms: *data* and *information*. When faced with this specific suggestion, computing people typically aver that it's far too late because the media and the public at large have come to accept the two terms as interchangeable. This is no excuse. If the profession promoted and practiced the distinction within the profession, it would gradually be adopted by others, starting with technical journalists. Indeed, the profession could press publishers to adopt the distinction in their style guides, and to enforce its use by their authors.

Anthropopathic language is rife in the computing industry, far in excess of this example. The next worst example, perhaps even more socially damaging, is the reckless use of *intelligent* in respect of machinery. There are many reasons to condemn this usage. Intelligence is essentially, and despite the misleading name of IQ tests, a collection of social skills. Machines have, in themselves, nothing at all to do with sociality, and it is grossly unethical to suggest that they do. The ugly initialism AI should be dropped, together with *artificial intelligence*, and replaced by neutral terminology; perhaps the old term *algoristics* could be revived. When faced with this suggestion, computing people typically say that, really, some programs do exhibit aspects of intelligence. This excuse misses the point. The computing profession has a moral responsibility to maintain a distinction between people and machines, and for this reason professionals should avoid any suggestion that machines possess intelligence, artificial or not.

The computing profession

Urging the computing profession to take a stand on ethical issues would be all very well if the profession were well defined, generally respected, and widely recognised. Sadly, it is not. Until 20 years or so ago, the profession was in good shape. Digital computers were costly and used almost entirely within large organisations that employed well-trained people to run them. These people had the straightforward tasks of writing programs, either to process numbers and text in bulk in the case of business data processing, or to carry out complex numeric calculations in the case of scientific data processing. Computer societies were set up to promote professional recognition for workers in computing and these grew rapidly. So did tertiary courses in computer science, of academic orientation, and information systems, of business orientation. A growing proportion of people working with computers had these tertiary qualifications and saw fit to join their local computer societies.

When computers became cheap enough to be used by small enterprises, or by small groups within large enterprises, however, fewer of their users saw themselves as computing professionals requiring tertiary qualifications to practice their profession. Computer society memberships levelled off. When computers became cheap enough to be owned personally, they proliferated, their coupling over the Internet became common, and the population of computer users grew rapidly. Increasing use of digital chips in everyday devices made many more people computer users.

That computer society memberships and tertiary student enrolments have declined recently implies that the professional societies are not in a position to apply governance to the generality of workers in the computing industry, and that the very idea of professional computing needs revision. To get some idea of what revision might be appropriate, the role of digital technology in society needs to be examined.

Digital technology

Digital technology pervades our society, but then, from an overall point of view, it always has. Our social development has only been made possible by the development of digital technology. Understanding this depends on appreciating the basic nature of digital technology, and its contrast to analogue technology.

> **Digit**. 3. a. Each of the numerals below ten … expressed in the Arabic notation by one figure
>
> **Digital**. 4. Of, or pertaining to, or using digits; spec. applied to a computer which operates on data in the form of digits or similar discrete elements
>
> *OED*

Digital technology is based on being able to discriminate between a few discrete elements. Analogue technology is not. The abacus is a digital calculator; the slide rule is an analogue calculator.

Early telegraphy used digital technology to allow semaphores and Morse signals to reliably transmit messages over long distances because the encoding was limited to only a few discrete elements. More recently, telephony was an analogue technology transmitting an electrical signal, analogous to the original acoustic waveform, and thus allowing it to be regenerated, providing too much noise did not accumulate during the often many stages of transmission.

The impressive success of modern digital technology springs from its binary encoding since, having only the two distinct elements — conventionally named *zero* and *one* — allows large amounts of noise to be cleaned off a signal at every stage. Binary encoding also allows systematic redundancy to be added when heavy noise is expected so that the received binary signal can be corrected. More complex binary encodings, such as alphabets, are represented as binary digit (bit) combinations.

Language

The basis of human society is language, originally only spoken or sung. Vocal language depends on distinguishing between words common to the vocabularies of the speaker and the listener. To distinguish between words is relatively easy, despite the large number of different words in a typical vocabulary, because spoken words are encoded from a limited number of phonemes. In English, the word *hit* has three phonemes, *hint* four, the second word being distinguished by the extra phoneme. Spoken language is thus a digital technology, a skill to be displayed and applied by its users. Speech, or an equivalent such as signed language, is a skill necessary to the functioning of a complex society. While speech is built from a digital vocabulary, it also transmits non-digital information to a listener, information such as the identity and mood of the speaker. Such extra information is the lifeblood of professional acting and political oratory.

 Speech is transient. Written language was developed as a durable representation of speech to allow tyrants and bureaucrats to control people over a widespread area. More recently, written language has in many countries been made available to the general populace. Such languages are digital in being represented by combinations of relatively few discrete elements. There are many writing systems, though there are many more spoken languages. The elements of alphabetic systems, for example, in principle represent phonemes of the spoken language, though extra elements such as the decimal digits represent words, and punctuation symbols suggest non-digital content. Written language used administratively ideally expresses rules unambiguously. If written language is used personally it needs to have the nuances of its meaning added by the reader. Reading is a thoughtful and imaginative activity, of richness and variety resulting from the mood and experience of the reader.

The digital computers of 50 years ago were used for administration and research. Business and government used them on transaction records, and scientists and engineers used them for numerical computation. As they became more capable, in particular when they grew beyond a character set consisting mostly of digits and upper case letters, they found further uses. In the 1980s, the personal computer and its printer began replacing the typewriter, a digital machine that had held great social significance. Connecting personal computers to the Internet added significant scope to the use of written language through user-group and personal email messages and, later, through the World Wide Web of documents.

The computing profession

Over the last 20 years, the development of binary digital machinery and communication has spread beyond traditional computing. The old telephonic cycle of speech being encoded, transmitted, and decoded by medium of an analogue electrical signal has been reversed. Analogue representations of material such as sounds and images are now being directly encoded digitally for storage and transmission, before being used to recreate the original sounds and images in a close approximation of their original analogue representation. While this has had a profound social effect, discussed below, it has also had a very significant professional effect. It has led to the use of computers being spread throughout the formal professions, even to the control of theatrical performance.

Though they are helped by relatively well-recognised software engineers, professionals of many kinds are using digital technology without using other computing professionals. And, in business and government, management prefers to use digital technology to drive employees through well-defined routines by deploying large, general-purpose software packages that have been built and installed by software engineers. Thus, the computing profession is faced with two problems.

The first problem is that the software engineering profession is becoming well recognised and accepted. Sooner or later it will become a formal branch of engineering, and it is arguable that this is only being delayed by the peculiar refusal to acknowledge the need to distinguish the trade of programming and to employ technicians skilled in that trade in the same way that other branches of engineering employ appropriate technicians. When this happens, the computing profession will have to adapt in some way to the loss of what many in the profession see as their most important component.

The second problem is that neither the personal computer user, nor the professional computer user, sees a need for computing professionals beyond software engineering. This is leading to a very significant reduction in tertiary computing enrolments. A factor here is that secondary education is diverting much of its focus to developing skills in the use of computers, a mistaken diversion in my opinion. Consequently, graduates from secondary school see neither a need for further education in computing nor a prospect for getting rewarding employment as a graduate in computing or information systems.

These problems argue for a revision of the computing profession by the professional body aiming to administer that profession. In my opinion, the way forward must be to recognise and promote the trade of programming. This would free up computing professionals to focus on the people using computers rather than the computers themselves. After all, it is said to be a characteristic of learned professions that they look primarily to the interests of people, and are responsible for how their technology benefits those people. Such is the ethical aim of a profession. However the computing profession is reformed, its professional body cannot prescribe ethical behaviour and enforce that prescription until its professional scope is clearly and restrictively defined. Reform is thus an ethical imperative.

Computing and people

The development of human societies has long depended on the use of language of one kind or another. Linguistic communication has made change possible in both human relationships and human activities. Socially and ethically, change in human relationships is of prime significance, and modern digital technology is having its most important effect on these relationships. To understand what is happening to our society, the role of language in past changes needs to be understood. While these changes have been both varied and complex, there do seem to be significant generalities, which are especially evident when looking at present-day society.

Oral society

In oral societies, social relationships were necessarily established by conversation. Cooperative activities depended on debate, and social hierarchies came from some people in a community having the authority to issue commands to others. In some languages the social structure is built into the vocabulary, for example, with different pronouns used for people of higher social status and people of lower. Of special significance are names for family relationships because they declare important responsibilities. The family is by far the most significant component in oral societies. Childhood is used by families to ensure that the next generation is capable of sustaining the social structure.

Literate society

Written language has been, for almost all of its history, confined to a privileged administrative stratum of society. This is most clearly seen in the role of the book as the main material source of authority in early religious society and in bureaucratic society. Quite recently, the digital technology of printing and the burgeoning of urban society made popular books affordable, widespread literacy desirable, and general education practical. This development created a discourse both social and technical at a level above the family and village. Without this discourse our modern society would not have happened.

The development of technology of any kind depends on the accumulation of knowledge. To accumulate steadily, knowledge must be durably represented, examined widely, and extended incrementally. The context for the development both of technology and of prosperity is the enlargement of society. Going far beyond the farm, the village, and the town, thousands upon thousands of people must be able to live and work together.

There have been several factors underlying whatever success modern society may claim to have won. First, is the oral underpinning of social and occupational everyday life. This has been created by the society of the family followed by and supporting the society of the school. The further that education goes, the richer the social and occupational intercourse. Second, is the literacy and numeracy that underpins life in wider society, recreational life in associations and meetings, and vocational life in the workplace and office. Literacy and numeracy make such extraneous personal linkages possible. Third, is the widening of social contact fostered by the development of personal transport and communication. Interaction with a variety of people makes life interesting and develops sociality. Fourth, is the personal enrichment fostered by the enlargement of social and vocational life. A richer life is not only a matter of personal experience, but, perhaps more importantly, of self knowledge through contemplation and the need for social sensitivity.

Binary digital society

This review of the social background of modern society has an old-fashioned flavour, and would be probably be looked at askance by young people today. Does this invalidate it? Well, validity depends on ethical values, and these seem to have changed over recent years. I have become increasingly aware of this since I moved from a semirural location to an inner suburb of a regional city. Being able to walk to shop or dine out, or just for the pleasure of wandering through tree-lined streets, is a delight. But the delight is tempered by evidence of values that are alien and abhorrent to me.

A few examples, then. School students who have just bought a hamburger drop the paper bags at their feet; a rubbish bin that is located a couple of paces away remains unused. By late afternoon, the plaza outside the fast-food shop is littered with discarded food and wrappings. Footpaths and nature strips near shops and hotels are spread with shards of broken glass. The young driver of a loud, speeding car and his three young passengers give me a one-finger salute and shout obscenities at me when I look wonderingly in their direction as they corner in front of me, though their shouts are softer than the thumping in the car that passes for music.

But these are just personal examples of what seems to me to be a drastic change in values and ethics from those that applied in my own youth. These are more evident in things I read about so frequently in the newspapers — violence, murders, pornography, rape, binge drinking, dangerous driving, robberies — all too often being committed by adolescents. And internationally, …. Maybe the seeming prevalence of these distressing events is just a result of what the media chooses to publish, but this would be, in any case, a change in values. Maybe I'm just an old-fashioned dodderer with old-fashioned values. If so, then I think the old-fashioned values should be brought back.

My observations, however, suggest to me that the problem is deeper than that, and that society has been badly weakened by the use of modern binary digital technology. The tragically misnamed social media seem likely to continue this weakening, perhaps even accelerate it.

Binary technology now gives priority to sensation and perception. Old-fashioned literacy and numeracy gave priority to imagination and contemplation. Popular music is rhythmic to excite, old-time music is melodic to soothe and charm. Modern games involve the player in rampant conflict, old-time games like chess and bridge are social and thoughtful. Nowadays, idle time, which used to be spent reading a book or chatting, is spent in excited watching of a screen. In short, modern culture is based on the satisfaction of individual desires, whereas old-time culture was based on family and other group sociality.

A great many influences have effected this change, and most of these influences have some technological aspects. It could be argued, for example, that production technologies have made consumer goods cheaper and that cheaper transport has made them more varied. Then, popular media has promoted individual desires for the consumption of these goods. Such technologies aren't basically digital, but digital technology has played a significant part in their recent development, as well as greatly amplifying the variety of consumable items.

The ethical imperative

The goodness or badness of the dominance of consumerism in our modern society is an ethical judgement. The diminution of sociality, however, whatever its cause, cannot be considered a good thing. That the computing profession and its binary digital technology has a significant rôle in that diminution is certain, however uncertain the exact nature of that rôle may be. The profession, therefore, has an ethical imperative to work to reverse the diminution. Some possibilities are obvious.

Basic skills

Consider in today's adolescent culture the four factors given above as underlying the success of modern society. First is the oral underpinning of everyday life, starting in the family. This is declining. A current problem in early schooling is that too many children starting school cannot converse. They understand spoken language but they cannot use it themselves. Second, is the literary and numerary underpinning of recreational and vocational life in wider society. Literacy and numeracy are both declining, at least in 'developed' societies like Australia. Both the vocabulary and the arithmetic of school leavers have, on average, become greatly reduced. Third, is the widening of social contact through better personal transport and communication. Some commentators consider that social contact among many young people is being narrowed by developments such as the prevalence of single-parent and working-parents families, by protective measures such as driving children to school and removing social activities like physical education from the curriculum, and by gluing students to Internet terminals during school time. Fourth, is the personal enrichment coming from enlarged social and vocational life. Tragically, social and vocational life is not enlarging for young people, who spend more and more time playing with their computers rather than with their peers. Their vocational life is becoming poorer because employers prefer to use complex software to enable them to use more low-paid and easily replaced unskilled workers. Worse, young people are being given the idea that they don't need to extend their general knowledge because they can find out anything they want to know from the Web.

These comments reflect the opinions of an earlier generation. Regardless of how well founded or how subjective they are, the underlying problem is that many young people are being inculcated with a sensual priority, and are not being given the possibility of knowing and preferring a richer, contemplative culture. The irony is that digital technology could be used in family life and formal schooling to give the average child the possibility of a far richer cultural life than all but the most gifted present-day children. The method is by using computers to develop basic skills.

Automatic drill

Advanced skills and accomplishments depend on the use of basic skills. The more basic skills that a person has, the more advanced their accomplishments can be. Basic skills are those skills that are applied without conscious thought. There was a time when skills such as reading, spelling, grammar and mental arithmetic were the normal result of elementary schooling. Nowadays, there are complaints at high bureaucratic levels that English spelling is too difficult for school children to learn; certainly there is plenty of evidence in public notices that they aren't learning it. What is the problem?

The problem is that basic skills come from persistent drill and practice. Ask any music teacher or athletics trainer. Fifty years ago, drill and practice was standard in primary schooling, even though it must have been very boring for the teachers. Today, it is dismissed with disgust by teachers as 'drill and kill'. The shame of this is that the digital computer can be vastly more effective at administering drill than any human primary classroom teacher. Automatic drill is intensely personal. The software can adapt the stored training to individual talents and capabilities, to individual moods and interests, and even to overcome defects such as dyslexia. It can be used to report to a teacher on each student's progress so that social classroom activities can be designed and corrective activities can be imposed.

Automatic drill is extremely versatile. With speech recognition it can be used with very young children. The customary primary skills of reading, writing and arithmetic can be developed way beyond present-day possibilities and, for quite young pupils, in different languages and writing systems. Less usual skills can be developed, such as singing, recitation, elocution, reading and composing music, recognising tunes and instruments, history, local and international geography, and many, many others. With haptic interfaces, like Nintendo's Wii for example, basic physical skills such as musical instrument performance and sign language could be developed.

Automatic drill is potentially revolutionary. In education, it would change the rôle of the teacher from autocrat to helpmate, eliminating fixed classes (every student would develop their own spectrum of individual skills) and making traditional tests and periodic examinations pointless (the software continually records basic skill levels, and advanced skills would be encouraged differently for different students). Outside the education system, it would empower people to develop basic skills personally and swiftly, once they identified their need and prepared to commit their time to drill. The introduction of pervasive automatic drill into schools would, of course, need to be done in stages. But the difficulty is to get it done at all.

Teachers have been thoroughly taught that drill is inhuman and, in any case, they would need themselves to be re-educated and trained to best exploit this technology. Politicians and demagogues would need to be convinced, or bypassed, to generate acceptance of a change in approach to education. Nevertheless, it is ethically imperative that the computing profession takes the lead in getting automatic drill accepted and adopted. This not to impose a particular culture on society but to give people in our society the potential to extend themselves in whatever culture they choose to adopt. It is also imperative because it will extend the capability of society in almost every way, an extension that is increasingly important because of the threats and challenges of a global nature that are becoming evident.

Data standard

While there is a general ethical imperative to get automatic drill widely accepted and practised, there is a customary danger to be avoided. In the past, new technologies such as radio and television have been seen as of enormous potential in improving education and community standards. As it happens, the educational and community potential has been swamped by commercial and marketing needs and influence. The same danger looms over automatic drill. Videogames, often of a murderous flavour, dominate the computing field at present, even at the level of chip manufacture. While these sensually absorbing computer applications are spreading ever wider, some commercial interests see some profit to be made from adapting these to mind-training use. This threatens the same fate for educational automatic drill as radio and television — overwhelming use for marketing.

Software for automatic drill does, of course, need to be developed commercially. The computing profession, however, needs to press for this development to be versatile enough for teachers and parents to be able to develop and adapt their own drill packages, rather than being forced to accept whatever commercial enterprises see as most beneficial to their profits. The only way to do this is to establish, and to compel acceptance of, an international standard for the representation of the data used by the drill packages. This would mean that commercial interests, and open-source software developers, would be forced to develop software to best use the drill data that is being generated free from dependence on particular software packages.

The standard needs to define formats for two different categories of data: the subject matter data and the individual learner's data. These would necessarily be linked, but the subject matter data of a drill package would be general in nature and be incrementally improved, while the learner data would be generated through the use of the drill package by each learner.

Conclusion

The computing profession has some very important ethical responsibilities in our present day society. The first is to get binary digital technology seen as a technology that people use, and for which use people must bear the responsibility. Computing professionals should see themselves as bearing a special responsibility, and as needing to see that the public at large understands that technology itself is not responsible for problems caused by shortcomings of digital technology.

The second is to get the computing profession respected as a learned profession that is capable of improving the use of computers in society. This cannot be done now because, given that computers are so widely used, a clear and reasonable public image of the profession no longer exists. Such an image depends on the profession revising its scope and responsibilities.

The third is to see that the *social* benefits of binary digital technology are broadened and amplified, rather than the *commercial* and *political* benefits. The most important way to achieve this is to get the technology used in early education to increase the potential of students, rather than to make it cheaper for government and less demanding for teachers.

References

Holmes, WN, 2006, *Computers and People*, Wiley–IEEE Computer Society Press, Hoboken, New Jersey (ISBN-13: 978-0-470-00859-1, xxii+324pp).

Simpson, JA, Weiner, ESC (edd.), 1991, *The Compact Oxford English Dictionary, Second Edition*, Clarendon Press, Oxford (ISBN: 0-19-861258-3, xvii+2386pp).

Tootill, GC (ch.), 1968, *IFIP-ICC Vocabulary of Information Processing*, North-Holland Publishing Company, Amsterdam (xii+2088pp).

4. The uncertainty of ethics in IT[1]

Mark Haughey
Fair Work Ombudsman, Australian Government

Introduction

This paper[2] focuses on ethical issues that are unique to Information Technology (IT) practice, not 'ethics in the workplace' issues or 'sales ethics' issues. These latter areas, while very relevant to IT (for example monitoring the workplace or promoting a product), are generic and existed before IT. As a discipline or profession, IT derives from engineering to a degree (the common job title of hardware engineer or the software engineering concept) and, therefore, some of the ethical issues faced by engineering will translate to IT. This is particularly the case with the development of projects and there will be many an ethical issue (although probably not considered in that light) in getting a project completed on time, on budget and in scope.

Fear, uncertainty, doubt (FUD)

The IT industry, in common with many human endeavours (just think of the recent debate on climate change) is skilled at generating fear, uncertainty and doubt. While most prevalent in IT sales, especially when considering switching suppliers, it also exists in the IT workplace when it comes time to consider extending people's contracts. This is not unique to IT, although the 'lock-in' that exists with a vendor or reliance on specific skill sets (not only product or tool related, but often unique to the particular business utilisation of a product

---

1 I have no formal training in ethics. My majors were in pure mathematics and computing science as part of my Bachelor of Science (Mathematics) degree from the University of Adelaide in 1979. Since then, I have worked exclusively for the Australian Public Service in the field of IT, mainly in senior executive roles. While I consider myself a public servant first, being an IT professional comes a close second. As a public servant, I have attended ethics training courses, principally in support of procurement activities. While I have a general interest in the subject I want to stress that this paper is not about ethics as a discipline.

2 Disclaimer: This paper represents my own opinions. It does not represent my current or past employers; and it does not represent the views of the Australian Public Service or the Australian Government, past, present or future. This paper is based on my observations of the IT industry: firsthand, observed, read about and speculated on. As such, it has been referred to as 'a view from the trenches'; in which case, be prepared for a bit of mud.

or what has been built with a tool) makes them particularly acute. The reliance on a software engineer easily surpasses that placed on a skilled tradesperson or a medical specialist (you can always get a second medical opinion).

But, what of the behaviour of the software engineer? What ethical challenges do they face? Should they be providing skills transfer? This assumes that there is salary or staffing available to provide an understudy, and that the understudy has the same skill potential. This concept of skill transfer is not common in other professions or trades. The best that is offered is advice on how to avoid the problem in the future (how to conduct necessary maintenance or preventative medicine, perhaps). What the IT professional faces is the belief that they can somehow transfer their skills. This is not true for all recipients of IT services, particularly in the hardware area or packaged software, but, in the development and maintenance of application systems, it is generally accepted that in-house staff can support the resultant system. In the absence of in-house IT staff, the solution is to contract the developer to provide ongoing support. A key component of this transfer of knowledge is through documentation, which of course is one of the first things that is negatively impacted if there are pressures on the project.

Should the IT professional be promoting solutions that have a transition path, which can be supported by a lesser skill base? Often this was a business requirement to start with, but invalidated as complexities increase (and any successful system is sure to be extended in innovative ways and even unsuccessful systems will have a series of fixes applied to remedy shortcomings). As this tends to happen after the initial budget has been consumed, the upkeep of documentation and other maintenance artefacts will often suffer. This might not be a problem if the people supporting were involved in the development and if they have the support as their primary focus.

Permanence

What makes IT workers a permanent fixture in a business? Unlike an engineering project that builds something concrete, there is often a less tangible outcome resulting from an IT project, but a much more flexible product. When a tangible product does not work it can be a project in its own right to rectify the malfunction (for example, product recalls, or even a return to the drawing board). With IT, it is often a simple fix, often as easy as restarting the computer or downloading a patch.

Is this a question of maintenance? Apart from operation and administration, the maintenance of a software system requires at least the same level of skill that was used in developing it (especially given the issues that can exist around the system documentation). When enhancements are added to the maintenance work, the value of IT as part of the business quickly becomes apparent (in the way that elite sports teams will have their own medical support). The ability of any significant business to function, to enter new markets or deploy new products or services, to meet changes to government regulations or gain insight into its business from the data it has collected in doing business, is directly tied to its IT systems and the ability of its IT workers to safely deliver enhancements and maintenance. This reliance makes the application of fear, uncertainty and doubt so effective in the IT industry.

The move to commercial, off-the-shelf software lessens this, if the customisation trap can be avoided and the vendor and any underpinning technology is sound. Often, however, it is just a deferral of the challenge, which is only alleviated by IT becoming a commodity item. Indeed, this has happened across the IT industry for a variety of things, most obviously those items in the end-user consumer space where scale provides the volumes for commoditisation to thrive. But what happens when the software is unique to a business?

Precision

As with the precision of mathematics, in computing there is the ability to programmatically represent real-life systems. As such, IT is about provable facts: the system did exactly what it was programmed to do. There is no room for ethical dilemmas as there is no choice. Bugs aside, an IT system does what it was programmed to do. Given that premise, there is no ethical issue with an IT system: it is black and white. An IT system treats everyone with the same level of access in the same way. It behaves identically each time that the IT system is used. There are no 'What ought I to do?' conscience challenges. While I have no data on it, I have heard many IT people described as black and white as well. This polarity can lead to the problem of a finished IT system behaving in the way that it was specified to function, not in the way it was necessarily intended to operate.

Ethical problems certainly arise in the creation and the subsequent use of IT systems — this is where the non-IT people get involved. What ethical challenges, however, might be presented to the IT people during the creation of an IT system?

The journey

Once initiated, the creation of an IT system has three aspects:

- A set of specifications
- A timeframe
- A set of resources, funding or treasure

To use the analogy of a journey: destination, time of arrival, and transport. What is really wanted is the IT system, created according to the specification, delivered on time and under budget. This is where the uncertainty starts to creep in.

Heisenberg uncertainty principle

To quote Wikipedia 'the uncertainty is an effect caused not just by observation, but by any entanglement with the environment'. And, in building an IT system, we have three potential moving surfaces: specifications, time and resources. Closely observing one of these three in an IT development — the journey — can cause the other two to distort unpredictably. Just as a constantly whining 'are we there yet' can produce an apparently much longer journey, locking in the delivery date for an IT system can see the specifications not being met and the costs going up. It is also true that the specifications are only final when the system is delivered and that the ultimate documentation of the specifications is the system's code. Of course, a skilled IT workforce is required to be able to read the code.

The dilemma

This uncertainty gives rise to unethical practices:

- 'We don't understand the specifications or the timeframe but will just ask for lots of money in the hope we can work something out.'
- 'We can never meet the timeframe so let's cut some corners to come in on budget.'
- 'They want it now so it's going to cost them dearly.'
- 'For this amount of money, this is as good as it gets.'

There is nothing unique to IT in these practices; they would be faced by most construction projects. In taking short cuts, the risk is not that certain things will be incomplete (for example, the composition of helpful error messages), but that the absence of such components will be acceptable because of the assumption that they can be easily added later. Indeed, this may not be an issue at all, but something that was negotiated with the business and a joint decision made. But, if the decision process is obscured in the project details, and the programmer chooses to reuse code, for instance, in multiple parts of an application rather than create it as a callable subroutine, is this what they ought to have done? Given these challenges — and with IT projects there are many more beyond these fundamental risks — how do we guard against them?

Educate the observer

This was a strategy that was employed widely and, possibly, successfully early on, before everyone became an expert when they got a PC or their nephew went to college and became a whiz at churning out systems in some small scale development tool. What is needed is a level of IT-maturity in business: experience with similar-sized implementations and requirements; the possibility of having worked with the same development team before, using the same tools and methodologies; and, knowing what tolerances should be built into the funding to cater for reasonable potential delays.

While IT awareness in business has increased, the focus is more on the use of good project management methodologies that allow for informed and reasonable observation. Such an approach provides a more holistic view of the project, the ability to manage multiple developments comprising different teams and tools (project management providing the common language) and to allow for changes in the business people engaged with the project. Failures of course, still occur, for example, when:

- The (absent) project manager is told a story different to that which is unfolding: 'We ought to tell the manager we're running behind, but that will only require us to write reports on why and put us further behind so we'll say everything is okay and work a bit harder to catch up.'
- The methodology employed was unsuitable for the size of the development (sledgehammer to crack a walnut or, more frequently, a bike to take the extended family on an African safari): 'I just need an issues list.'
- Or, just plain misunderstandings or poor communication creates confusion and delays (standard human failings).

Chunking

A good protection against these failures is chunking: breaking a large activity down into more manageable pieces. To use the journey analogy, it is to have frequent rest breaks. Breaking up IT developments into phases is an art form. Often, it is sufficiently tricky simply to identify meaningful milestones. It can also make funding harder in a variety of ways (business might be looking for a big investment or conversely, having made a small investment, may not want to make further ones). It all comes back to the maturity of business. Have we bitten off more than we can chew? Is a system like this appropriate for an organisation of our size? What are other, similar businesses doing? How can I prove the concept without betting the business? IT practitioners can help, but are often at risk of being blinded by business enthusiasm, or being lost withing a big project or, and this is more likely, excited at the prospect of using some new technology.

Value for money

This leads to IT governance — the accounting for the business value of the money spent on IT. It also needs to cover live systems and not just those under development. As an aside, it is much easier to commission a new system than terminate one that has got into trouble; and, for some obscure reason, it often seems that it is far harder to actually turn a system off that is no longer in use.

But how is business to know if they will get value for money? It can be hard getting accurate costs for an IT system. Even when it goes live, there is no certainly that all costs have been identified, especially those associated with maintenance. This represents further uncertainty for businesses, especially given the pace of technology change. Is it possible to relate the IT cost to a business transaction? If the cost of the IT system stays static, wild swings in business volumes can alter this figure dramatically — the dollar per transaction approach.

Expectation management

The business has requested an IT system and has a range of expectations about the system and its delivery. Many of these expectations will be associated with risks, but often hidden by assumptions, some obvious and some unstated.

In order to manage expectations, effective communication between the IT professional and the business is vital: it must be timely, relevant and regular.

There needs to be attention to the medium used for the message — how ought news to be conveyed? Face to face or a lengthy written report with the bad news hiding somewhere near the last page? Or, perhaps, a post-it note left on the door (or on the minute outcomes in the file so it can be removed later to preserve reputations)? Again, should I try email, SMS or a phone call? This all requires careful thought.

The reporting process needs to have been established in advance — a regular communication or governance arrangement put in place. It has to fit the culture of the organisation and the personality of the managers (it is no good sending the boss an email if they never check their inbox). The message needs to be composed in the language of the business, which is not easy if the IT shop is isolated from the business. Along with effective communication, the next two items also help with expectation management.

Engagement

Engagement by IT professionals is vital — they must be part of the business. Most businesses require that individuals perform their tasks without harming anyone in the process. Engagement shouldn't be that hard, although, to be fair, communicating with some people can be a challenge. Where the systems required are central to the business and are the things that make that business unique, the developers need to be associated with the business because, for one reason, the system will encode much of the intellectual property of the business processes and procedures. Heavily outsourced systems create a distance between the developers and the business, where one should not exist.

Being part of the business removes some of the uncertainty from when it is time just to say 'no' to an unreasonable or poorly conceived request, versus just getting on with the job. Being part of the business lessens the 'little knowledge of the business' problem; it at least alerts the developers to potential scalability requirements and it also allows an informed decision on how long to go on with a struggling project or system (unlike building a bridge, there are often simple fixes that can be made to an IT system to make two disparate parts meet).

Packaging

Unique challenges in IT relate to its often-intangible nature. This is why there tends to be a lot of packaging around off-the-shelf software. The old programs that I wrote using punched cards used to look impressive, but even that could be misleading if it was padded out with embedded comments or, more worryingly, highly embedded IF THEN statements when a single CASE statement would have been more elegant. The other challenge is the rapid creation of the IT industry and profession and the extent to which it now touches everyday essential activities. These facts further cause expectations to overflow and put pressure on people when they are thinking of what they ought to do.

Conclusion

What is to be made of my journey? As a chief information officer, with over 30-years experience in IT, working for what I would say is a very ethically aware employer — the Australian Public Service — I have seen that the demands on IT have become increasingly complex. Complexity breeds ethical challenges and IT is just as much at risk of ethical issues as any other profession or human endeavour. What can be a right or fair answer can often distort under a quest for accuracy — the uncertainty principle. To reply to a question with 'Yes and No' might be truthful, but it is generally not very helpful. It might not even be fair.

Section III

Professionalism

Professions, professionals, and professionalism

In this section, the concepts of professions, professionals and professionalism will be examined in more detail.

Someone can be a professional in one or both of two different senses, one broad and the other narrow. Likewise, 'profession' can have a broad or a narrow sense. In the broad sense, a profession is anything that is done for a living: a professional golfer or carpenter is one who makes a living out of playing golf or doing carpentry, as distinct from the amateur, who seeks no such reward but pursues the activity just for the love of it. In the narrow sense, a profession — such as law or medicine — is said to be a particular kind of occupation that satisfies a number of criteria (for typical sets of these criteria, see the following papers by Clive Boughton and by John Weckert and Douglas Adeney). The notion of a professional can have two different senses even when used in the narrow sense. A professional might be a member of a profession or a professional might be someone who has a particular attitude to his or her work. In the extract below by George Bernard Shaw, the two doctors talking are professionals in that they are both members of the medical profession, but neither possesses the attitude expected of professionals, so are not professionals in that sense.

Schutzmacher has dropped in on Ridgeon and is telling him how he, Schutzmacher, managed to make so much money as a doctor:

SCHUTZMACHER. Oh, in my case the secret was simple enough, though I suppose I should have got into trouble if it had attracted any notice. And I'm afraid that you will think it rather infra dig.

RIDGEON. Oh, I have an open mind. What was the secret?

SCHUTZMACHER. Well, the secret was just two words.

RIDGEON. Not Consultation Free, was it?

SCHUTZMACHER. [*shocked*] No, no. Really!

…

SCHUTZMACHER. My two words were simply Cure guaranteed!

RIDGEON. [*admiring*] Cure Guaranteed!

SCHUTZMACHER. Guaranteed. After all, that's what everybody wants from a doctor, isn't it?

...

RIDGEON. And the guaranteed proved sound nine times out of ten, eh?

SCHUTZMACHER. [*rather hurt at so moderate an estimate*] Oh, much oftener than that. You see, most people get well all right if they are careful and you give them a little sensible advice ...

(George Bernard Shaw, *The doctor's dilemma*, 1906)

Both of these members of the medical profession are behaving in an unseemly manner. They rate financial gain more highly than caring about, or being truthful with, their patients.

The first two chapters in this section focus on the distinction between the notions of a *professional* and that of a *profession*, although the perspectives are rather different. For Boughton, an information and communications technology (ICT) academic with vast industry experience, a professional satisfies criteria very similar to commonly accepted criteria for a profession. A profession, in his view, is 'usually exemplified in the form of an association/society/body' and has responsibilities to the professionals who are its members. Core components of a profession are its code of ethics and body of knowledge (BOK); both of which are essential in maintaining professional behaviour. Boughton explores BOKs in ICT and concludes that more work needs to be done in this area before ICT can be a real profession. ICT also still lacks a professional body that can enforce professional behaviour. He suggests that perhaps it is currently a 'semi-profession'. Weckert and Adeney, both philosophers, also distinguish *professional* from *profession* but, unlike Boughton, do not see the traditional concept of a profession as useful: what is important is that an ICT professional has a *professional attitude*, such an attitude being defined as 'taking pride in one's work, trying to do it as well as possible and actively considering ways in which it might be done better, looking at it in the broader context of society as a whole, and taking responsibility for what does.' On this account, professionalism is broader than on Boughton's account. For him, a professional is more closely related to a traditional profession than for Weckert and Adeney. For them, anyone can be a professional if they have the right attitude, whether they be a doctor, plumber or fruit picker. Like Boughton, they also doubt that ICT is a profession but not because it is as yet not mature enough but, rather, because there are no professions at all in any interesting or useful sense.

In the third paper, Richard Volkman, also a philosopher, presents a different perspective on ICT professionalism, one that is related to professional attitudes and the account of professionalism outlined by Boughton. Volkman argues that virtue ethics provides a much better foundation for ethics in computing than do either of the more common approaches of utilitarianism or deontology.

Utilitarianism, put simply, is the theory that the morality of actions is determined purely by their consequences, particularly to the extent that they create more happiness than the alternatives. Deontological accounts specify morality in terms of rules. Virtue ethics focuses on the character of the person and is interested in developing a 'good person'. Given the problems in regulating ICT, as the Hon Michael Kirby points out in his article, ICT professionals who behave well without external regulations are essential, if the industry is to maximise its benefit to society. Virtue ethics as presented by Volkman is an attractive option.

Finally in this section, Catherine Flick, a lecturer in computing and social responsibility, examines end user licence agreements (EULAs) in what is, in effect, a case study of responsibility in the ICT industry. It was noted earlier that responsibility is an important part of professionalism and Flick shows convincingly that serious problems exist in this area in ICT. Neither the content nor the presentation of standard EULAs, are, in most cases, conducive to users giving informed consent to the software that they are installing. This does not reflect well on the industry's sense of responsibility to its customers. The situation, however, is not all bleak. She also examines a variety of suggested solutions to remedy the situation: this shows that many in the industry are concerned with the current situation. Lastly, Flick outlines her own solution, which is based on a different account of informed consent from the one presupposed by most EULAs.

Reference

Shaw, George Bernard, *The doctor's dilemma*, 1906.

Biographies

Dr Douglas Adeney (MA Monash, PhD St Andrews) is a senior fellow in philosophy at the University of Melbourne. He taught for a number of years at Melbourne State College, and then at the University of Melbourne, retiring in 2007. He currently teaches short courses at the Centre for Adult Education in Melbourne, and is involved in various ways in the development and delivery of the VCE Philosophy course. Among his publications is *Computer and information ethics* (co-written with John Weckert, Greenwood, 1997).

Dr Clive Boughton has enjoyed a diverse career ranging from soil physics research and engineering, water pollution engineering, superconducting, plasma and molecular physics research and, ultimately, software engineering as developer, expert advisor, manager, educator and researcher.

His background has enabled him to observe and experience the affects of both professional and non-professional behaviour in many forms. Today, he believes that there is greater focus on defining and promoting professionalism, but that consistency and dedication to improvement is lacking. As chair of Australian Safety Critical Systems Association, Clive sees competence and professionalism as essential for people developing safety-related systems.

Dr Catherine Flick is a lecturer in computing and social responsibility at the Centre for Computing and Social Responsibility, De Montfort University, United Kingdom. Her research interests lie in the area of ethics in ICTs, particularly in social media, user experiences, and emerging technologies. Apart from her research into informed consent, some of her previous work has been in the areas of online child protection, ethical governance of emerging technologies, anonymous computing, and trusted computing. She has a computer science and computer ethics background, and gets excited about developing ethical, value-centred technologies from the ground up. She is an active member of the British Computing Society and the International Federation of Information Processing Working Group 9.2 and SIG 9.2.2.

Dr Richard Volkman is professor of philosophy at Southern Connecticut State University and associate director of the Research Center on Computing and Society. Dr Volkman's research evaluates the impact of information technologies on our abilities to lead the good life. Since the relevant information is decentralised, tacit, and local, this project involves articulating individualist moral and political philosophy for the information age and addressing associated issues, such as intellectual property, identity, privacy, and digital culture.

Prof John Weckert is professorial fellow at the Centre for Applied Philosophy and Public Ethics (CAPPE) and professor of computer ethics in the School of Humanities and Social Sciences at Charles Sturt University. He is editor-in-chief of the Springer journal *Nanoethics: Ethics for Technologies that Converge at the Nanoscale*. He is manager of the CAPPE program in technology and has published widely in the ethics of ICT and more recently on ethical issues in nanotechnology. His previous appointment was professor of information technology at Charles Sturt University and recently he was a visiting professor in philosophy at Dartmouth College and an Erasmus Mundus scholar in Norway and Sweden.

5. What is an ICT professional anyway?[1]

Clive Boughton
The Australian National University

Introduction

The intention of this paper is to instigate ongoing discussion surrounding the connected topics of information and communications technology (ICT) professionalism and the ICT profession. Part of that discussion needs to include suggestions for 'the way forward' for the development and recognition of an ICT professional body and the way it should govern/support/protect the professionals within it.

I want to start off by describing some of the types of people that I have met/ observed during my time in the systems/software industry and also within academia. The type classifications are mine, are not formal, and carry no essential meaning other than what I have opted to give them. They are like a loosely typed language: they only possess meaning in a chosen context, and I can change the meaning and context whenever I like. Hence, the classifications are not to be interpreted as authoritative.

The point in presenting these ICT types is to help to identify different angles to the accepted norms on the subject of professionalism, and to use them in a discussion on the state of professionalism and, indeed, the current state of the ICT profession.

To me, individual professionalism (as a norm), requires a combination of several necessary characteristics:

- competency
- ethicity
- knowledgeability
- learning ability
- care
- pride.

---

1 A version of this chapter was published in *The Australasian Journal of Information Systems*, 2009, vol 16, pp 149–63. It is published here with permission.

An individual cannot be a professional in the absence of even one of these characteristics. It is common, however, for ICT people to attach to themselves the label of 'professional' whilst lacking one or more of the characteristics listed above.

Aside from professionalism in ICT, it is not clear what the ICT profession is. Is it a combination of two (or more) different professions? Is it merely a label given to a broad range of computer, software, communication networks, electronics, and other practices and sciences? I will draw on more formal definitions to discuss the state of play of the ICT profession later in the paper.

Information and communications technologies types

The following descriptions are provided in the hope that readers, who have perhaps met people with many of the same traits, ponder the connection between the traits and 'professionalism'. The labels given to the 'types' are intended to summarise overall capability and social interaction.

The 'highly broadly technical and you know it' type

They know their stuff and make sure everyone else knows they know their stuff. They have an excellent grasp of what they know, and they ususaly interest themselves in highly technical/detailed information. Presumably it provides a degree of comfort/control to know a whole heap of stuff in which few others are interested. They always have something to say when discussions move to their area of expertise and, on such occasions, they nearly always make everyone feel inadequate or irritated. Such discussions don't last long, unless there's another 'highly technical and you know it' sort of person participating — but then they tend to be the last people standing, when everyone else has gone. These people are usually competent and ethical, and often perform best under pressure — they love challenges. They typically don't make successful team players/ managers because their organisational abilities only apply to themselves. When praised for their performance their attitude is typically one of: 'What did you expect?'. Diplomacy is not one of their strong points.

The 'highly broadly technical and you don't know it' type

They are quiet/unassuming and know their stuff but, generally, others don't know that they know their stuff. They have an excellent grasp of lots of things, not just their ICT knowledge — but few ever know this. They perform their work ethically with a high level of competence and diligence — they hate pressure, though. They are typically self-critical, even when told that they have performed excellently. These people often squirm when praised — but, in the end, they do appreciate being wanted. Typically they will not volunteer their expertise and they're just as happy to occupy themselves learning new stuff if no-one bothers to include them in a team/task. These people cannot manage anyone other than themselves, but at least they know it. They also know that there's a lot that they don't know and they often worry about this.

The 'highly narrowly technical and you know it' type

A little like the 'highly technical and you know it' types, but they are usually newbies to the game and/or immature. Every new problem put to them is solved using the 'one' technique that they know really, really well. These people are often seen as a pain. They usually don't perform well, even when they are given the opportunity to use their best knowledge assets — probably because they are not particularly good at solving complex problems. If these people resist learning new ideas, they become increasingly incompetent.

The 'highly narrowly technical and you don't know it' type

As with the 'highly narrowly technical and you know it' type, every new problem put to them is solved using the 'one' technique that they know really, really well. The trouble is, others don't necessarily know this is happening during a project and, if not discovered early, it can be disastrous. These people are a pain because they stubbornly insist that they are right and that everyone else is wrong. Again, just like the 'highly narrowly technical and you know it' type, they usually don't perform well, even when they are given the opportunity to use their best knowledge assets, simply because they are not good at solving complex problems. These people typically resist learning new ideas and so they remain incompetent.

The 'big-picture broadly technical' type

Tend to be the visionaries. They possess/understand enough/plentiful technical concepts/information to be able to provide effective plans of attack for solving large/complex problems without being swamped with detail. These people are typically the engineers of the ICT industry and they usually possess a wealth of experience. They are competent and always prepared to learn more, but not usually open to taking on detailed knowledge. They can be good leaders and managers, but they are best at analysis, architecture and design. Verification and validation of ideas are not their strong point, but they are very happy for others to provide that detail. These people know what they know and don't know, and are typically prepared to inform others of the same. Sometimes their ethics are questionable, especially when they produce solutions to large/complex problems that do not recognise, or simply ignore, the broader stakeholder community.

The 'big-picture narrowly technical' type

The industry is, unfortunately, riddled with these people. They try hard to emulate the 'big-picture broadly technical' type, but usually fail because they lack experience, or they have only worked within a narrow field of ICT, or within a single organisation. These people typically provide the same, conceptually limiting solutions to all large problems with which they are presented. Put them outside their safety net and they are flummoxed. They can behave unethically when under pressure, or when others have great expectations of them — but this is often caused by them having neglected to communicate what they don't know.

The 'concepts only' type

There doesn't seem to be too many of these people about these days. They typically possess very limited knowledge and understanding of the technical, process, quality, and people elements of the ICT industry. They try and make themselves seem important by producing a million (mostly crazy) ideas a minute. They do try and concentrate on human usage and efficacy aspects of ICT. This is not to suggest that good concepts cannot be produced by people who have never been trained in ICT, but they are usually experts within other domains.

The 'I can do it' type

Usually new graduates who are very enthusiastic and keen to make their mark. Often smart, but they tend to be over-confident. While these are good qualities for new professionals to possess, they do have to be controlled/directed so that the individual builds on good professional behaviour. If uncontrolled, all sorts of disasters can occur not only for the individual but also for teams and organisations. If these individuals are newbies, then their expectations are typically not well aligned with reality — and this example is not limited to ICT. There is the potential, however, to assume that 'I can do it' ICT graduates are somehow smarter and more correct than engineers, scientists, lawyers or artists, and their bosses. These people can possess new ideas and new technology, but this does not mean that they should be in control.

The 'I can't do it' type

Perhaps these people have chosen the wrong vocation. They lack confidence. They struggle to remember/understand almost anything, and they always need training. These people are not self motivated and they lack initiative. They don't want to be responsible for anything or anyone. Under pressure, they usually fail to perform. These people probably have other problems that override their professional responsibilities. Given the right opportunity and environment, however, they can change.

The 'other professional' type

There is a broad group of this type of person. From a professional viewpoint, however, they are often the people who both criticise the ICT profession and contribute to the apparently parlous state of the profession, typically by undertaking ICT-related work themselves (when they are not competent to do so) or making serious decisions on behalf of ICT professionals without seeking any advice/input from those professionals.

What have these different types to do with professionalism and professions? Well, for a start, it makes it clear that there is a diverse range of people in the ICT industry, possessing a diverse (not necessarily adequate) range of knowledge, abilities and personalities. On the one hand, many of the people working in the ICT industry and academia belong to no professional body, such as the ACS, ACM or IEEE. Many do work professionally, however, and demonstrate ethical behaviour — often to a greater degree than any of the professional bodies might expect. On the other hand, a number of people in the ICT industry do belong to a professional society, and yet they possess minimal knowledge of any part of the ICT domain, display little competence, and even occupy positions superior to others who are far more knowledgeable and competent. Often, for the latter 'professionals', it seems that personality is the most valued characteristic. Interestingly, and in contrast, while working as a scientific professional within the more traditional engineering and scientific domains, I observed that positions of importance/influence belonged to those with appropriate knowledge and competence — their personality traits were a less important factor in gaining such positions. Of course, this situation has another set of problems.

I have had about 25 years experience as a software engineer at various levels. In all that time, although I have identified lots of different work practices, ethical attitudes and personalities, I have not been able to establish precisely what comprises an IT or ICT professional. Hence the question associated with this paper: 'What is an ICT professional anyway?'

Since I am not sure of the answer to this question, it seems appropriate to start with one or more definitions of a professional.

The professional

There are numerous definitions of 'professional'. To begin with, the typical dictionary (in this instance *The Collins English Dictionary*) contains definitions for the word *professional* such as:

- of, or relating to, suitable for, or engaged in a profession
- engaging in an activity for gain or as means of livelihood
- extremely competent in a job, a person who engages in an activity with great competence
- undertaken or performed for gain or by people who are paid
- a person who belongs to, or engages in one of the professions.

More definitive views of professionalism include reference to profession, professional development and professional bodies. Such a view might include the following (Texas Library Service, nd):

- Professionals are considered experts in their chosen vocation/field.
- Professionals possess a broad range of (systematic) knowledge with a theoretical base.
- Professionals are responsible to the public and/or community.
- Professionals possess a high degree of autonomy regarding their decision-making and behavior.
- Professionals are governed by a code of ethics, which:

    - is a statement of rules and values

    - intends to ensure a high quality of service

    - intends to guarantee competency of membership, honor and integrity

    - is an expression of a professions' principles and what it expects of its members

    - emphasises no personal gain at cost to others (co-professionals, clients, community etc).

- The professional's system of rewards is primarily recognition for building up appropriate knowledge and experience, and furthering the respect of the profession.
- There is a system for testing the competence of members.

David Maister provides a more humanist-oriented view of professionalism, by de-emphasising the knowledge/skill aspects. To paraphrase Maister slightly, he believes that '... real professionalism has little to do with which business you are in, what role in that business you perform, or how many degrees you have. Rather it implies a pride in work, a commitment to quality, a dedication to the interests of clients, and a sincere desire to help' (2001). An important indication of the nature of professionalism made by Maister is that it is '... predominantly an attitude, not a set of competencies. ... skills you can teach, attitudes and character are inherent'.

The key traits of a professional are shown in Table 1, where they are presented as a set of criteria for professionalism. All but the last of these have been taken from the Texas Library Service tutorial on professionalism (nd). The last trait has been added on the basis of Maister's views on professionalism.

Table 1: Criteria for professionalism

Training	There is an extensive period of training, often after a combination of formal education, training and apprenticeship; this training is usually undertaken in a higher education environment.
Intellectualism	The intellectual component is dominant.
Autonomy	Professionals usually have autonomy in their work.
Judgement	Professionals, because of their training, education, knowledge and experience, may use their own judgement in determining the appropriate approach to their clients or customers.
Independence	Professionals can work independently and charge fees, or they can be part of an organisation.
Service	Professionals possess abilities to provide a valuable service to society and operate with little or no self-interest.
Dedication	Professionals are dedicated to service and institutions.
Pride	Professionals take pride in the quality of their work.
Honest & trustworthy	Professionals can be trusted to behave honestly.

Source: Texas Library Service

It is interesting to apply the criteria in Table 1 to some of the ICT types described earlier. These are shown in Tables 2 and 3 where, again, the resulting summaries are drawn from my own experience and observations.

Table 2: 'Highly broadly technical and you know it' professionalism

Training	Usually possess a technical degree. Also do much self-education.
Intellectualism	Not usually dominant.
Autonomy	Usually.
Judgement	Whenever allowed or are asked to advise, but more likely to criticise other peoples' judgements.
Independence	Not usually. Confidence in own abilities does not usually extend to taking on the responsibilities of being independent.
Service	Usually.
Dedication	Not usually.
Pride	Usually.
Honest & trustworthy	Usually.

Source: Author's research

Table 3: 'Big-picture narrowly technical' professionalism

Training	Usually possess a non-technical degree. Don't do much self-education.
Intellectualism	Very little.
Autonomy	Not usually but do try.
Judgement	Whenever the opportunity arises. These people make the judgements that the ICT types in Table 2 criticise.
Independence	Not usually.
Service	Inconclusive observation.
Dedication	Inconclusive observation.
Pride	Sometimes.
Honest & trustworthy	Often not honest about their knowledge/capability.

Source: Author's research

The reason for choosing the two particular ICT types in Tables 2 and 3 is simply that they are contrasting but, nonetheless, represent a significant percentage of the ICT-industry membership.

Slight variations within Table 2 would result when applying the criteria to the other 'broadly technical' types described earlier. Similarly, variations within

Table 3 would result when applying the criteria to the other 'narrowly technical' types. The 'concepts only', 'I can't do it', and 'other professional' types should probably be disqualified as professionals, let alone ICT professionals.

One might conclude that there are no real ICT professionals. I doubt that's true. Because there is a confusion of diverse backgrounds and training of people working in the ICT industry, it seems appropriate to create an effective method for identifying the various professionals — the ICT professionals specifically.

How do we distinguish an ICT professional from other professionals working within the ICT industry? Perhaps the answer can be found by examining the profession that supports the ICT professional.

The profession

Professions, usually exemplified in the form of an association/society/body of like-minded colleagues, are typically created to represent and support those who are accepted as accredited members ('qualified' professionals), and to maintain professional integrity, values and recognition. Hence, real professions have responsibilities to the professionals that they purport to support. For example they:

- create structures within professional subcultures
- provide legal reinforcement for the activities of professionals
- strive to provide environments of public acceptance
- promote ethical practices
- define penalties for professionals who work against the tenets and practices of the profession.

The traits of a profession are summarised in Table 4.

Table 4: Traits of a profession

Exclusive control — esoteric and systematic BOK (body of knowledge)	Members have a monopoly on understanding and applying the BOK.
Autonomy of practice	Members control the standards of the society.
Norm of altruism	Members act in the best interests of clients.
Authority over clients	Professionals control the client/practitioner relationship.
Distinctive occupational culture	Occupation is set apart by a distinctive set of norms, values, and symbols.
Recognition	Usually legal requirement for specific training and preparation prior to practice.

Source: Dinsmore & Cabanis-Brewin, 2006

The *Software engineering body of knowledge* (SWEBOK, nd) provides a characterisation of an engineering profession as supporting the follow components:

- an initial *professional education* in a curriculum validated by society through *accreditation*
- registration of fitness to practice via voluntary *certification* or mandatory *licensing*
- specialised *skill development* and *continuing professional education*
- communal support via a *professional society*
- a commitment to norms of conduct often prescribed in a *code of ethics*.

According to Andrew Abbot (1998), professions begin when people recognise that they are undertaking a vocation/occupation that is not covered by any existing profession. *Perhaps this is the current state of the ICT industry*. The creation of a professional association/society defines the extent of practices and necessary competencies to perform those practices — thus defining a 'competence territory' (see Dinsmore & Cabanis-Brewin, 2006). The process, or life cycle, of professionalisation typically consists of the following steps/phases (not in strict order):

- create a full-time occupation/vocation
- establish professional association
- create and enforce code of ethics
- control use of the (profession's) name and the body of knowledge (BOK)
- develop recognised training procedures
- win political, social, and legal recognition.

A profession provides comfort and security to people who fulfil its requirements and obey its tenets. Professionalisation, however, can be a torturous and drawn-out process simply because of other existing professions. Even when the first five steps of professionalisation are largely fulfilled, gaining political, social and legal recognition can be a serious problem for the associated professionals — *software engineering is a good example of such a profession*.

The status of a profession can change significantly with time, especially when it becomes too protective of its members and doesn't respond appropriately to serious challenges — of (say) malpractice. Giving any impression to society of cover-ups and lack of interest in upholding codes of ethics can only damage a profession. Charges of malpractice need to be investigated without bias or conflicts of interest by professions, and the processes of investigation need to be transparent.

With time, professions are also affected by changing political, social and legal environments and, if they are too slow to move with those types of external changes, they may suffer diminished recognition.

The *AMA handbook on project management* (Dinsmore & Cabanis-Brewin, 2006) presents the concept of semi-professions by examining the state of the occupation of project management using the traits of a profession from Table 4. Table 5 represents a summary of that examination.

Table 5: The status of project management as a profession

Exclusive control — esoteric and systematic BOK (body of knowledge)	No — *BOKs are beginning to be recognised but still highly contested.*
Autonomy of practice	No — *members contribute to the standards of practice.*
Norm of altruism	Not usually — *societal impact of failed projects not recognised.*
Authority over clients	Not usually — *project managers tend to work within corporations.*
Distinctive occupational culture	Possibly — *certain aspects exist.*
Recognition	Not yet — *project management not legally recognised as a profession in any jurisdiction.*

Source: Based on Dinsmore & Cabinis-Brewin, 2006

After having (myself) attempted to elicit (without success) a specific ICT BOK it would seem that the state of the ICT profession is similar to that of project management — it is a semi-profession because, as summarised in Table 6, it possesses only elements of the 'required' traits, described in Table 4. Certainly ICT 'practice' is at the stage in the path of professionalisation of being a full-time occupation/vocation. By splitting ICT into the components of IT and communications, it is evident that there is at least one core/common IT BOK (Australian Computer Society) but there seems to be no obvious communications BOK — although there is a WEBOK (Wireless Engineering BOK within the IEEE Communications Society). As an aside, the term 'body of knowledge' seems relatively new, and so, many older professional societies may either use an alternative term or not actually refer to any specific knowledge base. The latter is probably due to an acceptance that (tertiary) educational institutions will provide the primary body of knowledge to support a society and its membership. Thus, identifying bodies of knowledge that might underpin a professional society may not always be straightforward.

A BOK is important to a professional society, not only in defining/creating boundaries to the essential knowledge of members, but also in helping to define required education/training programs for basic/extra qualifications for members at different levels of the society. The BOK may, in fact, enable the identification of several related occupations/vocations for which different certification programs be defined. It is one of the essential elements that a profession's membership should control, and be prepared to consider, according to changing technology and societal concerns.

One of the important duties of a professional society is to ensure the disambiguation of the name of the profession. A case in point is the profession of software engineering within Australia. Aside from the fact that computer scientists possess a different view of what software engineering is, there is also the issue that, despite Engineers Australia accrediting software engineering degrees, there is no apparent support to disambiguate either the typical industry or the computer scientific usage of the name. Ambiguity of a profession's name can be damaging to the professionals who enter the profession, only to find themselves competing for positions that are apparently advertised for them, but open to those who are either less qualified or from different professions. Recognition of the profession externally becomes more difficult when recognition internal to a professional society is not obvious.

A professional society's code of ethics is intended to ensure its members maintain competency and appropriate behaviour and consideration toward other professionals, clients and the community. Together, the BOK and the code of ethics are important in both the mounting of, and defence against, charges of unprofessional behaviour. Any charge of misconduct against a professional can only be successful if it is proved that best practices have not been followed and that codes of ethics have been contravened. Defence of such charges follows the opposite view. If unprofessional behaviour is proved, then appropriate penalties should be applied.

This is well and good, but I have only ever heard of one person in the Australian ICT industry being punished for misconduct — and this was undertaken independent of any professional body. Is it the case that there are no charges concerning professional misconduct in ICT? Or, are the associated professional bodies defending charges so successfully that very few individuals have been punished for professional misconduct? Is it that legal processes are more likely to ascertain charges of professional misconduct and also deliver penalties — perhaps using associated ICT professions' BOKs and codes of ethics to do so?

Table 6: The possible status of the ICT profession

Exclusive control — esoteric and systematic BOK (body of knowledge)	Not fully apparent — *there are contributing BOKs such as SWEBOK and possibly the ACS IT Core BOK, but there doesn't seem to be a specific BOK for ICT. University degrees confuse the situation because of little consistency on base knowledge/courses.*
Autonomy of practice	No — *members sometimes contribute to the standards of practice, but not necessarily for community interests.*
Norm of altruism	Not usually — *still a general lack of best fit for overall approaches and technological decisions, and many ICT projects still fail on cost and achievement of deadlines.*
Authority over clients	Not usually — *ICT professionals tend to work to other (non)professionals who control the client relationship.*
Distinctive occupational culture	Possibly — *certain aspects exist, but the norms, values and symbols are confused. Partly because ICT is a multi-profession profession.*
Recognition	Seemingly — *ICT is assumed to be a profession, or at least, it is often referred to in that way.*

Note: This is the opinion of the author and is based on a combination of long-term observation and the seemingly confused state of an ICT BOK.

Don Gotterbarn (2008) relates stories of people working in the ICT/software industry who having been fired because of unethical practices, but with no intervention by appropriate professional societies to either condone the company that instigated the act, nor impede/prevent the perpetrators of the unethical act(s) in simply moving to similar jobs at other companies.

Is it any wonder that there are sceptics concerning professionalism and professions? For example, according to John Quiggin (2003):

> Professionalism is both an individual characteristic and an ideological position. The primary definition is that of individual professionalism: the idea that membership of a profession carries with it a set of internalised values that will be reflected in the way in which work is carried out and the ethical standards that are adhered to.

This statement appears to support the concept of professionalism, however, Quiggin goes on to say that *the idea of professionalism as an individual characteristic is false* — given the chance to pursue their own self-interest (narrowly defined to exclude such items as the approval of professional peers) individuals will do so regardless of any attempts to instill professional values. In elaborating on the words of George Bernard Shaw, 'The professions are a conspiracy against the laity', Quiggin writes, 'That is, adherence to professional values and ethics serves to advance the collective interests of the profession at the expense of society as a whole. Exactly the same critique is made of trade unions.'

When it comes to the ICT profession/professionals, I don't wholly agree with Quiggin. I rather think that the ICT profession is nebulous in terms of BOK and lacks authority, or the right sort of recognition, to ensure the professional behaviour of anyone (society members and/or non-members) practising in the ICT industry.

Some conclusions about the state of ICT professions/ professionalism

I believe it *is* appropriate for professions to set expectations of knowledge, competence and behaviour. Professions and the professionals within them should encourage continuous learning, sharing of knowledge, informing of experiences, and reflection on episodes of good (and perhaps bad) ethical behaviour and the benefits (or negatives) to others and themselves.

Establishing and maintaining a BOK and a code of ethics are the easiest activities for a profession to undertake. Because humans are the enablers and disablers of professionalism, there is always going to be a spectrum of acquired knowledge, ethical behaviours and personalities, ranging from competent, ethical, empathetic, and earnest individual professionals down to those who just barely meet those standards and who should, perhaps change their profession.

The most difficult activities for a professional society to undertake seem to be related to maintaining the integrity of its membership (using the criteria of Table 1 and the traits of Table 4 as guides). I believe there are ICT workers, and indeed have met several people working in the ICT industry, who rate as professionals according to Table 1. However, I don't currently know of any ICT type professional body, that maintains any but a few of the traits listed in Table 4.

So far I have concentrated on trying to establish what a professional is (independent of vocation) and what a profession is. I have used definitions that establish that (really) the ICT profession is young and is evolving slowly.

In answering the question: 'What is an ICT professional anyway?', I have to admit that I'm still not sure, but I think I have a reasonable idea of the sort(s) of knowledge, competence and ethics that someone who is labelled as an ICT professional, should possess. I do not, however, have a definitive answer to the question.

The ACS Core BOK for IT professionals does not significantly assist me, either, in answering the question. The ACS BOK contains the following diverse areas of knowlcdgc:

- computer organisation and architecture
- conceptual modelling
- database management
- data communications and networks
- data structures and algorithms
- discrete mathematics
- ethics/social implications/professional practice*
- interpersonal communications*
- program design and implementation
- project management and quality assurance*
- security
- software engineering and methodologies
- systems analysis and design
- systems software.

Those items that are marked by an asterix are designated as mandatory knowledge within all professional-level IT courses accredited by the ACS. Any one of the listed areas of knowledge could constitute a lifetime of professional activity/work, and indeed such professionals exist. For example, within the software engineering and methodologies area of knowledge, there is another recognised BOK (SWEBOK, nd) and it contains the following knowledge areas:

- software requirements
- software design
- software construction
- software testing
- software maintenance
- software configuration management
- software engineering management
- software engineering process
- software engineering tools and methods
- software quality.

Again, any one of these knowledge areas could occupy a lifetime of professional activity/work and, again, such professionals exist.

As indicated earlier, it is appropriate to define a BOK. It is also appropriate, however, to define levels of practice. The SWEBOK (nd) includes descriptions of practices and processes for each of its knowledge areas, but it does not define levels of practice. At this stage, the ACS Core BOK for IT professionals contains little in the way of appropriate practices/processes for each of the areas of knowledge, nor anything about levels of practice. This is not intended to be a criticism, the SWEBOK took six to seven years and the help of well over 500 professionals to define. It does show, however, that the ACS BOK is not yet complete.

It is appropriate that anyone possessing significant capability surrounding any one of the knowledge areas of the SWEBOK might be an eligible ICT professional. They would, however, have to possess significant understanding of other (related) knowledge areas in order to do their jobs comprehensively. So, what minimum depth of overall knowledge would anyone need to reach some base level of practice enabling professional status in (say) software testing? Also, what other higher levels of practice might these base level professionals aspire to, which are both defined and supported by an ICT professional society?

I could conclude by simply stating that I do not believe the ICT profession, as it stands today, is, or is even capable of, providing adequate support to ICT professionals and non-professionals. Until that happens I doubt I, or anyone, else will be able to answer the question — 'What is an ICT professional, anyway?' While this conclusion may be correct, it neither helps in maturing the ICT profession further, nor in providing less confusion for those working as professionals and non-professionals in the ICT industry.

Further work

It is apparent that significant work needs to be done to fully establish an appropriate ICT BOK and associated practices. To that extent, it should be noted that the ACS is in the process of establishing a more comprehensive BOK than is currently available. This endeavour is important for the ACS because (as a professional body) it accredits tertiary ICT programs throughout the country. In this process, existing BOKs, such as SWEBOK — which provides a very good basis for describing base knowledge, methods and practices for each of its knowledge areas — need to be taken into account.

In this and other related BOK endeavours, it is necessary to ensure input from a broad range of stakeholders — not, for example, academics only. The mixture of stakeholders needs to include people and organisations who are not only creators of ICT-related knowledge, products or services, but who are the users/ receivers and/or who are impacted in some way by ICT activities.

Regarding the matter of codes of ethics, it seems that professional societies provide these mostly as guidelines of acceptable behaviour and also as a basis for examination of accusations of unprofessional behaviour. On the one hand, codes of ethics can be used as blunt instruments, only to be made apparent when a professional member has committed some (alleged) malpractice. On the other hand, codes of ethics should be used to actively encourage greater, unwavering commitment to integrity, honesty and trustworthiness. Obviously, to establish a profession containing professional members requires education, not just to learn the wording of codes of ethics but also how to use the codes of ethics as a framework to clearly promote what is expected of professionals and how they should behave when ethical behaviour is at risk. All this is easily stated but hard to do. Nonetheless, ethical behaviour needs to be imparted and governed effectively in order to raise the bar on expectations of all ICT stakeholders in regard to ICT activities.

It seems that ICT professional bodies need an injection of governance practices, especially in being proactive about supporting current members and expanding membership to help ensure wider consistency of professional behaviours, practices and project outcomes.

Aside from dealing with the BOK, ethical behaviour and expectations, there is the matter of determining different levels of practice/skills regarding the broad range of capabilities among people working in the ICT industry. Some of the ICT types I described at the beginning of this paper may better fit the job description of 'technician' rather than (say) 'engineer' or 'scientist'. All of these roles should, of course, follow professional practices and abide by ethical standards. The purpose in providing a range of levels of practice, however, is to set in place a reasonable framework to establish minimum requirements for being able to practice as a particular type of ICT professional, and also determine what 'endorsing' criteria need to be met to advance to higher levels of practice.

The concept of levels of practice is not new, and has the same advantages within other more established professions of protecting all stakeholders. For example, it should not be possible for a new graduate of the ICT profession to be placed in a managerial role requiring significant technical and people management experience, to be handling significant numbers of staff (including senior staff), and to also be expected to make technical decisions on topics about which they possess only superficial knowledge. I have seen this occur on several occasions, and the result is unacceptable because it affects a significant body of stakeholders, including the junior professional and their employer. The junior can be 'scarred' for life professionally, and the organisation for which they work rarely takes the blame for the inevitable failure(s). This sort of situation is caused through the combination of lack of ethics (of both parties) and no limitation concerning a role in which a junior ICT professional can be involved. Ethically, junior professionals must be able to both admit their inability and to resist the temptation/encouragement to be involved in a role for which they do not have the requisite knowledge or skill. Ethically, the organisation should not be able to advise a junior professional to undertake a role for which they do not possess the appropriate capability.

So, if the ICT profession is to gain traction as a true profession that supports true professionals, then steps need to be made now along the lines described above. The magnitude of such a task may seem insurmountable just now and, indeed, it will take some time to evolve. The point is that the first steps need to be undertaken as soon as possible so that ICT comes to earn its recognition as a profession, which it currently does not deserve.

References

Abbot, A, 1988, *The system of professions*, University of Chicago Press.

Dinsmore, P & Cabanis-Brewin, J, 2006, *The AMA handbook of project management* (2nd edn).

Gotterbarn, D, 2008, 'Software engineering ethics research institute', paper presented to EtGovICT, Canberra, 1–2 May.

IEEE Computer Society, nd, 'SWEBOK', IEEE Computer Society, <http://www.swebok.org>

Maister, DH, 2001, *True professionalism*, Touchstone.

Quiggin, J, 2003, Word for Wednesday Professionalism: Definition <http://johnquiggin.com/2003/09/10/word-for-wednesday-professionalism-definition>

Ridge, J, 2008, 'Australian Computer Society', paper presented to EtGovICT, Canberra, 1–2 May.

Texas Library Service Training Program, nd, <http://www.tsl.state.tx.us/ld/tutorials/professionalism/prof.html>

6. ICT is not a profession: So what?

John Weckert and Douglas Adeney
Charles Sturt University and University of Melbourne

What is a profession?

Information and communications technology (ICT) is not a profession in any significant sense, but this is not a slight on ICT. It is, of course, a profession in an insignificant sense. If a person develops software for a living as opposed to doing it as a hobby, that person is a professional software developer. Some people sing for a living and others just for fun. The former are professional singers while the latter are amateurs. In this sense any occupation is a profession insofar as those engaged in it are making a living from it, but this is not what is meant in discussions about whether or not ICT is a profession. What is in contention in those discussions is whether or not ICT is *merely* an occupation or whether it has characteristics, like medicine and law are claimed to have, that set it apart as a profession (this is the sense in which 'profession' will be used in the rest of this chapter).

What are the characteristics that distinguish a (so-called) profession from other occupations? The following, commonly suggested criteria for a profession, based on Deborah Johnson's list (2001), are:[1]

(a) The need to master a special, esoteric body of knowledge. This body of knowledge is usually obtained through higher education — for example, a degree in law — and is required in order to practise in the legal profession.

(b) Some amount of autonomy for both the profession as a whole and the individuals in it. Professionals have the autonomy to make decisions in their daily work, justified by the fact that they are experts and this is

---

1 The Australian Council of Professions definition is this:

> A profession is a disciplined group of individuals who adhere to ethical standards and who hold themselves out as, and are accepted by the public as possessing special knowledge and skills in a widely recognised body of learning derived from research, education and training at a high level, and who are prepared to apply this knowledge and exercise these skills in the interest of others. It is inherent in the definition of a profession that a code of ethics governs the activities of each profession. Such codes require behaviour and practice beyond the personal moral obligations of an individual. They define and demand high standards of behaviour in respect to the services provided to the public and in dealing with professional colleagues. Further, these codes are enforced by the profession and are acknowledged and accepted by the community.

based on their training. So, for instance, a lawyer will have the autonomy to decide on the best way to defend his or her client, as opposed to taking orders from some higher authority. This autonomy also extends to the profession as a whole. The profession regulates itself, making decisions about who is able to enter it and to what standards they should be held. It will generally be able to do this with little government interference.

(c) Connected to the autonomy of professions as a whole is the existence of a formal organisation. A profession will usually be governed by a unifying body that is recognised by the government — for example, the Australian Medical Association (AMA). This body may have several roles, such as setting standards for certification and having the authority to expel members from the profession.

(d) A profession will also generally have a code of ethics that both sets the standards of behaviour for its members and signals to members of the public what to expect when they engage the services of a professional. For example, medical professionals are usually taken to be governed by the general principles of beneficence, non-maleficence, autonomy and justice.

(e) Finally, professions are typically understood to fulfil a social function. This will often be connected to a social good, such as health or justice, or concerned with more general social benefits, such as providing the technologies and infrastructures to enable society to function effectively. In general, the idea is that professionals serve the interests of the public. So, for instance, the role of the lawyer is to promote justice, the role of police is taken to be the provision of security, and the role of engineers is the provision of technologies that enable things such as communication and transportation (Johnson, 2001: 60–61).

Doubts about ICT as a profession

We have three reasons for doubting that ICT is a profession. The first is the diverse nature of ICT, the second relates to the debate over licensing and certification, and the third is that there are no compelling reasons to believe that there are any professions in any significant sense at all.

The diverse nature of ICT

One difficulty in calling ICT a profession should be noted, and that is the difficulty of knowing exactly what ICT is. ICT covers a wide variety of activities and is, to a large extent, an umbrella term. These activities include those of:

- information technology managers
- system managers, designers and analysts
- programmers
- software designers and engineers
- technical support
- ICT trainers
- database and systems administrators
- multimedia specialists and web developers
- computer network professionals.

(see also ACS, 2011)

Computers play a role in all of these occupations, but that is not enough to justify calling ICT as a whole, a profession. It does not follow from this, however, that some parts of ICT could not be professions; for example, software engineering. If there is to be serious discussion of ICT and professions, it should probably begin with an examination of which parts of ICT most resemble recognised professions (see Holmes in this volume for a similar position). Currently software engineering is probably the most likely candidate and most of the discussion concerning licensing has been in this field. But it is even doubtful whether software engineering can be considered a profession.

Certification and licensing

Software engineering (SE), as was just suggested, has the most potential to be considered a profession. The vigorous discussion regarding licensing and certification in the industry suggests, however, that SE (and by implication ICT in general as well) is not considered a profession by a substantial number of those employed in the industry. Licensing has frequently been advocated, but just as frequently attacked, and implemented only in very few places. The ACM, for example, after giving the idea consideration, came out against it in 1999. Before proceeding, it is worth distinguishing between licensing and certification. A person is certified as being a qualified member of an occupation if, typically, he or she has satisfied certain requirements considered necessary for that occupation, for example, has passed particular examinations. While certification indicates that a person has certain skills, someone without certification can still practise.

Licensing carries legal weight. In occupations where there is licensing, a licence is necessary to practise legally. In the ICT industry, discussion of licensing has generally been limited to software engineering, but possible certification has been raised for ICT more generally.

The most common objection raised against licensing seems to be that, because SE is relatively new, there is as yet no recognised body of standard professional knowledge that all practitioners must have. This is something that is one of the definitive aspects of other professions (condition (a) above). According to John White and Barbara Simons, the ACM council believed that the 'state of knowledge and practice' was too immature to give assurances that 'the licensed engineer is capable of producing software systems of consistent reliability, dependability, and usability' (2002: 91). Tony Healy (2004) and John Knight and Nancy Leveson (2002) also argue that no generally accepted body of knowledge exists. Others, however, see it differently. Danielle Boykin (2007) discusses the introduction of examinations for SE licensing, something that clearly assumes an accepted body of knowledge. Don Gotterbarn, a well-known campaigner for SE professionalism, has suggested (in conversation) that there is something absurd in maintaining the argument that, because knowledge is still developing in some field, that field cannot be a profession. On that basis, not even medicine would qualify as a profession, given that medical knowledge is still advancing. Some background to this debate can be found in Nancy R Mead (2009).

Another concern is that licensing 'would expose software developers to malpractice suits' (Healy, 2004). This, in itself, is no objection to licensing. Most professionals are in this position and software developers are the odd ones out. This objection to licensing is expressed also by Knight and Leveson (2002), who believe that currently (or in 2002) professional standards for SE are not yet clear enough to make 'best practice' easily discernable. If indeed best practice and body of knowledge are so poorly defined in SE then, perhaps, the objection does have teeth but, as has just been seen, not all believe that the state of SE is so parlous.

While certification is weaker than licensing, as we noted, some argue that many of the same problems arise in ascertaining the criteria for certification (Knight and Leveson 2002).

The argument here is not about whether there should be licensing and/or certification in the ICT industry. It is about whether ICT is mature enough yet to be considered a profession. The doubt within the industry does not of course show decisively that ICT is not a profession.

Scepticism regarding professions

A stronger reason for being sceptical of any claim that ICT is a profession is that good reasons exist to doubt that there are professions at all in any significant sense. Two problems become apparent when attempting to give an account of a profession.

First, standard accounts do not pick out only what are currently considered professions; they clearly encompass occupations that are not normally considered professions. On those criteria, plumbing should be a profession. Plumbers clearly have an esoteric body of knowledge not acquired by the rest of us. The skills of a plumber might not be obtained through a higher education degree, but their achievement does require a number of years of training on top of normal secondary education. Plumbers also have a degree of autonomy regarding how they undertake their work and how they solve particular problems; and they must belong to a formal organisation in order to practise, and this formal organisation has a code of ethics. Finally, plumbers clearly fulfil a social function (imagine a city without proper plumbing). They appear, therefore, to satisfy the criteria for a profession.

Consider the following extract from the Master Plumbers' Association of Queensland (2002):

> CODE OF ETHICS, TRADE CUSTOMS AND PRACTICES:
>
> 1. It is considered that the health and well being of the community as it is affected by all Branches of the Plumbing Trade is of prime importance and shall be protected to the best of the member's ability. The task of raising the living standards of the community is one entrusted to the members of the Association, …
>
> 2. A member shall observe the highest standards of honesty in all of his transactions, …

It calls for the health and safety of the community and high ethical standards, and assumes special knowledge and skills when it calls for 'the highest quality of service and true value for money'. This is something that could not be provided without that knowledge, and its inclusion acknowledges a clear social function for the profession. It also acknowledges the vulnerability of their clientele. This is true not only regarding the plumber's work, but also because frequently, clients must give plumbers access to their homes at times when they are not present.

The only real difference from members of recognised professions, say doctors and lawyers, seems to be that plumbers get dirty, with real dirt; but adding a criterion of cleanliness to the defintion of a profession seems odd. In any case, some scientists, for example those working in agriculture and biology, often get dirty, with real dirt, too; so, either they are not members of a profession or plumbing is no less a profession if getting dirty is allowed.

An additional reason for denying that plumbing is a profession, from Gotterbarn, is that members of professions have responsibilities to advise customers 'of potential negative impacts that a new system may have on the customer's business, quality of life, or the environment' (2004: 116). Members of professions do have this responsibility, but so do plumbers. When a plumber is hired, more is wanted than a technically good job; advice is normally required on whether what is asked for is, in the plumber's opinion, a good idea. Again, this does not distinguish plumbers from accepted professions.

The second problem is that the criteria offered seem arbitrary. Picking an occupation, say medicine or law, and then constructing a list of its characteristics and using that list as the criteria for a profession is hardly convincing. Why not choose plumbing or fruit picking? Historically a profession was a vocation, a calling from God. The clergy had a vocation, artisans an occupation, but this too has the ring of arbitrariness.

The considerations so far suggest that the notion of a profession is not useful in discussions of ICT, for four reasons: first, ICT is too diverse to be considered a 'profession'; second, doubt exists amongst ITC practitioners that ITC is ready to be called a profession; third, accounts of a 'profession' do not foreshadow only those pursuits that are commonly called 'professions'; and, finally, the criteria seem arbitrary. Before rejecting the notion altogether, though, it should be noted that it may have a use in encouraging good behaviour. Those who consider that they have a profession, rather than merely a job, may have extra motivation to carry out their work in a more conscientious and moral fashion, in order to uphold the good reputation of the so-called profession and because they are bound by a code of ethics. But, as we have just seen, plumbers too are bound by such codes, so it is not clear that the notion of a profession is important for providing motivation for good conduct. In any case, there does not seem to be much empirical evidence that members of recognised professions behave more ethically in their work than others.

A professional attitude

The notion of a profession may not amount to much, but that of a *professional* is important. The view being advocated here is that someone is not a professional on the basis of being a member of a certain occupation, a 'profession'. Rather, professionalism comes from having a *professional attitude*. The fictional mechanic, Mr JLB Matekoni, has such an attitude:

> Mechanics could be the conveyers of the most serious news ... on those occasions a suitably grave expression was required; one should not give bad mechanical news lightly, as Mr JLB Matekoni had felt obliged to tell his apprentices. He had overheard Charlie telling a woman that her car was *finished*, and on another occasion the young man had told a client that his brakes were the worst brakes in Botswana, adding, *And I've seen some pretty bad brakes in my time!* No, that was not the professional way, not that those young men understood what professionalism was all about (McCall Smith, 2010: 32).

For Mr JLB Matekoni, being a motor mechanic involves much more than being able to fix cars; equally important is how the mechanic deals with customers. He has a *professional attitude*. A professional attitude involves, at least, taking pride in one's work, trying to do it as well as possible and actively considering ways in which it might be done better, looking at it in the broader context of society as a whole, and taking responsibility for what does. Admittedly, there is more scope for the exercise of such an attitude in some fields of employment than in others: a worker doing repetitive work on a factory production line may not have much scope for reflection on how he or she might do his or her job better, or for pride in doing it well. But in a great many types of work, for example, law, medicine, plumbing, and ICT, there is ample scope for a *professional attitude*, and such an attitude should be encouraged and appreciated.

A professional attitude is particularly important in occupations where the person is an expert in the sense of having significant knowledge and skills not had by those outside the occupation. Special responsibilities, over and above those that we all have as human beings, are attached to this expertise, and a professional attitude involves taking these responsibilities seriously. Before elaborating on this, we look a little more closely at the concept of responsibility because of its central role in discussions of ICT professionalism.

The word 'responsibility' has three distinct senses that are important for our purposes. First, there is the *causal* sense: if it is said that lightning was responsible for a fire, what is meant is that lightning caused the fire. It is ambiguous however to say that someone was responsible for a fire. This might just mean that the person indicated caused the fire, or it could mean that he or she is blameworthy (or praiseworthy, depending on the circumstances) for lighting it. This second sense is *moral* responsibility. Someone is morally responsible for an action if he or she played a causal role in the action and if it is appropriate to attribute praise or blame for causing it. A person who is responsible in the causal sense is not always responsible in the moral sense. The third relevant sense is *role* responsibility. Certain responsibilities go with particular roles. The project manager's role carries the responsibility of overseeing the whole project, while the programmer's role has the responsibility to produce good code.

Someone may not be morally responsible for an event even when that person is causally responsible. This is an important point, especially in the context of professional attitudes in the ICT industry. When is a person (P) morally responsible for some event (E)? Three conditions must be satisfied: (1) P must have caused E, or knowingly allowed E to happen when he or she could have prevented it; (2) P must have intended to cause E, or allowed it through neglect or carelessness; (3) P must have been free to choose to, or not to, cause E. Condition (1) says only that P must have done whatever it was that caused E or not done what could have prevented it. If P rather than some other person, Q developed faulty software, then P is responsible, not Q. The second condition is satisfied if P intended to produce faulty software, or if it resulted from neglect or carelessness, for example, through neglecting to test the software thoroughly. Condition (3) shows the importance of autonomy. If P is insane or has no choice but to develop the software, and within an unreasonable amount of time, then moral responsibility is diminished or absent altogether.

In summary, then, P is morally responsible for faulty software if he or she

- developed it
- intended it to have faults or (more likely) was careless or negligent in some way
- was acting autonomously and not under duress.

Moral responsibility is related to *accountability*, but is not identical with it. Accountability can be taken as moral responsibility but it is not always used in this sense. Someone might be morally responsible for an action but not held accountable. This is commonly the situation in the computer software industry, where software developers are commonly not held accountable for malfunctioning software, even though a good case can be made that they ought to be because they are morally responsible. They are morally responsible because they developed the software (as well as meeting the other two conditions above). Frequently of course, where people are not held accountable, it would be argued that this is simply because they are not morally responsible. It is one thing, then, to urge workers in a given field to develop a professional attitude, or a more professional one, for the sake of their clients. It is another thing to seek to protect those clients by making the workers accountable. Some system of certification and/or licensing can support this move.

It might be objected, reasonably, that this picture is overly simplistic because software is not usually developed by just one person. Normally, the software development will be undertaken by a team, possibly a large one. In such cases, the argument goes, nobody can be held morally responsible. This is the problem of collective responsibility, sometimes called the 'many hands' problem.

Because many people are involved in typical software development projects, when something goes wrong, it is not always easy to say who is morally responsible, and who, if anyone, ought to be held accountable and liable for any damages. One solution is just to say that the group, or organisation, is responsible. In everyday talk we do this frequently, for example, when a large company is responsible for environmental damage. Two problems are worth pointing out here. One is a degree of unfairness. Not everyone in a group is equally responsible or even responsible at all. Not all employees will have played a role in the damage that was caused. The second problem is that moral responsibility is something that can only be attributed to autonomous human beings, not organisations.

While it is probably true that, in most cases where there is collective responsibility, no individual will bear the entire blame, it does not follow that moral responsibility cannot be attributed to many individuals. Many can, therefore, be held accountable to varying degrees. Not everyone in the development team will necessarily bear the same level of responsibility for the faulty software and some, perhaps, will have no responsibility at all. A careless programmer can bear responsibility, as can someone who is negligent in the testing of the product. In most cases, ultimate responsibility must be borne by the project manager whose role is to oversee the project. Responsibility cannot be avoided simply because a large team is involved in software development.

A further aspect of collective responsibility must be noted. While not all individuals are at fault, and those that are do not share equal responsibility, there is a sense in which all, with only several exceptions, can plausibly be held accountable. If I do nothing to try to change the situation of my team's or company's carelessness, I am helping to perpetuate a climate in which faults, mistakes or accidents are more likely to occur. So I can and ought to be held accountable to some extent, even though I did not cause the events, or intend them to happen. In Larry May's terms, I can be morally tainted even if I cannot be blamed for the event itself (May and Hoffman, 1991, chpt 15). The exceptions are where I have protested or attempted to change the situation, or am not in a position of influence; for example, if I could not reasonably be expected to know of the situation (See Feinberg 1970, chpt 9 for more discussion of these exceptions). Given this, all, or most, members of a group, team or company can be held collectively responsible for faulty software. Software developers can no more avoid moral responsibility for their products than developers and builders of aeroplanes, cars or bridges.

A central component of a professional attitude, it was stated earlier, was taking seriously one's responsibilities and acknowledging that one is morally responsible for one's actions. Three senses of responsibility were distinguished: causal, moral and role. The discussion so far has primarily concerned the first two, but role responsibility is also important in professional attitudes. The practitioners of many different occupations have responsibilities to others — because of their roles — over and above the responsibilities that they have as human beings. An electrician has special responsibilities with respect to the wiring of houses, just as a medical practitioner has special responsibilities towards patients, and software developers do towards their clients. Doctors, lawyers, computer scientists and others have professional bodies with codes of ethics to regulate or guide their members. Builders and plumbers have the same, even if they are not called professional bodies. A professional attitude involves a moral stance. This includes the ability to see one's work in the context of the larger society, how it contributes to that society, and taking moral responsibility for one's work. Codes of ethics attempt to clarify and partly formalise this stance.

A professional attitude in ICT

At the beginning of this chapter we made the twin claims that ICT was not a profession in any significant sense and that this did not matter. The argument has been that there are no professions, in the sense that many claim that there are, but that nothing much follows from that because what is important are *professional attitudes*. These professional attitudes are closely related to the special responsibilities that come with particular occupations, and this set of occupations is much broader than the set of the so-called professions. People in many occupations have special responsibilities over and above the responsibilities that we all have. A medical doctor, for example, has the responsibility to provide health care, and 'the role of the lawyer is to promote justice', as we saw in e) above. We all have some responsibility as citizens to assist others with respect to health care and justice, but not to the degree that doctors and lawyers respectively do. ICT practitioners have special responsibilities, too, with respect to society because of their expertise. Not only their direct clients but also the general public are vulnerable to the systems developed and maintained by these workers. It does not matter if ICT is not a profession and is not seen as one, even by its practitioners, but it is vitally important that those in the industry have professional attitudes toward their work, especially those who develop and maintain computer systems, including networks. This is the area in which professionalism in ICT has been most discussed and with good reason. These systems have a profound impact on modern life and system failures or malfunctions can vary from annoying to catastrophic. While there have been many calls for greater professionalism, it is not clear that these calls have been heeded. A recent report suggested that professional attitudes in the industry are not as prevalent as is desirable, and the industry as a whole has been reticent to recognise moral responsibility for software systems. The industry clearly is not devoid of people with professional attitudes (see Section Two of this book), but such attitudes do not always come to the fore as much as they should. For example, in a recent study, some of the interview comments suggested that there is not a strong commitment from all ICT workers to professionalism. While only one comment states explicitly that 'Most IT people do not want to be professional', other statements, such as 'Lots of money in the industry', 'Computing has been seen as a way to make money and then leave', 'People don't take responsibility', 'Gap between what the customer wants and what they need. Often they are given what they want with full knowledge that this is not what they need', 'No encouragement to be responsible', and 'Outsourcing is a problem because of lack of responsibility' indicate that many practitioners do have professional attitudes and they are concerned about the substantial number of their colleagues who do not share their attitude (Lucas and Weckert, 2008).

Typical licensing agreements for shrink-wrapped software also indicate a lack of professional attitudes in the industry:

8. Disclaimer of Warranties.

A. You expressly acknowledge and agree that use of the Apple software and services is at your sole risk …, the Apple software and services are provided 'as is', with all faults and without warranty of any kind, and Apple and Apple's licensors … hereby disclaim all warranties and conditions with respect to the Apple software and services, …, including … satisfactory quality, …. Apple does not warrant … that the operation of the Apple software or services will be uninterrupted or error-free, or that defects in the Apple software or services will be corrected (the text has been changed from uppercase in the original to make it more readable) (APPLE INC, 2008).

The above disclaimer makes it clear that no responsibility is taken for faults in software. It is common for new cars, vacuum cleaners and so on to be faulty, but we expect that the warranty will cover their replacement or repair. Responsibility is taken for those products. It is true, of course, that thorough software testing is difficult. It is true, too, that where a problem arises it could be caused by the operating system or the hardware on which the purchased software is running. Nonetheless, these disclaimers, which are the norm rather than the exception, do suggest that the ICT industry as a whole does not take responsibility as seriously as it could, and taking responsibility is, as we have argued, part of professionalism. The commitment to professionalism in the sense of having a professional attitude as defined earlier, does not appear to be as strong as is desirable.

Conclusions

We have argued here that ICT is not a profession, in any significant sense, for a number of reasons, but, primarily, because the notion of a profession itself is not important or useful. What is important is having a professional attitude and all that that entails, particularly a willingness to take responsibility for one's actions. Encouraging such attitudes should take precedence over efforts to make ICT a recognised profession. ICT practitioners play an essential role in many aspects of modern life — their work has major impacts on the economy and, in the case of safety-critical systems, their decisions can have life and death implications. Having a professional attitude in the ICT industry is equally important as in medicine and law.

References

Apple Inc, 2008, 'iwork software license agreement', <http://images.apple.com/legal/sla/docs/iwork09.pdf> 10 November.

Australian Council of Professions (ACP), 1997, 'Definition of a profession', <http://www.professions.com.au/defineprofession.html>

ACS, 2010, *Australian ICT statistical compendium 2011*. <http://www.acs.org.au/__data/assets/pdf_file/0019/9307/Australian-ICT-Statistical-Compendium-2011.pdf>

Boykin, D, 2007, 'Is it time to license software engineers?', *PE Magazine*, December, <http://www.nspe.org/PEmagazine/pe_1207_Software_License.html>

Feinberg, J, 1970, *Doing and deserving: essays in the theory of responsibility*, Princeton University Press.

Gotterbarn, D, 2004, 'Informatics and professional responsibility', in TW Bynum & S Rogerson (eds), *Computer ethics and professional responsibility*, Wiley-Blackwell, pp 107–18.

Healy, T, 2004, 'Licensing developers won't work', *TechRepublic*, 24 November, <http://www.techrepublic.com/article/licensing-developers-wont-work/6310377>

Johnson, D, 2001, *Computer ethics*, 3rd edn, Prentice-Hall, Upper Saddle River, NJ.

Knight, JC, & Leveson, N, 2002, 'Should software engineers be licensed?', *Communications of the ACM*, November, pp 87–90.

Lucas, R & Weckert, J, 2008, *Ethics and regulation in the ICT industry*, report for the Australian Computer Society, Centre for Applied Philosophy and Public Ethics, Charles Sturt University, Canberra.

McCall Smith, Alexander, 2010, *The double comfort safari club*, Little, Brown Book Group Limited.

Master Plumbers' Association of Queensland, 2002, 'Code of ethics, trade customs and practices', <http://www.thisplace.com.au/sponsors/mpaq.htm>

May, L, 1991, 'Metaphysical guilt and moral taint', in L May & S Hoffman (eds), *Collective responsibility: five decades of debate in theoretical and applied ethics*, Rowman & Littlefield Publishers, Savage, Maryland. pp 239–54

Mead, NR 2009, 'Software engineering education: how far we've come and how far we have to go', *The Journal of Systems and Software*, vol 82, pp 571–75.

White, J, and Simons, B, 2002, 'ACM's position on licensing of software engineers', *Communications of the ACM*, November 2002, p 91.

7. Being a good computer professional: The advantages of virtue ethics in computing

Richard Volkman
Southern Connecticut State University

Introduction

When smart and well-educated professionals misbehave, ethicists have to wonder if we could have done anything to prevent it. After all, while it may be morally satisfying to simply assign full blame for the woes of Enron, Tyco, Worldcom, and others, to corrupt corporate leaders, such an analysis begs the further questions: Why did morally deficient actors rise to such prominent positions in the first place? Why were the prevailing standards, policies, and practices of professional ethics — embodied in implicit and explicit ethical controls — so unable to regulate conduct that in hindsight seems obviously beyond the pale? One plausible answer has to do with the impotence of standard utilitarian and deontological styles of reasoning, in the face of prevailing uncertainty about values and outcomes in the context of the 'new economy.' These concerns are especially acute in computer ethics; while change and uncertainty is a problem in business, it is at the very heart of computing and the circumstance that gives rise to computer ethics (Moor, 1985). In general, if one cannot rationally determine the outcomes of action or how to evaluate those outcomes, then utilitarian and deontological reasoning cannot guide action, and this leaves ample room for rationalisation, and this undermines the efficacy of these approaches to computer ethics.

In contrast, principles of good character do not derive from statements of eternal, universal values or fortuitous outcomes; a virtue is a trait of character that is good for the person who has it, where the value of good character derives from the agent's own commitments. It follows that virtue ethics is not susceptible to rationalisations based on extrinsic rewards, which are made uncertain by a prevailing relativism about values or by a volatile economic environment. An analysis of the culture of computing grounds the classical virtues of integrity, honesty, courage, and good judgement in the antecedent beliefs and values typical of those entering the profession. Virtue helps to flesh out the spirit of the profession, as this stands behind professional codes and other artefacts of professional ethics.

The principle aim of this paper is to introduce the educated lay reader, especially current and future computer professionals, to the basic advantages and strategies associated with the use of virtue terms in the description and prescription of ethical conduct. In light of this goal, the essay seeks to capture the main ideas that motivate a virtue- or character-centred approach to ethics, while remaining uncommitted with respect to a number of interesting theoretical and philosophical matters, including even the question of whether the thoughtful application of virtue terms to practical reasoning should be conceived as any kind of theory at all. An applied ethics of virtue is found in thinkers as diverse as Aristotle to Nietzsche, and it is not the aim of this essay to advance one or another particular approach or to address the metaphysical or epistemological doctrines that distinguish them. Virtue discourse works as a guide to action, no matter how these foundational questions should turn out, and that may be the principle advantage of the virtue ethics approach to applied ethics. If this essay succeeds in showing that there are good reasons to become fluent in the language of virtue, then a further examination of the particular accounts of virtues, and whether virtue ethics is an alternative ethical theory or an alternative *to* ethical theory, can be fruitfully explored.

Problems in applying an ethics of rules and outcomes

At least when applied to the contexts of professional ethics, utilitarian and deontological ethics generally boils down to a concern for good outcomes, on the one hand, or obedience to good rules on the other. This way of thinking about ethics is problematic, however, especially in a cultural and economic context characterised by relativism and uncertainty. Relativism about values and uncertainty about outcomes straightforwardly undermine any utilitarian calculus as a guide to action. One cannot choose the course of action that will generate the best outcome if one has no standard for evaluating outcomes and no means of predicting outcomes. The situation is little better for deontology. If morality is ultimately about acting on the right rules of conduct, then one cannot figure out how to behave morally unless one is able to determine what the right rules are and how they apply to a given circumstance. But conditions of change call established rules into question and can blur the lines between appropriate and inappropriate application of the rules. To offer a concrete example, no deontologist (except perhaps for Kant himself) would maintain that it is *always* wrong to lie. In the context of a perceived 'new economy', it is a short leap to the view that 'creative' accounting is not really lying, or, if it is lying, then it is not a case where lying is wrong. After all, the stock just keeps going up.

Of course, this is plainly mere rationalisation. Utilitarian and deontological styles of reasoning, however, invite this sort of rationalisation. In both cases, an illusion of ethical precision combined with the notion that ethics is a constraint on one's self-interest creates an incentive to find ethical loopholes. Even if professional philosophers can discern something amiss in this characterisation (and I do mean *if*), it is plain enough that business leaders and other professionals could get this impression. Whatever one thinks about the ultimate grounds of ethics, the point here is that ethicists need to talk about and emphasise something more than rules of conduct and calculations of the overall good. If morality seems to be nothing but a bunch of rules or calculations that operate against what I would really like to do, and if certain clever considerations seem to show that a given rule does not apply in this particular case, then there is a strong incentive to be overly clever. This encourages a lawyerly approach to ethics that is quite at odds with the spirit of utilitarian or deontological morality, but entirely predictable if morality is reduced to good outcomes and right rules. If one tinkers with the calculation or evaluation of outcomes, or if one fuzzes the rules or their application, then just about anything goes. And the tinkering and fuzzing do not require willful fudging by the agent. In our current cultural, economic, and technological circumstance, transformative change does this all by itself.

Of course, all this applies to the world of computing even more plainly than the world of business. Jim Moor's seminal 'What is computer ethics?' identifies the field with 'conceptual vacuums' that give rise to 'policy vacuums' as a consequence of 'the transformation of social institutions' attributable to the introduction of computing (Moor, 1985). Against this, Don Gotterbarn has argued that an overbroad conception of computer ethics is 'unmanageable' and that it makes for 'unsolvable problems', while appealing to the uniqueness or transforming effect of computing distracts the field from its proper focus on professional ethics (Gotterbarn, 1991). In this connection, Gotterbarn and others have emphasised the importance of articulating the nature and morality of computing as a profession, culminating in the drafting of the 'Software Engineering Code of Ethics and Professional Practice' by a joint task force of the IEEE and the Association for Computing Machinery (ACM).

But it has to be acknowledged that the difficulties noted above for utilitarian and deontological ethics apply to the creation and application of a code of ethics (which is, after all, an essentially deontological enterprise). If a code of ethics encourages professionals to think of ethics as reduced to a collection of rules, then lawyerly rationalisations are encouraged, especially insofar as relativism and change undermine the authority of any particular rule or its application. In this vein, John Ladd worries that, 'a code of ethics can be used as a cover-up for what might be called basically "unethical" or "irresponsible" conduct' (Ladd, 1980). In anticipation of this concern, the IEEE/ACM code explicitly rejects reading the code as a moral checklist. The preamble asserts:

> It is not intended that the individual parts of the Code be used in isolation to justify errors of omission or commission. The list of Principles and Clauses is not exhaustive. The Clauses should not be read as separating the acceptable from the unacceptable in professional conduct in all practical situations. The Code is not a simple ethical algorithm that generates ethical decisions. In some situations, standards may be in tension with each other or with standards from other sources. These situations require the software engineer to use ethical judgment to act in a manner that is most consistent with the spirit of the Code of Ethics and Professional Practice, given the circumstances.

Of course, this leads one to wonder how to identify the 'spirit' of the code. The answer seems to lie in the genealogy of the code itself: 'As this Code expresses the consensus of the profession on ethical issues, it is a means to educate both the public and aspiring professionals about the ethical obligations of all software engineers' (IEEE/ACM Joint Task Force, 1999).

This suggests that the spirit of the IEEE/ACM code can be understood as the consensus of the profession on ethical matters. Before joining the task force, Michael Davis had argued 'a code of ethics is primarily a convention between professionals. According to this explanation, a profession is a group of persons who want to cooperate in serving the same ideal better than they could if they did not cooperate' (Davis, 1992). The moral force of the code, in this view, comes from its being the articulation of the considered judgement of those in the profession about the best way to achieve the values that define the profession. Engineering, for example, is about 'the efficient design, construction, and maintenance of safe and useful objects,' and the values embedded in this description of what engineering is about characterise what it means to be 'thinking like an engineer' (Davis, 1992). The code of ethics expresses the culture of the profession, and the process whereby that culture is articulated in a written form is normatively more important than the precise articulation of the rules or values in the artefact itself. The culture that animates the code identifies the 'spirit' of the code.

This account goes a long way towards resolving the main problems, noted above, for utilitarian and deontological ethical approaches. But one needs to see that the solution is the result of abandoning stereotypical deontological and utilitarian modes of thinking in favour of 'thinking like an engineer'. Instead of specifying *the* rules of right conduct, the antecedent culture identifies values that generate reasons for action, even in the absence of clear rules of conduct. Moreover, the authority of the culture need not be asserted as expressing categorical judgements of value. Relative assertions will do just fine. The normative force of the code is settled, at least within the practice of engineers, by correctly noting: 'This is who we are; this is what we value; this is how we achieve our goals.'

This, however, raises a number of questions. Some have to do with making the account more clear: Where the particular principles and clauses of the code break down, how does one guide action? That is, how can one articulate the spirit of the code when its deontological formulation is questioned? What are the units for such an analysis? What are the terms of debate when the spirit of the code is itself in dispute? Other questions have to do with the evaluation of the written code: Does this code actually express the culture of the profession? Is that culture worthy of expression in a code? Some of these issues turn on an empirical description of the culture, while others demand normative and conceptual clarification. Both sorts of concerns will be addressed below with respect to the profession of computing. What we find is that the space 'between the lines' of the code is populated by virtue, and that virtue terms can adjudicate and articulate moral debate without the problems identified for utilitarianism and deontology.

The culture of computing

By a culture, I mean 'the values, attitudes, beliefs, orientations, and underlying assumptions prevalent among people in a society' (Huntington, 2000). The relevant 'society' in this context consists especially of those 'involved in the design and development of computer artifacts' (Gotterbarn, 1991), but it also extends to those consumers of computer artefacts who identify with technology as an integral part of their lives. In short, I hope to identify something of the culture of 'computer geeks.' That there is such a culture, antecedent to philosophical theorising about professional ethics, is indicated by the fact that everyone has a sense of who the 'computer geeks' are, just as surely as one has a sense of who is meant by 'hippies' or 'Bible-thumpers'. It is significant that there is no similar sense of who is picked out by 'telephone enthusiasts' or 'pea-lovers.' While plenty of people like telephones, no-one seems much interested in evangelising for them, and people who like telephones probably have very little else in common. This explains why there is no such thing as 'telephone ethics', despite the transforming effect of telephones on society. Telephones do not represent a way of life. Things differ in important ways in computing.

Computer technology does have evangelists, and those who embrace computing sufficiently share values and beliefs such that they represent a coherent social unit.

On the other hand, as with hippies or Bible-thumpers, there are a host of subcultures and competing assumptions among computing enthusiasts. Furthermore, social categories routinely overlap. Some computer geeks are Bible-thumpers, and some computer geeks are hippies. Fortunately, it turns out that these complications have very little impact with respect to the main thesis of this essay. If an appeal to the core values of geek culture is sufficient to ground the virtues, then it is proved that geek culture combined with Evangelical Christianity or tree-hugging environmentalism is sufficient to ground the virtues. It is left open whether these combinations can be coherent and sustained. While I am sure there is good reason to have other commitments as well, the purpose of this essay is simply to see how much ethical content can be gleaned from the culture of computing by itself. For that purpose, what matters is whether the spirit of professional ethics can be articulated in terms any computer enthusiast would have to grant simply in virtue of being a computer enthusiast.

The reason for focusing on the culture of the enthusiast instead of what might be the more enlightened or elevated culture of mature computer professionals stems from the educational function of professional ethics, and especially the IEEE/ACM code. (IEEE/ACM, 1999) Appealing to the judgements of mature members of the profession amounts to little more than preaching to the choir, unless it can be shown that these judgements are implied by the antecedent aspirations of the intended audience. Insofar as professional ethics is for the enculturation of new members to the profession, arguments for more mature attitudes must appeal to attitudes already consonant with those of prospective professionals. Students who aspire to become computer professionals are typically motivated by either the promise of a lucrative career in computing, or an antecedent enthusiasm for technology (or both). Enthusiasm for money has little to do with computing, per se; enthusiasm for technology, however, obviously does. As we shall see, an analysis of the virtues (especially integrity) shows that mere enthusiasm for wealth is problematic as a way of life, even on its own terms; but if it is demonstrated that the values of the profession can be conceived as the mature expression of enthusiasm for technology, then one important category of serious students can be motivated to think professionally. The bottom line is that a successful education needs to start where the students are, and an important category of students has already embraced geek culture.

It may seem that mere enthusiasm for technology is too squalid as a way of life to ground ethical judgements. A closer look at the cultural artefacts and self-reflecting essays of geek culture, however, reveals values that amply serve as raw material for a more substantive ethics. The assertion that 'Technology is cool' expresses an evaluation that connects with a wide range of other values and ultimately grounds a coherent conception of the good life.

Wired magazine features a monthly department titled, 'Fetish: technolust'. In this department, cool new gadgets and technologies are described alongside full-colour glossy photos. Sometimes the descriptions make it clear why the editors believe a featured device is desirable enough to be compared to the erotic, but most of the time it suffices to simply describe what it does. The audience of *Wired* does not need to be told what makes something cool. However, such judgements need not be inarticulate. In fact, just the opposite is the norm. Product reviews are surely among the most ubiquitous of written artefacts in geek culture. They are found in magazines like *Wired*, *MacWorld*, *SysAdmin*, *PlanetPDA*, and *PCWorld*; they are the foundation of huge and famous websites like *Tom's Hardware Guide*, and they are regular subjects of debate on *Slashdot*. For the ethicist or cultural anthropologist, what is significant about product reviews is that they express evaluations based on the common values that help to define the culture. In this context, the fact that *Wired* can simply display certain items as obviously worthy of technolust demonstrates the extent to which these values can be taken for granted, at least among members of the culture. The coolness of some objects does not need to be demonstrated; it is *observed*.

Judging that a gadget is cool is in large measure an aesthetic judgement. This aesthetic is exemplified in the behaviour of a high-school friend who used to compete in computer programming contests at a nearby engineering college. These events were typically followed by a small trade show where technology and engineering companies would display their latest work in the hopes of attracting future employees. (Note, they displayed their works; they did not advertise salaries). My friend discovered at one of these shows a mini-switch that completely captivated him. These switches were nothing special to look at, but he loved the solid, sturdy feel in such a small device, and especially the tactile and auditory feedback it gave the user. When one clicked the button, one knew instantly whether the thing had switched. *Clickety-click*. He handed one to me, beaming with the same expression a singer friend wore at his first Broadway musical. *Clickety-click*. 'This is a *nice* switch,' he gushed. *Clickety-click*. For weeks after, he carried a pocketful of these switches with him, absentmindedly clicking them in satisfaction as he worked and played. Like most aesthetic judgements, it is hard to convey the meaning of this experience to anyone who does not already get it; it is not merely to recognise but truly to *appreciate* quality craftsmanship and clever technical solutions to ordinary problems, the way one might appreciate a painting in a museum or a music performance.

I submit that this appreciation reflects a judgement — experienced by the initiated as an immediate observation — that the object of appreciation is a clever, elegant, well-designed, efficient, or otherwise powerful solution to some technical puzzle. It is in these terms that product reviews tend to extol or disparage their objects, and it is in these terms that debate rages about the merits and demerits of competing operating systems, chip architectures, or industrial design solutions. To be a computer geek is to be fluent in this language, and it is a language of values with implications beyond what kind of computer to buy.

Many of these enthusiasts aspire to go beyond being connoisseurs of technology and to design and create their own solutions. These are the students we are trying to reach in courses about professionalism in computing, and it is fortunate for us that they are already embedded in a way of life receptive to talk of objective evaluations. From this commitment stem the other main hallmarks of geek culture, especially their commitment to reasoning and justification. In the world of technology, reasoning and know-how matter; mere appeals to authority are worthless. This is evident in any serious debate on *Slashdot*, where the argument is not settled until detailed references are provided. Usually, this takes the form of links to other sites, so each participant can be satisfied about the validity of the claims. As in academic debate in other venues, however, these discussions usually are not settled but are met with competing arguments and data, similarly referenced.

This commitment to reasoning and justification in debate also indicates that members of the community are generally self-navigating and resistant to leadership or indoctrination. To substantiate one's case by linking to external documentation so anyone can examine the data for herself only makes sense in a community whose members insist on thinking for themselves. As Jon Katz has amply documented, the online community does not tolerate self-proclaimed leaders and does not promote leaders of its own. For geeks, mainstream symbols of authority and prestige are suspect, while scientists, technologists, and others who routinely defend their claims against invited criticism, are revered (Katz, 2000). Instead of appealing to the authority of tradition or position, arguments are expected to stand or fall on their own merits, with the further expectation that the very best arguments will withstand withering criticism and scrutiny from the community at large and ultimately emerge as the prevailing view. Geek culture thus embodies an instinct for Popperian scientific method and the marketplace of ideas. In contrast to following authoritative leaders, the geek community sees itself as tracking truth, guided by the invisible hand of reason. There is a palpable and sometimes snide impatience with those who will not or cannot justify their claims, as one *Slashdot* article makes explicit: 'This is the story of one computer professional's explorations in the world of postmodern literary criticism. Wouldn't it be nice to work in a field where nobody can say you're wrong?'

The idea that there is a best way of doing things and that we can learn what it is through a process that rationally filters and evaluates distributed knowledge grounds an optimistic belief in technological progress. The notion of progress is controversial among some social scientists, but progress is obvious to tech enthusiasts because they experience it daily. It is manifest that the computers, cell phones, cameras, and other gadgets they work and play with today are much better than the ones they worked and played with just a few years ago. Everything has gotten faster, more powerful, smaller, cleverer, more efficient, etc. *Clickety-click*. If things outside of technology and engineering have not been improving at the same rate, it is because 'they' (the non-geeks who dominate mainstream society) are too stupid, cowardly and sheeplike to submit their ideas to the processes of rational scrutiny and justification that generate real progress. This commitment to progress implies an embrace of change that is nicely captured by the conclusion of *Wired*'s montage, 'Change is good': next to a photo of burning man reads the caption, 'The future is in Beta' (Rossetto et. al., 1998).

This commitment to progress helps to ground a coherent and ethical way of life in geek culture because it permits members of the community to conceive of their lives in the manner of a 'quest.' Practical optimism and progress imply that things can be made better and that technologists are on the front lines of making it happen. They are the solution to whatever might be the problem. Moreover, those who participate in the tech community are each playing a role in making things better simply by actively participating in the process that generates and regulates technological change. This notion that embracing the process of technological change can ground a coherent life plan is perhaps best illustrated by Steve Jobs' legendary wooing of Pepsi CEO John Sculley to serve as CEO of Apple: 'Do you want to sell sugar water for the rest of your life, or do you want to change the world?' Geek culture wants to change the world, or at least surf the wave of technological change to a better life. This is the quest of geek culture and, as Alasdair MacIntyre notes, making sense of one's life in terms of a quest is the hallmark of a way of life that can ground the virtues (MacIntyre, 1984).

The analysis of character

To speak of one's character is to talk about who one is. Describing a person's character identifies him or her in terms of the values, beliefs, projects, commitments, and ordinary ways of reacting to the world that explain and justify his or her behaviour. We routinely rely on traits of character to predict behaviour and to thereby justify reliance. For example, we say, 'He is a hard-working fellow. I'm sure he'll get the job done.' And to be sure, the fact that he is a 'hard-working fellow' is among the reasons he was entrusted with the task in the first place. Evaluations of character are standard practice in everyday life — probably more common and useful than evaluations of action. Both positive terms like 'sweety', 'cool', and 'upbeat'; and epithets like 'creep', 'jerk', 'putz', or 'bonehead' are all evaluations of character in the sense that they speak to the being of the person evaluated rather than his or her actions. The ethical terms of everyday life — in contrast to the stuffy and stilted evaluations of academic philosophy — are aretaic, not deontic. Notice that these evaluations and our reliance on them to predict and explain behaviour assume that traits of character are relatively immutable and deep. For this reason, any agent will have to take evaluations of her own character very seriously, whatever her particular projects and other pursuits might be.

Character is typically analysed and evaluated in terms of the virtues. By a virtue is meant any trait of character that is good for the agent who manifests it, while a vice is a character trait that is bad for the agent. To say that a person manifests or lacks a virtue is both to describe and evaluate that person at the same time, and the description and evaluation cannot be separated without distorting the meaning of the term. As Philippa Foot notes, virtue terms are not unique in this respect. For example, one cannot understand what an 'injury' describes without also understanding that it is bad to be injured. The surgeon does not 'injure' a patient in the course of performing needed surgery, even if her actions are accurately described as 'cutting someone with a sharp knife, leaving a wound' (Foot, 1978). Similarly, one cannot gloss the descriptive meaning of a virtue or vice without understanding the normative context that would make the designation appropriate or inappropriate. For example, while honesty is the virtue related to truth-telling, it would be wrong to say a person is dishonest because he told a client that last minute changes to a project will be 'no problem'. It is surely not a mark of bad character that a person takes challenging clients in stride. Even though his utterance is not the whole truth, dishonesty does not describe everyone who makes an untrue statement. For proper use of the term, the evaluation matters as much as the description.

So we need to get clear about the evaluative content of the virtues as well as their descriptive components. In the Aristotelian tradition, when thinking of the virtues it is helpful to put things in terms of the Doctrine of the Golden Mean, which is a sort of heuristic for understanding the way virtue terms typically work. It must be emphasised that the Doctrine of the Golden Mean is not a principle for deducing the nature of a virtue, nor is it the last word in an analysis of virtue. Indeed, it is hardly the first word. It is, however, a good first step to take in organising one's thoughts about the content of our virtue terms. On the Aristotelian view, virtues describe the character of one with just the right amount of concern for some good, where the evaluative component of the virtue is unpacked in terms of the 'right amount' of concern for that good. Failures of character (vices) typically result from having either too much or too little concern for some good, and virtue is the mean between these vices of excess or deficiency. To pursue the example of honesty, truth-telling is a good. Too much concern for truth-telling, however, makes one a stickler. Being a stickler will make it hard to get along with others, thereby undermining one's pursuit of other goods like friendship or wealth. Even worse than being a stickler, however, is being dishonest, which is an even greater threat to friendships and trade relations. Even if one could *be* dishonest but *appear* honest (which is unlikely in any event), such a character could never *be* a good friend. Thus, anyone who recognises the value of real friendship and the relative immutability of character will have to admit the value of honesty.

While relativism about value and uncertainty about outcomes can undermine utilitarianism and deontology as guides to action, virtue terms do not derive their normative force from asserting that any particular state of affairs is categorically good. Instead, virtues relate to things that are more-or-less obviously good for me (and good for you). While there is endless debate about what is 'really' good, and what rules if any 'really' count as the categorical command of reason (or even whether there are any such rules), and while these debates are exacerbated by the various transformations wrought by the computer revolution, there is nothing mysterious or controversial about the claim that my own safety is good for me; at least, no-one generally has to be persuaded of such a claim. If I admit that safety is good for me, then I have a clear reason to manifest some concern for my own safety, and if I admit that there are other things that are also good for me, then I have a reason to manifest just the right amount of concern for safety — no more and no less. That is, I have a reason to manifest courage and to avoid being rash or cowardly. Instead of endless debate about categorical goods or uncertain outcomes, a substantive ethics emerges once one identifies the ends to which one is actually committed. Hence the Socratic dictum, 'Know Thyself!' and the Nietzschean imperative, 'Become Who You Are!'.

At a more practical level, an emphasis on character that starts from one's present commitments does not invite overly clever or lawyerly rationalisations of actions that are contrary to the 'spirit' of morality. Only a fool would expect (or try to provide) a rigorous refutation of the whole of deontological or consequentialist thought in a single paper, but it should be plain that an ethics of character enjoys very real advantages in the sphere of applied ethics. The virtue ethicist is not trying to convince anyone to abandon or limit the pursuit of her commitments in the name of some abstract notion of the 'moral law' or 'maximum overall net utility'. Rather, virtues are an indispensable means to success in one's own pursuits. Who does one intend to convince with lawyerly and overly clever defences? Talking one's self into bad character does the gravest harm to one's self, on one's own terms. When the ultimate judge and jury consist in the values to which one is already personally committed, the notion of 'getting away with it' loses all meaning. Getting away from whom? Yourself? How could that be a victory? Once this is understood, the more abstract and theoretical concerns of moral philosophy no longer stand in the way of applied ethics. This is especially valuable in the context of professional ethics, since an analysis of the culture and values that define the profession reveals the relevant commitments (and the ongoing terms of discussion about them within the tradition), such that 'Know Thyself!' becomes a tractable imperative.

As for guiding action, one must not misunderstand the sense in which attention to one's own character serves as a guide to action. Sometimes, utilitarian and deontological ethics have encouraged scholars to mistakenly think that a good guide to action will churn out a precise list of dos and don'ts. Against this, and echoing the IEEE/ACM code cited above, Rosalind Hursthouse notes the absurdity of thinking that ethics can be reduced to a simple algorithm that can be mastered by 'any clever adolescent' (Hursthouse, 1991). Ethics involves reflecting on the nature and meaning of life itself. It is preposterous to suppose this can be accomplished by simply doing the maths, or that it can be reduced to one or several simple rules.

Virtue ethics 'guides action' by focusing one's attention on one's striving to be the sort of person one wants to be. It asks one to reflect on who one is, and to evaluate one's own way of life in light of the sum total of one's commitments and values. Of course, not all ways of life can withstand this scrutiny. While desirable ways of life will ground virtues that form a coherent whole and confer direction and meaning to one's life, ways of life that do not imply the classical virtues are generally incoherent or otherwise undesirable. Acknowledging this presents one with an opportunity to revise one's conception of the good life and to become the sort of person one wishes to become — a person of good character, on one's own terms. When a person manifests good character, her actions flow immediately from her character in a manner akin to instinct or habit. The 'spirit' of ethics guides her action, not the letter of the law.

Good character matters in the culture of computing

Virtue ethics is illustrated by showing how the way of life associated with the culture of computing generates the virtues while close alternatives do not. As we shall see, this analysis makes it clear that those living by the close alternatives have good reason to adopt the culture of computing instead. As it turns out, the failure to support virtuous conduct creates serious obstacles to leading the good life, even if the good life is conceived solely on one's own terms. In contrast, the culture of computing does support virtuous conduct leading to a coherent and defensible conception of the good life. In short, advocates of the values that define the technological enthusiast will be better able to succeed in the pursuit of these values if they manifest the virtues of integrity, honesty, courage, and good judgement.

Integrity is the virtue associated with the right amount of steadfastness in one's defining values, goals, and other commitments. Just as the integrity of a building is given by its structural soundness, the person of integrity stands firm and true. Being pig-headed, narrow, or stubborn are vices of excess with respect to steadfastness, while one who is flighty, unserious, or inconstant is deficient in this good. In contrast, the person of integrity is open to criticism and will change her course as necessary, but only in light of good reasons and due deliberation. The person of integrity can be relied on to be who she is in spite of temptations or distractions. As one slogan has it, 'Integrity is doing the right thing even when no-one is looking.' The person of integrity does what she does because that is what she is all about, not because somebody might or might not be looking.

For the culture of computing, integrity involves conceiving one's life as a quest in pursuit of technological progress and embracing the practices and social institutions necessary for that pursuit. One does not manifest integrity if one behaves in a manner that will undermine the success of one's core projects. In light of this, teaching new members of the profession to value the profession itself can be cast in terms of integrity. As scholars of technology routinely emphasise, and as contributors to *Slashdot* often assert, technology is a social endeavour. 'Technology consists not only of artifacts and the tools and processes needed to produce them, but also of the entire social organisation of people and materials that permits the acquisition of the knowledge and skills needed to design, manufacture, distribute, use, repair, and eventually dispose of these artifacts' (Winston, 2003). Commitment to the technological project therefore issues in a commitment to the whole set of social institutions presupposed by that project. Thus, even for a computer professional who starts with nothing but an appreciation of cool technology, being true to his conception of the good requires a commitment to, among other things, the profession itself. The virtue of personal integrity hereby stands behind the IEEE/ACM code's declaration that, 'Software engineers shall advance the integrity and reputation of the profession consistent with the public interest.'

Less obviously, the integrity of a computer professional also requires commitment to the wellbeing of end users. After all, they are an indispensable part of the social milieu of technological progress. This responds to Gotterbarn's worry that a puzzle-solving culture does not sufficiently take users into account (Gotterbarn, 1991). As conceived here, concern for the user is not an extra burden for the software engineer but an integral part of the whole pursuit. Gotterbarn is surely correct to note that those who focus on puzzle-solving are prone to overlook this. Moreover, the elitism of computer culture tends to exacerbate the problem by inspiring contempt for the ignorance and irrationality of the lowly end user. At least for those already embedded in the culture of computing, however, demonstration of the integral relation between end user satisfaction and technological progress will do more to address these concerns than any amount of finger wagging about 'service' or 'public interest.'

Honesty is closely related to integrity. As noted above, it is associated with truth-telling, and it is necessary for trade relationships and friendships of all sorts. The dishonest computer professional not only brings shame to the profession, thereby revealing his failure of integrity, but also undermines the general social circumstances necessary for the creation, implementation, and dissemination of technological solutions. Hype and marketing are not necessarily dishonest, even where such communications are not entirely frank. However, when the vague boundary between hype and fraud is crossed, progress suffers. Similarly, disseminating FUD (fear, uncertainty, doubt) about a competing product or announcing vaporware in the hopes that a potential competitor will never come to market, plainly stands in the way of technological progress. The whole point of these practices is to protect or gain a market advantage that is out of proportion to the merits of one's technology, and that plainly subverts the market's ability to adjudicate competing technologies in favour of the best solution. This analysis helps the new professional to understand why such dishonest actions are rightly denounced in the IEEE/ACM code, helping to flesh out the 'spirit' of the code.

Courage is also a virtue of the good computer professional. The willingness to embrace innovation and change is not characteristic of the coward; nor does one who is over concerned with safety submit beliefs and ideas to the rough and tumble of rational debate. On the other hand, the rash inventor or engineer will tend to make sometimes-tragic mistakes and, therefore, fail to advance technology. In geek culture, success requires real courage. In the day-to-day affairs of the good computer professional, courage reveals itself in standing up for one's ideas while submitting them to rational scrutiny. In more extreme cases, the computer professional may need the courage to blow the whistle or walk away from a project that directly or indirectly conflicts with the integrity of the profession. For example, a database of medical records that is deficient in security undermines the advance of technology, since any harm to the public will engender fear of similar projects in the future. Especially when money is tight or when managers do not comprehend the possible consequences of shoddy work, it can take real courage to maintain one's integrity and the integrity of the profession.

After integrity, good judgement may be the most important of the virtues since, without good judgement, the other virtues cannot be relied on to secure one's goals. If one mistakes technology for the gadgets it produces, or if one does not accurately foresee how a project might advance or undermine the goal of technological progress, then one's bad judgement will undermine one's integrity. Good judgement is difficult to characterise without circularity; one cannot merely describe good judgement in terms of getting right results, since that would leave evaluations of character hostage to fortuitous outcomes. Instead, good judgement needs to be conceived in terms of a method for arriving at right answers, and this suggests having the right amount of concern for both rational rigour and educated perception in fixing belief. Reasoning that is shallow, narrow, or short-sighted will not advance technological solutions, and may actually retard the progress of technology. On the other hand, pessimistic scepticism is directly contrary to the ethos of innovation and progress. To avoid either extreme, the good computer professional needs to develop a fine aesthetic sensibility for good design and a habit of working out problems through rational debate within a wider community. Fortunately, the resources of good judgement are at the very heart of geek culture's embrace of rationality situated in a critical community of debate and justification.

So it seems that the classical virtues can be grounded in the values of geek culture. While our discussion has only addressed a handful of the classical virtues, and only in a cursory way, it should be clear how further discussion might proceed. In general, once one has uncovered a set of values and beliefs sufficient to ground a way of life as a quest, showing how the virtues emerge from such a way of life is a trivial matter. Success in any quest will require integrity, and integrity will demand coherence and richness in one's conception of the good such that honesty, courage, good judgement, and the other virtues more or less immediately follow.

To see this, it helps to contrast the way of life of the good computer professional with close alternatives. For example, consider the way of life of the 'script kiddie'. Folks who get their kicks out of exploiting online security vulnerabilities may have a lot in common with computer geeks, and their actions may be dimly inspired by the same cultural background. It is clear, however, that script kiddies lack integrity. As the term 'kiddie' suggests, these people do not have the technological sophistication to really understand what they are exploiting. This reveals that they are not motivated by the technological aesthetic, at least not in any mature form. Instead, they seem to be operating on the same base motives as common vandals. Getting kicks from destroying things fails, not only in terms of geek culture, but on its own terms. The purely negative aim of destruction depends for its meaning on the work of those who create what is to be destroyed. But without some positive agenda, it is impossible to conceive one's life as a quest for something, or to evaluate one's progress with respect to what is valuable. Because their actions are entirely parasitic on the discoveries and creations of the geek community, the way of life of the script kiddie has no intrinsic meaning, and will fail as a way of life. Much the same can be said of 'crackers', who make it their business to defeat intellectual property protections and distribute expensive software packages without compensation for the authors. While this may require some technical knowhow and sophistication, 'crackers' are nonetheless parasites on the body of good computer professionals.

These failed ways of life might be contrasted with the info-terrorist or technological civil disobedient. Those who conceive themselves as fighting a guerrilla war against the perceived injustices of intellectual property rights, dangerously sloppy computer artefacts, or even the whole technological worldview might very well be on a quest. Indeed, they might even defend their actions in terms drawn straight from computer culture. 'Information wants to be free,' they might say, or, 'We exploit security flaws in technology in order to reveal them and make the technology better'. These attempted justifications of unethical conduct reveal a commitment to the values that guide the computer profession itself, and the debate is now simply about means. The consensus view among computer professionals is that civil disobedience is not the best means for advancing technology. For those who oppose the hoarding of intellectual property, participation in the creation of Free Software alternatives like Linux or OpenOffice is a more constructive and rewarding path. Security flaws can be revealed without handing the exploits over to script kiddies. The justifications of technological civil disobedience are either short-sighted or they reveal an arrogance with respect to the considered judgements of the profession and computing culture. Either way, they reflect bad judgement.

More dangerous and troubling are those who do not embrace technological culture, but who use technology as a means for other ends. Technology creates powerful tools that may be put to immoral or selfish purposes. With technical skills, con artists and other thieves can perpetrate crimes of a scale and breadth that was previously unthinkable, while maintaining an invisibility or anonymity that conceals the perpetrator or even the crime. Clearly, such crimes constitute real obstacles to the advance of technology, and it is the responsibility of the computing profession to mitigate the risks posed by those who would misuse technology. Clearly, these behaviours are not an expression of geek culture. Showing the ethical bankruptcy of the base hedonism that usually motivates these alternatives to computing culture is relatively straightforward. How can the hedonist, the con artist, or parasite conceive of life as a quest? Answering this question is beyond the scope of this essay, but the analysis here should indicate how reflecting on the role of character and integrity in pursuit of the good life already sketches the case to be made against those who would misuse technology, even as it fleshes out the spirit of those who create and discover it.

References

Davis, M, 1991, 'Thinking like an engineer: the place of a code of ethics in the practice of a profession', *Philosophy and Public Affairs*, Spring, pp 150–67.

Foot, P, 1978, 'Moral beliefs', in P Foot (ed), *Virtues and vices*, University of California Press, pp 110–31.

Gotterbarn, D, 1991, 'Computer ethics: responsibility regained', *National Forum*, Summer, pp 26–31.

Huntington, S, 2000, 'Cultures count', in L Harrison & S Huntington (eds), *Culture matters*, Basic Books, pp xiii–xvi.

Hursthouse, R, 1991, 'Virtue theory and abortion', *Philosophy and Public Affairs*, Summer, pp 223–46.

IEEE-CS/ACE Joint Task Force, 1999, 'Software engineering code of ethics and professional practice,' online at <http://www.computer.org/cms/Computer. org/Publications/code-of-ethics.pdf>

Katz, J, 2000, *Geeks*, Broadway Books.

Ladd, J, 1980, 'The quest for a code of ethics: an intellectual and moral confusion', in R Chalk, M Frankel, & S Chafer (eds), *AAAS professional ethics project: professional ethics activities in the scientific and engineering societies*, American Association for the Advancement of Science, pp 154–59.

MacIntyre, A, 1984, *After virtue*, University of Notre Dame Press.

Moor, J, 1985, 'What is computer ethics', *Metaphilosophy 16*, pp 266–75.

Rossetto, L, et al, 1998, 'Change is good', *Wired*, January, pp 163–207.

Winston, M, 2003, 'Children of invention,' in M Winston & R Edelbach (eds), *Society, ethics, and technology*, Wadsworth, pp 1–19.

8. Informed consent in information technology: Improving end user licence agreements

Catherine Flick

De Montfort University

Information technology suffers from a distinct lack of care with respect to adequate informed consent procedures. Computer users are commonly asked to consent to various things that could threaten their personal identity, privacy, and property, yet little care is taken in assessing whether the consent is truly informed. Some software even takes advantage of the confusion rife in informed consent procedures in order to install otherwise unwanted software on users' computers (such as adware or spyware). End user licence agreements (EULAs) are a common example of these poorinformed consent procedures, which have their basis in the inappropriate use of medical informed consent procedures, instead of being based on concepts that have been developed specifically for IT. In this paper, I outline some of the problems that need to be overcome in informed consent in IT, and present a more appropriate theory for informed consent (based on the Manson and O'Neill waiver of normative expectations theory). I will then look at some current and recent suggestions for improvements or practical implementations of mechanisms to deal with informed consent. Then, in the third section, I will firstly identify the normative expectations, establish a communication framework, and then suggest an agreement mechanism that greatly improves the potential for informed consent decisions for EULAs. Examples will be used to illustrate the process, but will not cover the details of the actual legal agreement. It will, instead, focus on delivering the content of this agreement in a way that improves the user experience and focuses on changing the approach for informed consent in EULAs. Finally, I demonstrate two models that show examples of practical implementation of these concepts and make some conclusions about their feasibility.

The problems

The problems we wish to solve in this chapter are those of informed consent in EULAs. Currently they focus on a disclosure and effective consent approach to informed consent: an EULA will (usually) be presented as a text box containing a large amount of legal text, with a scroll function to enable reading the entire document. Typically, a check box is placed at the bottom of the window that the user is required to tick if they have read and understood the agreement,[1] after which, if the response is positive, the software will install. An example is given in Figure 1.

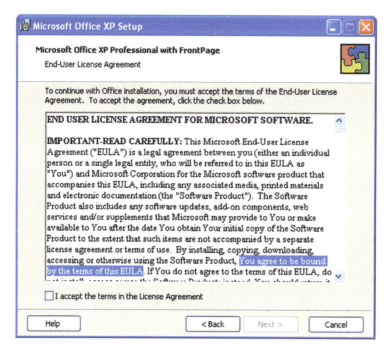

Figure 1: Microsoft Windows XP Service Pack 2 EULA screenshot

Source: http://www.microsoft.com

This disclosure approach makes no attempt to check to see if the user understands the agreement, often contains 'hidden' text (text deep within a long agreement that users would not generally read) that allows the software to perform tasks that may not be acceptable to the user, such as pop-up advertising, or which would violate the expectations of the user, such as privacy or security issues. These sorts of agreements are also prevalent in everyday use of computers, since each piece of software installed will generally require consent to an agreement similar to the one illustrated above. Typically, computers will have a number of pieces of software installed, which means that users have to consent to many such agreements. This process is repeated, often to the point where users know that it will be tedious to read all the material and skip through the stages without concentrating on the detail of the agreements. This 'numbness' is a serious problem that needs overcoming before we can claim to have a good informed consent procedure.

---

1   Whether or not they have read the agreement, ticking the box entitles the requester to assume it has been read and understood.

Thus, we have two sets of problems: one being the content of the agreement, and the other the display of the agreement (and acceptance mechanism). The content of the agreement, mostly legal text, is difficult to read, difficult to understand and, in some cases, almost deliberately obscure in order to dissuade users from reading it or to make it ambiguous in its meaning. The display of the agreement does not encourage a thorough reading of the agreement, even if it were easy to read, with the tiny text box holding thousands of words of text, often in all-capital letters, which can be difficult to read and comprehend. The consent mechanism in place (sometimes ticking a box, many times simply clicking 'I agree') is rudimentary at best, but coupled with the poor quality of the licence display, it is insufficient for the consent process to be considered informed. In order for this to improve, either the agreement mechanism needs to be changed (the more difficult route, since it could require some sort of randomness such as changing button locations to be introduced so that a new 'numbness' doesn't eventuate), or the preceding communication framework (detailing the contents of the agreement) needs to be improved (a more attractive route, since it would more easily fit within the current paradigm, giving a much better cost-benefit ratio), or both.

The theory

The current theories used for informed consent in the medical and research fields comprise theories based on autonomy such as Faden and Beauchamp's autonomous authorisation and effective consent models, or the general autonomy-based theory of Appelbaum et al (1987), yet both are insufficient (as is obvious from the current practice) for use in IT due to the high levels of expertise and face-to-face discussion interactions that need to take place in order for fully informed consent to take place. The current approach used for informed consent in IT is more closely aligned to the Faden and Beauchamp effective consent model than an autonomous authorisation approach or that of Appelbaum's.

The effective consent model involves sets of policy rules that aim to regulate the consent-seeker in order to protect the consenter. It does this by utilising disclosure and competence tests that conform to a legal or generally accepted standard framework suitable for the context. It often uses appeals to autonomy as its justification, but requires little evaluation of any sort of autonomy on the part of the consenter (Faden & Beauchamp, 1986). This is particularly obvious in the case of EULAs which default to a de facto standard that was developed by the same set of people who also commonly act as the consent-seekers, or are closely aligned with other consent-seekers, and thus are more interested in providing the bare minimum for legally sufficient informed consent so that they are protected from liability. In doing this, they present a bare-boned version of effective consent, which centres on disclosure yet provides almost no competence testing. This is an unacceptable approach for informed consent in EULAs, since it offers almost no attempt to inform the user of simple disclosure, which, as discussed, is presented in such a way that it is almost always ignored. This definitely benefits the consent-seeker over the consenter, and shows the need for an overhauled informed consent system for EULAs. On a theoretical level, it shows the need for rethinking the approach to informed consent in IT, since the situations found in IT are very different from those in the medical and research fields. I have therefore adopted Manson and O'Neill's objections to the basis of informed consent on justifications of autonomy, and their revised theory that relies on the establishment of informed consent as a waiver of normative expectations.

Informed consent as waiver shows how the identification of normative expectations, coupled with a good communicative transaction framework, can improve informed consent procedures in computing situations. Choosing normative expectations needing to be waived restricts the scope of the informed consent procedure to the values considered important by the users of the software, rather than the more abstract and idealised values of the autonomy-based theories. A good communicative transaction framework allows for more direct disclosure of information that is relevant and accurate, rather than a general full disclosure that could easily flood the user with too much information, resulting in a confused or ignorant decision rather than an informed one.

The practical implementations and improvements discussed in this paper aim to not only provide an example for the application of the above theory, but also to demonstrate ways of overcoming developer and vendor apathy toward change. It does this by recommending a third-party mediation group that handles the development of new EULAs through an easy-to-use interactive interface. This group would provide a modular licence builder for simple and standard licence agreements waiving common normative expectations. These expectations would be standardised into language that can be easily understood, yet legally enforceable.

Some recent suggestions for improvements and practical implementations

The problem of informed consent in IT is not a new one. There have been several suggestions for improvements, including some user-tested implementations of these recommendations, such as with cookies in the Mozilla web browser (Friedman et al, 2000, 2002). Here I will look at some suggestions and implementations of solutions for the problem of EULAs, as well as some suggestions for how to encourage uptake and use of a revised system. These examples were found through an extensive Internet search for suggestions for solutions to this problem (such as editorials calling for changes to EULAs), as well as actual solutions that have been developed (software, development frameworks), along with tools used for similar areas (such as copyright and fair use, certification for privacy or security) with the analysis concentrating on the benefits and limitations of each proposal or solution to the particular problem of informed consent in EULAs. Any proposal or solution that showed any sign of motivating improvements to EULAs (such as EULAlyzer) or informed consent decisions by users (such as Creative Commons, TRUSTe) was examined. Also discussed are the potential ways in which a trusted, third-party medium could enter the picture in order to attempt to regulate and standardise a system of licence agreements to make it easier for users to give more informed consent decisions.

The 'Pure Software Labelling Act'

Simson Garfinkel compares the current situation with spyware and adware with the problems of patent medicines before the United States' *Pure Food and Drug Act* of 1906 was passed (Garfinkel, 2004a). With the introduction of the *Pure Food and Drug Act*, '… with the knowledge of what they were to ingest, consumers were able to identify and avoid (if they wished) consuming potions that were "habit forming"'. Garfinkel claims that this could be a useful way to approach the problems of 'overly- broad and turgid click-through licence agreements' (Garfinkel, 2004a). Garfinkel specifically suggests using icons on a label at the top of licence agreements, and in other visible places, such as on install screens and in the Windows 'Add or Remove Programs' area of the operating system user interface. Software that displays such icons is not necessarily engaging in user-hostile practices, but the aim is to make it easy for consumers to easily identify which software displays which behaviour.

The icons he suggests are shown in Figure 2. They cover a few possible issues that users encounter in software, and which could be considered violations of normative expectations.

Figure 2: Label icon suggestions by Garfinkel (2004a) for software licence agreements

Source: Garfinkle, 2004a

One of these issues is software that modifies the operating system, usually to access pieces of hardware by installing drivers, or to access some of the lower-level functions of the operating system, such as manipulating the system memory used by other applications. Another is the issue of monitoring, such as in the example of keystroke logging, where a piece of software keeps a record of every keystroke made on the keyboard (this is often used by malicious software aiming to access usernames and passwords to websites and other services and information contained in private files), or software that tracks what websites the user visits from their web browser. Pop-up windows are also addressed, as is software that downloads updates for itself 'that could change its behaviour'. Although diallers have become outdated, since many Internet connections are no longer directly through a phone line, they used to be a major issue, because the software would hijack the phone line and make expensive phone calls (often overseas!) without the knowledge of the user. Another issue Garfinkel recommends highlighting is the problem of software running at start-up. Software that loads its core functionality when the computer starts up appears to load faster when the application icon is later clicked by the user, but, ultimately, it makes the computer run slower, even when the application is not being used. When many applications use this functionality, the slowdown becomes noticeable, especially when the computer starts up. Also, software that allows others to control the computer remotely (I believe the intention of the author is to call that label 'Remote' rather than 'Remove'), such as with the popular software program called VNC, often used in company technical support situations, and software that does not allow itself to be easily uninstalled would get their own label under Garfinkel's scheme.

One requirement of Garfinkel's idea is that the labelling be mandatory and legislated by government. In this he appeals to the *Pure Food and Drug Act's* success in giving people the knowledge of a food product's ingredients in order to effect changes in peoples' decisions regarding their consumption of food items. Such mandatory labelling would also cause companies to change the inclusions in their products, in response to public acceptance or rejection of those inclusions. At any rate, Garfinkel's idea is worth considering, given not only the success of the *Pure Food and Drug Act*, but the appealing nature of labelling things simply.

The benefit of this sort of system lie in the fact that it would be immediately obvious what norms would potentially be violated should the user install the software. A standard set of icons for the label would be available for software manufacturers to describe the activities of their software, and an external group would have jurisdiction over the use (and misuse) of the labels. It would have the potential, if fully enacted as Garfinkel proposes, to change the behaviour of software manufacturers, that is, some mechanisms would no longer be used because they would be considered detrimental if obviously part of the software (such as pop-up advertising or hijacking the telephone line). Overall, users would have significantly more control over their software through market force, and would appreciate the easy way to determine what sort of behaviour is intrinsic to their software.

Of course, there are (somewhat unfortunately) obvious problems with this approach. For one, unlike the importation and buying and selling of food and drugs, software is not physically restricted within a particular jurisdiction. The availability of software on the Internet makes it difficult to require mandatory labelling, since software companies could easily set up in a location that doesn't require this standard, especially companies that rely on expectation violations to make money, such as advertising software companies. The market forces at work here mean that there is a lot of momentum required to change procedures, since markets only improve products under competitive conditions. This doesn't make the idea unviable, however, because ideally the companies that work out of that jurisdiction would be able to establish that use of the labelling system would have advertising, and therefore revenue, merit as well as inherent moral worth. This sort of marketing strategy could improve both the company brand and the view of the licensing scheme, with a labelling culture eventually seen as a competitive status symbol.

Garfinkel points out another issue, in that it would be difficult to decide what sorts of potential expectation violations should be labelled.

> The more information required on the label, the more expensive it will be to produce, and the less likely that consumers would be to actually pay attention to the information. Any regulatory body implementing this policy will need to avoid icon creep — having 23 different icons on each piece of software won't serve the needs of consumers, it will just cause confusion. (Garfinkel, 2004b)

This is an important point, and one that would have to be dealt with comprehensively, since it is clear that there would not only be potential creep but also the requirement to deal with old and outdated labels as well (such as in the case of the dialler).

This could be overcome, however, by developing icons that can encapsulate at least the essence of the issue, thereby allowing users to acquire a reasonable idea of the software contents. This would still be preferable to the computer user having to wade through 5000 words of legalese, as is the current expectation.

Overall, Garfinkel's proposal is an interesting approach to solving the problem of difficult-to-read licence agreements. Unfortunately, it is unlikely to work in its current incarnation, since there is little momentum behind it (little active advocacy or any real attempt to institutionalise it) and it requires government legislation to be fully effective, which requires lobbying and probable industry backing, or other critical mass elements. It also ideally requires global acceptance and institution in order to become the dominant culture in the Internet-enabled market. This problem, as well as some of the other limitations discussed above shows that, although this idea does not fully solve the informed consent problem, it certainly is a step in the right direction.

StopBadware

StopBadware is a database of known 'badware' (such as Trojan horses,[2] or viruses that exploit web browser vulnerabilities) and badware-serving websites. Some large companies and popular software, such as Google and Mozilla Firefox, use this service to help protect their users against malicious software.

StopBadware is not specifically aimed at improving informed consent, but it provides a service that shows how dissemination of information about the importance of informed consent in IT situations can be achieved and, ultimately, identify improvements to the quality of software to help to improve informed consent. StopBadware's mission statement declares that they are '... committed to protecting Internet and computer users from the threats to privacy and security that are caused by bad software' (StopBadware, 2008b). They do this by maintaining a database of badware, that is, software that violates their guidelines for good software design. In general, their definition of badware is this:

An application is badware in one of two cases:

1. If the application acts deceptively or irreversibly.

2. If the application engages in potentially objectionable behaviour without:

• First, prominently disclosing to the user that it will engage in such behaviour, in clear and non-technical language, and

• Then, obtaining the user's affirmative consent to that aspect of the application.'

(StopBadware, 2008b)

There are specific references to informed consent within the guidelines, with StopBadware requiring applications to disclose the installation, origin, function, and any potential 'unexpected or unwelcome' behaviour of the software, and then seek consent for these. This, indeed, is a very specific example of Faden

---

2   Malicious software that 'hides' inside legitimate-looking software, installing invisibly when the legitimate-looking software is installed. The 'rootkit' is also a type of Trojan horse.

and Beauchamp's effective consent model, that is, the requirement of disclosure, and then the consent to the terms disclosed (although there are no tests for understanding on the user's behalf). It is with this in mind that I discuss this example because, although it has a large amount of focus on informed consent, it doesn't actually take any steps to improve the current flawed consent model. Instead, the StopBadware model simply requires an EULA and privacy policy to be present and 'written in as understandable a manner as possible'.

Thus, I use this example as a general illustration of how a better model of informed consent could be instituted: the StopBadware database of badware is highly regarded and successful, and the organisation has some powerful industry backing, with the likes of Google, Mozilla, PayPal, AOL, VeriSign, and Trend Micro among its partners. The reason it is so successful is because of its integration into Google and Mozilla Firefox. By accessing Google, the most popular Internet search engine, and Mozilla Firefox, one of the more popular web browsers, many users encounter the work of StopBadware. The StopBadware process works as follows: if a website is reported to StopBadware as distributing (knowingly, or unknowingly due to being hacked, for example) badware, its details are entered in a database which is checked by Google before it displays Internet search results. Google then displays a warning to the user before the user clicks through to the reported website, stating that 'this site may harm your computer' (Google, 2008). Firefox will also display a warning before loading the website (StopBadware, 2008a). Because companies rely on Google placement in search results, the detrimental effect of having a warning placed on their website impacts their business, and thus effects change in their behaviour. StopBadware has a 'clearinghouse' of many software applications and websites that have improved as a result of this (StopBadware, 2008b).

This shows that there is certainly a way to encourage software vendors and manufacturers to change their behaviour without legislative requirements or the need for a critical mass. Instead, social pressure can be placed on companies through negative publicity, with companies eager to be determined suitable for exclusion from the badware list. StopBadware, Google, and Firefox thus have a significant ability to effect change and, although they mostly target dangerous software, they also require that software display certain positive behaviours through disclosure of information through an EULA and/or a privacy policy. This could potentially go further by requiring certain ways of displaying agreements and policies, or including a labelling scheme similar to Garfinkel's. It should be noted, however, that excessive regulation could be detrimental to the project, or that this could be seen as an abuse of power (to negative effect) or as a 'slippery slope' to other requirements for warning-free display on Google or Mozilla Firefox. So long as it is simply a warning and not a complete restriction mechanism, the potential good should outweigh the slight inconvenience for non-conforming software websites. Overall, it is a considerably effective delivery mechanism for protection of users from potentially dangerous software on the Internet, and could be used for further improvement of informed consent procedures beyond immediate jurisdictions.

EULAlyzer

EULAlyzer (Javacool Software, 2007) is a piece of concept software that takes any EULA and analyses it for particular words and phrases considered 'interesting'. These include words and phrases involving advertising, third parties, changes that may be made without notifying the user, or privacy information. When a user installs a new piece of software, they can use the EULAlyzer software to copy an EULA into the analysis tool, which EULAlyzer scans for terms of interest, displaying the results in a window for the user to easily scan through the pertinent phrases and work out if they are important. A screenshot of its operation can be seen in Figure 3. In this figure, the software licence agreement for Download Accelerator Plus (SpeedBit, 2008) has been analysed. In the upper screen, a list of suspect text is shown ('Flagged Text'); this shows the word as well as the context of the word as it appears in the EULA itself. The user can then click on the green arrow beside it to go straight to the pertinent spot in the EULA on the bottom half of the screen, where they can read more of the surrounding text. The 'interest level' shows how likely it is that the user would be interested in those pieces of text. I found that gauge unhelpful, since I was interested in seeing what all of the flagged text was, rather than selectively looking at or ignoring other phrases. The user interface screen for choosing the EULA to read was a difficult to use, since the window of the EULAlyzer couldn't be resized easily and the program requires the user to drag an icon from the EULAlyzer program to the text box of the licence agreement on the installation screen of the software in question.

The difficulty of dragging the icon and completing the procedure is likely to be discouraging for users of the free version of the program. Since the point of such a program would, ideally, be to encourage use and therefore improve understanding and knowledge of the contents of EULAs, the free version of EULAlyzer is not the best solution to the problem of EULAs.

There is a more advanced version of the software (which requires payment; the simpler version is free) that runs continuously in the background, and analyses any licences that the user encounters when installing software. This is much easier to use, since the user doesn't have to open the program to run the EULA through it each time that they install software (which would mean they could easily become lazy and not bother to actually do it). It also only runs on one platform (Microsoft Windows) and this limits its usefulness. It would be beneficial if this, or a similar program, became standard on all operating system platforms, but, as it is, it is simply a niche market application that appeals to the more privacy and security-conscious computer users.

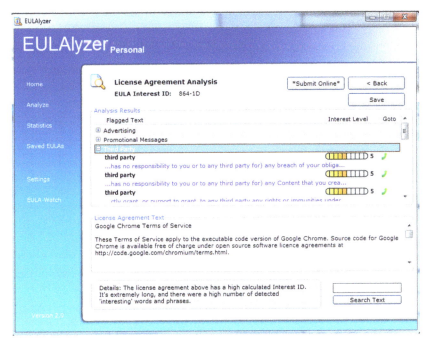

Figure 3: A screenshot of EULAlyzer, showing some analysis results

Source: http://www.javacoolsoftware.com/eulalyzer.html

EULAlyzer is essentially a patch approach to the systemic problem of informed consent in EULAs and, although it is useful, it doesn't attempt to change the poor behaviour of companies, but simply provides an option for users who are concerned about their computer and data's welfare. Since it is also analysing the EULA documents by scanning for certain words, it could also ultimately be a reasonable expectation that software wishing to avoid clauses being flagged by the analysis could rewrite their EULA to use words that do not trigger an alert in the software. EULAlyzer would then give a false sense of security, and could ultimately be more dangerous than not having the analysis tool at all.

TRUSTe

TRUSTe is an independent body which provides a service certifying that a website or other Internet service (including email lists), or software conforms to their best practices with regard to privacy. I am interested in the 'Trusted Download Program', since TRUSTe makes the claim that, with their certification, the software cleared by the program has been proved to be spyware free, and that they can 'build [their] reputation and increase consumer confidence', 'open doors to major advertisers and partners', 'reassure distribution partners and increase downloads', and 'join the only industry-recognised safe-software whitelist' (TRUSTe, 2008b). TRUSTe go on to state that the software certified by them 'directs software distribution opportunities and advertising dollars to programs and affiliate networks that respect consumer choice'. Essentially, it is a middleman between software companies and advertising companies, with a degree of respectability brought on by the standards that they require of the software, which provides a certain exclusivity. The requirements of software for compliance with the program are for 'transparent distribution practices', 'clear disclosure and informed consent prior to download', 'clean uninstall' and 'respect for the end-user' (TRUSTe, 2008b).

The organisation has close ties to advertising companies and download websites that subsist on advertisements (through both adware and website advertising) for their income, which raises questions as to their integrity. They do, however, have a very comprehensive list of specific policies that the software must abide by (and be subject to audits on), and the penalties are reasonable (consisting of being removed from the program, possibly with public advertisement). They also have strict requirements for EULAs and consent, including that on installation, the software displays a primary notice of functionality that could impact the user, and that an EULA (or opt-out mechanism) is 'insufficient for providing such notice or obtaining consent' (TRUSTe, 2008b). These are, for the most part, excellent policies, however, looking at the published 'whitelist' (TRUSTe, 2008a), there are a large number of questionable pieces of software, especially the advertising and tracking software listed, which includes some software with particularly questionable activities. One such example is the software WhenU Save/SaveNow, which is listed as generally unwanted adware by many anti-malware programs (Healan, 2005), and which is often packaged as a third-party piece of software along with other 'free' software, such as the aforementioned screensavers or movie players. Perhaps TRUSTe's for-profit status, and reliance on companies to pay subscription fees to display the certification seals, means that it needs to form working relationships with these companies in order to work out any potential difficulties rather than simply offering the service as a simple 'tick or cross' system. According to Hansell (2008), TRUSTe has a record of questionable toughness on its clients:

> ... it does not always tell the public if it discovers violations of its principles, even if the violations are so egregious that it kicks a site out of its program. Last year, it ejected three of the 1500 companies in its certification program, and three more chose not to continue because their business models no longer complied ... The organisation declined [...] to name those six companies, saying it only makes public cases of 'blatant violations with probable consumer harm'.

It also encountered image problems with its approval seal for certified websites being targeted in cross-site scripting attacks, which allowed spamming websites to illegally display the image in order to lull the consumer into false confidence about its activities. Worse, these attacks tricked the TRUSTe server into confirming these websites as legitimate, thus misleadingly approving them. This was more destructive than a simple re-using of the seal image to instil a false sense of security in users, because it returned results that said the use of the seal was legitimate (Wagstaff, 2004). This problem undermined the system for TRUSTe and, although the issue was resolved, it was damaging to their image as a trusted third party. A similar issue occurred when the lists of TRUSTe-certified websites was correlated with McAfee's SiteAdvisor, a service that tracks websites that have viruses or other problematic exploits, and there was a number of TRUSTe certified sites on the SiteAdvisor list. This, too, was problematic for TRUSTe's image, since its aim is to certify a website as free of these sorts of exploits and viruses (Leyden, 2006). While these problems are not unique to a service like TRUSTe, the fact that their auditing process was flawed is of note, since it would seem obvious that a website operator could set up a site that conforms to the TRUSTe guidelines then, using the respectability that comes with the TRUSTe label, gain trust of its own. The operator could then start to take advantage of that trust by introducing viruses or exploits on the site, taking as much advantage of it as possible before TRUSTe investigated the breach of its policies.

This sort of approach is attractive to software companies that are engaged in creating and promoting software that requires more credibility than usual: makers of advertising software or software that includes some other feature which may be considered unwanted or violating major expectations of a user, such as a keylogger or a client for a remote access server. These sorts of applications already have a certain stigma associated with them; recognition by a service such as TRUSTe is one way to gain respectability. Thus, it seems that it is unlikely that this sort of approach will eventually encourage a critical mass of software manufacturers, since many, like Google, like to at least claim to 'keep good company' (Google, Inc., 2006), and not taint their brand with association with these companies and their software.

This is not to say that all third-party certification is a bad thing: becoming certified as conforming to ISO standards or Hazard analysis and critical control points HACCP standards for food safety, for example, is widely regarded amongst the relevant industries. TRUSTe, unfortunately, does not have the reputation of these standards organisations. Ultimately, a reputable, third-party standards organisation can enhance the reputation of software, but, until standards are developed that are well-balanced and focus on identifying values and setting up frameworks for dealing with user and company expectations, this remains a problematic process. The guidelines discussed here could be used to develop such a framework for independent, third-party standards assessment, and could be used as the ethical basis for a legal regulatory framework that would support such a standards organisation.

Creative Commons

Creative Commons is a non-profit organisation that provides licences for use of creative works; licences that span a variety of different restrictions and freedoms in an easy-to-access model allowing users to easily licence their work. Their work is based on the idea of a 'commons', 'resources that are not divided into individual bits of property but rather are jointly held so that anyone may use them without special permission. Think of public streets, parks, waterways, outer space, and creative works in the public domain' (Creative Commons, 2008b). Thus the licences allow the holder to retain their copyright, but allow people to distribute, modify, and/or copy the work, providing that due credit is given (depending on the specifications chosen by the user).

The website allows users to choose a licence after answering a series of questions, such as whether to allow commercial uses or modifications of the work, and the jurisdiction of the licence, and then presents it simply, as is illustrated in Figure 4.

Note the use of icons, similar to the type Garfinkel proposes, to allow for an easier understanding and a more immediate recognition of the requirements of the licence. The standardisation of the licences, too, makes it easy for people wanting to use the work to know what requirements they need to fulfil. The third-party nature of the Creative Commons organisation means that it is not simply acting on the copyright holder's behalf, but has sought to make the licences and licence procedures understandable and fair to the user.

Another positive side to the Creative Commons approach is the easy-to-use form. Within a few minutes, a copyright holder can have a socially acceptable licence drawn up that has a solid legal foundation. This independent, non-profit, unbiased third party runs from donations and is part of a grassroots advocacy organisation with a large support network (Creative Commons, 2008a). In this way it has attracted many of the newer websites, particularly social networking and media sites such as Flickr, Google, Yahoo! (Creative Commons, 2008b) and, with them, many users of the Creative Commons licensing system. With this large industry backing, Creative Commons is a good case study for reasonable ways to approach change in culture within the IT industry.

Although there is the potential for misuse of the Creative Commons labels, similar to the issue face by TRUSTe, because the use is within copyright and contract law, misuse could be opposed by legal action. The situation is different with TRUSTe, which has no legal backing beyond striking a certificate from its list. What Creative Commons does differently is not simply act as a third-party auditor, but provide a service with which users and companies can create licences with which to claim their own legal rights.

It is with this in mind that I move on to my recommendations for practical implementation of informed consent reform within EULAs.

You are free:

to copy, distribute, display, and perform the work

to make derivative works

Under the following conditions:

 Attribution — You must give the original author credit.

 Non-Commercial — You may not use this work for commercial purposes.

With the understanding that:

Waiver — Any of the above conditions can be **waived** if you get permission from the copyright holder.

Public Domain — Where the work or any of its elements is in the **public domain** under applicable law, that status is in no way affected by the licence.

Other Rights — In no way are any of the following rights affected by the licence:

- Your fair dealing or **fair use** rights, or other applicable copyright exceptions and limitations;
- The author's **moral** rights;
- Rights other persons may have either in the work itself or in how the work is used, such as **publicity** or privacy rights.

Notice — For any reuse or distribution, you must make clear to others the licence terms of this work.

Figure 4: An example of a Creative Commons license, with simple icons and explanations: The licence illustrated here is the Attribution–Noncommercial–Share Alike 2.5 Australia licence

Source: Creative Commons, 2008

Recommendations for practical reform of EULA informed consent

Having analysed the current approaches to dealing with informed consent issues in IT, it is clear that there are several different ways of, firstly, appealing to the industry and, secondly, making these consent issues meaningful for both users and manufacturers, thereby improving the overall level of consent. TRUSTe and Creative Commons are (currently) both non-profit organisations, yet one attracts a particular market (software requiring certification in order to gain a certain amount of respectability), whereas the other has a much wider base, with a highly respected group of industry links that use their system, and considerable respectability attached to it. The difference between them is that one requires end-user trust in the label and protection of that label, whereas the other requires trust in the system in order to obtain a legally backed and protected label. It is obvious, also, that a common suggestion for improvement of ease-of-use for users of such systems is by employing icons to allow for easier identification of important information. Also important are ways for industry to incorporate the system into their own systems, such as StopBadware and Google's approach to improving consumer awareness of potentially dangerous websites. It is obvious, too, that there needs to be a more basic approach than the EULAlyzer, which works to reform the system rather than simply deal with it as it is.

In this way I propose an EULA system by which a genuinely independent, third-party, non-profit organisation establishes a set of standard agreements, in a similar way to the standard licences from Creative Commons, coupled with an iconic labelling system that identifies key aspects of the agreements that users would consider important, in a way that allows users to waive normative expectations in order to give proper informed consent.

In establishing this system, I will outline some basic normative expectations that I would expect to be the main focus of such a system, and reiterate the process by which to identify the expectations, discuss the merits and potential issues with a set of standard agreements, along with a process for adding additional clauses to a standard agreement under this system, and then discuss the method by which such licences would be distributed, through a Creative Commons-like, non-profit trusted mediation service.

Choosing normative expectations

The expectations users have of software vary considerably, but there are several important contraventions of these expectations that can easily be outlined as being potentially harmful to the user or having a negative impact on the performance of the computer, or are generally actively avoided by users, such as advertising.

Some potential activities of software that would generally violate a user's normative expectations include (but are not limited to) the following:

- monitoring of user behaviour

- collection of personal information
- sending personal information to a third party
- changing the behaviour of other software
- changing the behaviour of the operating system
- installation of additional software
- display of advertisements
- requiring extra computing skill to remove
- starting automatically on boot
- running in the background
- automatic upgrading.

A user would need to know if their behaviour is being monitored, such as through software that monitors keystrokes or captures screenshots, or which monitors traffic flow over the network. Display of advertisements is another problematic function. Other activities that affect the user could be the practice of directing the software to start up when the computer first boots up, and then running it in the background seemingly invisibly until the user actually wants to use the application, at which point it loads very quickly because it doesn't need to perform all of its start up routines. This is not a bad thing on its own, and can result in speedier loading times for programs. The problem is that with many programs performing this activity, the computer starts to take a long time to start up on boot. This could mean that the computer actually starts to perform much worse than usual because of all of the programs that are running in the background, taking up system resources such as memory and CPU cycles and thus encumbering the computer. Automatic upgrading is also a potential issue, since, although it is usually welcomed by most users and recommended for security issues, there are some upgrades that could be harmful to particular users, especially if the upgrade significantly changes the behaviour of the software. These sorts of changes could be problematic for users that rely on a particular feature of the software that is removed by the upgrade, so there should be at least some sort of notification and re-acceptance of the agreement before the upgrade is made.

Other things not directly related to software activity that should be considered for explicit waiving within an EULA include the normative expectations of merchantability and fitness for a particular purpose, complete termination and deletion of details of any account information on uninstallation or unsubscription, the ability to obtain a refund if the terms are not agreed to, and notification of changes to the terms of the agreement.

Sets of logos or icons would need to be developed in order to best represent the ideas for waiving. The sorts of things Garfinkel had for his set is a good start, but more would need to be developed. The problem of an excessive number of meaningless icons being displayed could be avoided in several ways. Accurate and representative graphics that respond with a mouse-over display summarising the icon's meaning would be a useful starting point.

Standard agreements

With the normative expectations identified, and mechanisms in place to allow users to confidently waive them with informed consent, it is important to look at the other end of the system, that is, the rights and responsibilities of the software writers and vendors, which were previously the sole focus of an EULA. To effectively deal with the legal aspects of EULAs, I propose that a set of standard agreements be drawn up, with the basic legal requirements available to be added in as 'modules', similar to the drop-down menu selection screens of the Creative Commons website (Creative Commons, 2008b). In this way, licencers could choose which normative expectations licencees need to waive in their agreement, with appropriate text and icons inserted into the created agreement. It would, by necessity, be a more complicated than the Creative Commons example, but, considering there would be fewer licences compiled and there are greater ramifications of poor informed consent at stake, this would be a necessary and reasonable trade-off to make.

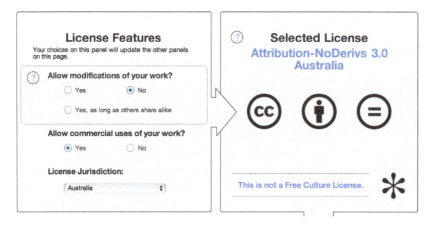

Figure 5: The Creative Commons licence choice screen

Source: Creative Commons, 2008

The use of modular standard agreements would be a convenient, clear, and effective way to communicate rights and responsibilities of both the user and consent-seeker, and to obtain waivers of a user's normative expectations in order to install and use the software. Not only would a user be able to predict the content of the agreement by the icons used to draw attention to potentially problematic areas of the agreement, but there would be a greater level of trust in there being a minimum of the previously identified 'hidden' or otherwise ambiguous clauses.

Should any conditions required by the licencer fall outside the provided sets of modules, a separate 'additional clauses' section could be provided, on a second screen, stating that they fall outside the conditions provided by the standard licence template. Generally speaking, these instances should be discouraged, or be considered for module status with appropriate standard wording devised, but, for more obscure software inclusions, this option should be available.

Trusted mediation

With these sorts of requirements for improvement of informed consent in EULAs, it would be necessary for an organisation, similar to that of the Creative Commons, that develops, maintains, and lobbies for the use of the above system. In particular, the organisation should be non-partisan, resulting in a legally supported and industry backed mediation group that is trusted by both users and industry to be working in the interest of all stakeholders.

Such an organisation would instil confidence in the users of software, in line with John Weckert's (2005) discussion of solutions to the problems with the establishment of online trust. In this, Weckert notes the use of certificatory third-party endorsements by way of 'seals', for which a company must satisfy certain requirements before they can display the seal (Weckert, 2005). Unlike the TRUSTe attempt, however, there would be no requirement for membership or licensing fees, thus eliminating the potential to favour one set of stakeholders over another. The lack of accountability to anyone but the organisation would only serve to strengthen the trust placed in the organisation by all stakeholders, since it would act as a mediation between groups with different sets of moral values. Also, since the licence is incorporated into the installation procedure of the software, there would be little room to doubt the authenticity of the licence, as it would have to read and perform the same way as the sample licences, or be held unconscionable if brought to legal action since outright deception would be taking place.

Thus, setting up a trusted mediation organisation that is non-profit and concerned with all stakeholders is the best way to develop, maintain, and lobby for these improvements to EULA policies.

Theory in practice: An example

The following example of an EULA, which puts the theory into practice, would require both legal and user interface design expertise for commercial application. The system that I propose assumes the oversight of a third party that is trusted to establish and govern the standard agreements. The situation that I outline could, however, be used outside of a third party too, given the consent-seeker is able to derive their own normative expectations from the guidelines and recommendations outlined previously.

Software background and purpose

The software in question, which is entirely fictional, will be named Admo. Admo is software that allows content providers, such as video or game developers/producers to generate an income from their work by way of advertisement support. For a computer user to view such videos or play games in this scheme, they will need to firstly install Admo, which will present them with targeted advertisements once an hour by way of a popup window. The advertising software will use information gathered from the computer user's web-browsing habits to work out what advertisements would be appropriate.

Standard agreement and modules

The trusted third party would ideally provide a basic standard agreement, which outlines the requirement for the user to accept the agreement before use, identify the jurisdiction under which the licence is legal, and other legal requirements that do not involve the software activity or function specifically (the exception to this is the module requirement for notification of policy changes). This section would incorporate the traditional first few paragraphs of EULAs. The standard agreement provided by the trusted third party would include modules for the waiving of the selected expectations, chosen for a particular jurisdiction. Each module should be assessed for inclusion depending on the software's requirements and purpose. The modules in italics below are the modules that Admo will need to seek explicit consent for waiving. All expectations it adheres to would be listed in positive language within the body of the agreement as well.

These modules include, for the purposes of this example:

- The software meets minimum standards of merchantability and fitness for the purpose for which it is made (according to law in that jurisdiction).
- *The software does not disrupt the existing operation of the computer.*

- *The software does not distract the computer user.*
- *The software does not track the behaviour of the user when using the software.*
- *The software does not track the behaviour of the user when using other software.*
- *The software does not collect personal information for its own use.*
- *The software does not collect personal information to sell or give to third parties.*
- The software allows a user to easily opt-out or uninstall, that decision will be final and persistent, and result in the removal of all traces of the software and the user being unsubscribed from the service.
- The software does not install additional software.
- The user may return the software to the place of purchase, unused, for a full
- refund if they do not agree to the terms of the agreement.
- The user will be notified of policy or specification changes and need to give explicit consent to these changes.

These modules (of the software Admo) would be developed by the trusted third party (not the software company!) from normative expectations about the particular type of software (downloaded from the Internet, requiring some Internet connectivity to function, as opposed to being bought from a shop and not requiring any Internet connectivity to function), based on industry guidelines and focus groups (or similar). They would have accompanying legally binding text, such as is seen in current agreements. Admo's company simply needs to decide which ones are applicable to their software, which will be highlighted and indicated plainly on the main window of the agreement.

Communication: EULA design

Once Admo has a set of normative expectations to waive associated with it, an EULA window for their software will be created to facilitate the communication transaction that needs to take place to fulfil the requirements of the theory. I outline here two possibilities for the design of the agreement panel, which would run on installation or re-request for consent. One is a two-window design, and the other displays multiple windows depending on how many modules are used (that is, how many normative expectations need to be waived). It could be that these are organised hierarchically, or kept separate, but determining this and which sort of model is suitable for the target audience is best left to further feasibility studies and trials.

The major problem with designing agreements is the need to balance the requirement for informed consent with an annoyance factor on the part of the user at having to perform several minor tasks before installation of the chosen software is complete. The more detailed the consent process, the more likely it is to irritate the user, which is why I propose two slightly different models, one of which is not as involved as the other. Ultimately, though, I suspect that the overwhelming desire to install the software would override the annoyance of the EULA windows, especially if there were obvious benefits to the requirement of such windows.

Two-step agreement

This model would use a similar style of screen (or window) to the current generic EULA window , but the user would have more control over the process. While the user could access the legal document in its entirety if they wish, the icon highlights would act as signifiers to the user of items for potential concern within the agreement. Hovering the mouse over the icon would bring up a tooltip with a short explanation of the behaviour requiring explicit consent. Clicking on the icon would take the user to the pertinent point in the formal document, which would be written in plain, yet legally binding, English (or have a plain English equivalent), and highlight the area of interest. A requirement to check off each expectation, as encapsulated by its icon, could be built into the screen. If one or more of the individual expectations are not waived by the user (that is, the checkbox under the icon is not checked), some more information about each one could be communicated to ensure that the omission is intentional, with an explanation that it will mean the software will not be installed. The agreement also asks for the user's date of birth and location, as a basic form of competence testing and jurisdiction identification. A mock up of the first example window for this style of agreement is displayed in Figure 6.

To address expectations needing to be waived but not present in the collection of modules, a second step can be optionally added which allows for additional expectation waiving. Here, any expectation that does not fit under the previously waived expectations should be listed. An example of the second step window is in Figure 7. This allows for an element of futureproofing and flexibility for the scheme. The danger is, however, that with technology changing, new types of expectations would start to become apparent, and this second window could become cluttered. It is because of this that the trusted third party should monitor the sorts of expectations that are outside the standard ones, and eventually include them in the list of standard expectations, while retiring those that are no longer necessary. This style of consent window, though, means that consent-seekers, who might otherwise only see the standard modules as being required for disclosure and consent, cannot modify the standard agreement to 'hide' other things in it, and so must have a visible second window that draws attention to additional requirements.

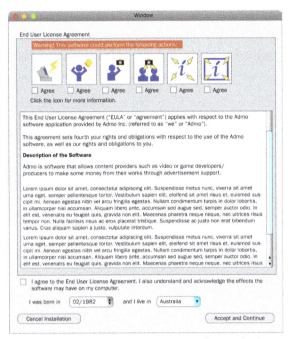

Figure 6: Two-step agreement, first window, showing main standard information and modules

Source: Author's research

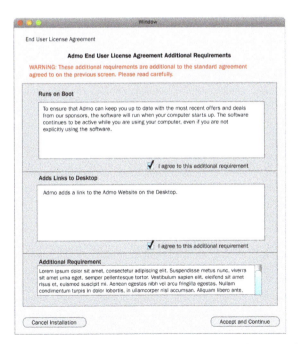

Figure 7: Two-step agreement, second step, showing extras

Source: Author's research

Multiple-step agreement

The second option I present is a more thorough attempt to inform the user before seeking consent. It starts with the standard basic agreement and then moves on to a single window per expectation module (or group of expectations).

Figure 8 illustrates a similar window to Figure 6, except it presents only the basic agreement. The same basic competence test is set out (asking for date of birth and jurisdiction), and the language is simple. Accepting this window of the agreement would bring the user to the next step in the agreement.

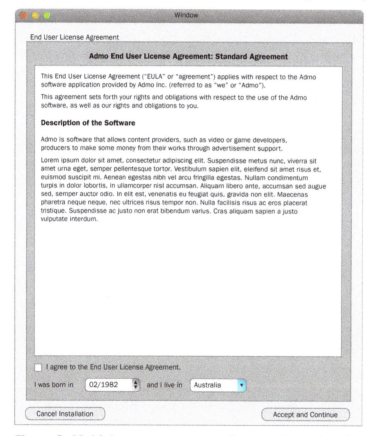

Figure 8: Multiple-step agreement, first window, showing standard information (no modules)

Source: Author's research

The following step is where the bulk of the expectation waiving is done. Each expectation module, possibly encapsulating related expectations, shows a set of icons identifying the expectations, and a summary of the effect of waiving those expectations (shown in Figure 9). Below the summary, the effects are explained in full, with a checkbox for the user to agree to each section, and identification of how many steps remain in the process. After agreeing to a module, the next module appears, until all modules have been agreed to. In this way it is much easier for the user to see more information about the potential issues that might affect them. Finally, following the module acceptance, an extras window, as shown in Figure 7, would be presented, with the same potential benefits and drawbacks as discussed in the two-step agreement, above. Only after accepting every step in the agreement would the installation of the software continue.

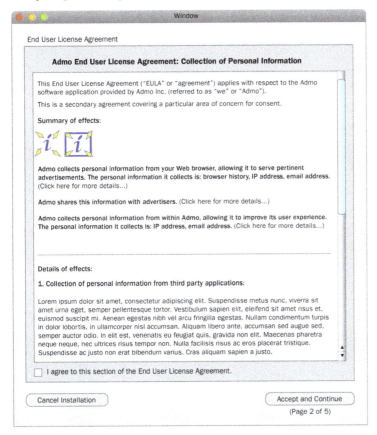

Figure 9: Multiple-step agreement, second window, showing information for each module

Source: Author's research

It is likely that the uncluttered appearance of the multiple-step example will meet with more approval. While it gives the user more useful information in easier to understand short statements, it is more likely to cause annoyance because of the number of windows requiring the attention of the user. The annoyance factor may be mitigated by the education of users in the importance of giving informed consent to EULAs.

Conclusion

In this paper I have outlined my recommendations for improvement of informed consent in IT by improving the worst offending area: that of EULAs. In doing this, I outlined the obvious problems plaguing EULAs, and then outlined some of the suggestions that have been put forward to improve the situation. Some of the suggestions were directly related, such as Garfinkel's labelling system, whereas others provided examples of trusted third party support mechanisms such as the Creative Commons and TRUSTe. I evaluated each one in terms of the problems of EULAs, and then established my own recommendations: a hybrid system of labelling and trusted third-party mediation and support that provided a foundation and design paradigm for modular standard license agreements. In keeping with the normative theory I have adopted for this task, I identified the normative expectations, discussing them in the context of the suggestions made in the previous chapter, and then established an effective communication framework: using trusted third party mediation to develop a set of modular standard agreements that are easy for users to understand, with labels identifying the particular normative expectations that would need to be waived for each piece of software. I discussed the merits and problems of the system, particularly some places that would need workable trade-offs, such as the potential problems of crowding of labels on the license window, lack of label recognition, and complication of license choosing. I also dismissed a few problems that have plagued similar projects, namely the problem with 'seals' in that they are easy to forge. Finally, I demonstrated a practical implementation of the theory proposed for the case study of EULAs. I offered two models, each with benefits and disadvantages, and suggested some further research for trialling these models.

In doing so I have outlined a serious practical application recommendation for improvement of an example of informed consent in information technology.

References

Appelbaum, PS, Lidz, CW & Meisel, A, 1987, *Informed consent: legal theory and clinical practice*, Oxford University Press.

Beauchamp, TL, & Childress, JF, 1994, *Principles of biomedical ethics*, 4th edn, Oxford University Press.

Creative Commons, 2008, 'Support Creative Commons', viewed 20 October 2008, <http://creativecommons.net>

Creative Commons, 2008, website, viewed 20 October 2009, <http://creativecommons.org>

Friedman, B, Millett, L & Felten, E, 2000, 'Informed consent online: a conceptual model and design principles', *Tech. rept.*, 00-12-2. UW-CSE.

——— & Howe, DC, 2002, 'Informed consent in the Mozilla browser: implementing value-sensitive design', in *Proceedings of the 35th Hawaii International Conference on System Sciences* (HICSS'02), vol 8, IEEE Computer Society, Washington.

Garfinkel, S, 2004a, 'The Pure Software Act: a proposal for mandatory software labeling', in S Egelman & P Kumaraguru (eds), *DI- MACS workshop and working group meeting on usable privacy and security software*, Carnegie Mellon University.

———, 2004b, 'The Pure Software Act of 2006', *Technology Review*, April.

Google, Inc., 2006, 'Software principles', viewed 10 September 2008, <http://web.archive.org/web/20080913202410/http://www.google.com/corporate/software_principles.html>

———, 2008, 'Google FAQ', viewed 14 October 2008, <http://web.archive.org/web-cdx/20120718191321/http://support.google.com/websearch/bin/answer.py?hl=en&answer=45449&rd=1>

Hansell, S, 2008, 'Will the profit motive undermine trust in TRUSTe?' *New York Times*, July, viewed 2 January 2009, <http://web.archive.org/web-cdx/20121105072940/http://bits.blogs.nytimes.com/2008/07/15/will-profit-motive-undermine-trust-in-truste//>

Healan, M, 2005, 'Funny business in antispywareland', *SpywareInfo*, March, viewed 2 January 2009, <http://web.archive.org/web/20060206125008/http://www.spywareinfo.com/newsletter/archives/2005/mar13.php>

Javacool Software, 2007, EULAlyzer, webpage, viewed 15 October 2008, <http://web.archive.org/web/20081005054651/http://www.javacoolsoftware.com/eulalyzer.html>

Leyden, J, 2006, 'Malware lurks behind safety seal', *Register*, September, viewed 22 October 2008, <http://web.archive.org/web-cdx/20121008085336/http://www.theregister.co.uk/2006/09/26/truste_privacy_seal_row/ />

SpeedBit, 2008, website, viewed 15 October 2008, <http://web.archive.org/web-cdx/20080930081941/http://www.speedbit.com/>

StopBadware, 2008, 'Mozilla partner to protect consumers' September, viewed 14 October 2008, <http://web.archive.org/web-cdx/20081009191105/http://www.stopbadware.org/home/pr_091508>

StopBadware, 2008, website, viewed 14 October 2008, <http://www.stopbadware.org/home>

TRUSTe, 2008, TRUSTe Software Whitelist, viewed 2 January 2009, <http://www.truste.org/pvr.php?page=td_licensees>

TRUSTe, 2008, 'TRUSTe trusted download program', viewed 16 October 2008, <http://web.archive.org/web-cdx/20120607062440/http://www.truste.com/products-and-services/enterprise_privacy/trusted_download_program>

Wagstaff, J, 2004, 'TRUSTe's own phishing hole', viewed 22 October 2008, <http://web.archive.org/web-cdx/20120701044650/http://www.loosewireblog.com/2004/11/trustes_own_phi.html>

Weckert, J, 2005, 'On-line trust', in R Cavalier (ed), *The impact of the Internet on our moral lives*, SUNY Press, pp 95–117.

Section IV

ICT governance

What is good governance?

The previous sections have considered aspects of regulation and professionalism. Both are important in information and communications technology (ICT) governance. Good governance requires some regulation but, as the Hon Michael Kirby pointed out, regulation has problems and hence the importance of professional behaviour in ICT.

ICT governance can be discussed at various levels, for example the global, national, industry, corporate and project level. These are not wholly distinct, but it is useful to separate them for clarity. Governance at the global level has chiefly been discussed in relation to the Internet, but its discussion is outside the scope of this book. National governance, rarely discussed as *governance*, is primarily an issue of national policy regarding ICT, often Internet policy and, again, this is outside the scope here (see Weckert and Al-Saggaf, 2008, for a discussion of Internet governance). Corporate governance, which has been noted by many as having come to the fore after some spectacular corporate collapses, generally refers to the running of the whole corporation, while corporate ICT governance is concerned with the role of ICT within the corporation (for a very useful background discussion of ICT governance see de Cruz (a)). Part of this is the governance of ICT projects, something that has a close link with professionalism, as discussed in this book. ICT industry governance is, again, little discussed as *governance*, but elements — for example, codes of ethics — are emphasised by the ACS and other professional bodies. In this section, the main emphasis will be on ICT corporate governance, although much of what is said will have relevance for the other levels as well.

What is ICT governance? Standards Australia has this definition (AS8015:2005):

> Corporate Governance of Information and Communication Technology (ICT) is the system by which the current and future use of ICT is directed and controlled. It involves evaluating and directing the plans for the use of ICT to support the organization and monitoring this use to achieve plans. It includes the strategy and policies for using ICT within an organization. (For a discussion see da Cruz (b))

This definition is clearly focussed on ICT in corporations and on how ICT can benefit the corporation.

The definition below, from the South Australian Government, contains the additional feature of encouraging 'desirable behaviour'.

> ICT Governance is a collection of principles and practices that guide the correct application and delivery of ICT. Components include;

- Strategy — clear direction statements.
- Policies — clear boundaries are set.
- Procedures & Guidelines — clear detail as to who, what, how.
- Accountabilities — clarity of roles and responsibilities.

> ICT Governance is defined as 'specifying the decision rights and accountability framework to encourage desirable behaviour in using IT'.

ICT governance is a subset of corporate governance, and models exist such as that published by the OECD, which defines corporate governance as 'providing the structure for determining organizational objectives and monitoring performance to ensure that objectives are attained' (DECS, nd).

Other definitions are discussed by Richard Lucas and Cecilia Ridgley in this section, with Ridgley providing critical analyses of a number of them.

A notion that is closely related to governance is that of an integrity system (see the discussion in the introduction to Section One). A governance system can be seen as a more formal part of an integrity system. Integrity systems are non-legal institutional mechanisms that are designed to reduce inappropriate behaviour and to promote and support an ethical climate. According to Seumas Miller, an integrity system 'is an assemblage of institutional entities, roles, mechanisms and procedures, the purpose of which is to ensure compliance with minimum ethical standards and promote the pursuit of ethical goals' (Miller, 2010: 354). These mechanisms might include behavioural codes, unofficial sanctions, meetings in which ethical topics are raised, education on ethical issues, mechanisms of accountability and so on. As this suggests 'integrity 'systems' are a messy assemblage of formal and informal devices and processes' (Alexandra and Miller, 2010: 39), which are focused on developing and maintaining the individual integrity of the members of the organisation. For example, codes of ethics include standards of conduct and occupational ideals to guide members; a statement of the fundamental goals of the occupation; and the principal rights of members of the occupation in relation to groups such as employers, peers and clients. These codes clarify the kinds of behaviours that are expected from members. They can be used as a resource to help make decisions and guide behaviour (especially in situations which are not covered by legal regulations), as well as a tool for raising awareness about particular ethical issues that may not otherwise be obvious. Unofficial sanctions are a means of punishing inappropriate behaviours. These may include cautions, reasons for promotion or demotion or other such means of encouraging or discouraging particular behaviours. They operate as clear feedback on behaviours. The role of ethical discussions is to raise awareness of ethical issues and encourage industry members to engage with them. Such discussions could be structured or informal, and take the form of discussing real cases in the industry or introducing hypothetical ethical scenarios. As can be seen, there is much in integrity systems that is not normally thought of as governance, nevertheless it is useful to see governance in the context of integrity systems.[1]

The two papers in this section take rather different stances on ICT governance. In the first, Lucas, an academic with backgrounds in both ICT and philosophy and long industry experience, focuses on the professional, claiming that governance is a proper subset of professionalism, meaning that full ICT professionalism includes ICT governance as well as knowledge while, in the second, Ridgley, an ICT practitioner, argues that an approach that focuses on the individual is inadequate. Lucas takes as his starting point the relationship between the professional and virtue (see also Richard Volkman's paper on virtue ethics). In general, a virtuous professional is one who promotes human flourishing. But, he argues, there is an inherent problem with seeing an ICT professional (Lucas uses IT rather than ICT) in this light because ICT *in itself* does not contribute to human flourishing in the way that, say, medicine does. ICT contributes through its contributions to those other fields. Ridgley agrees that values play a central role in governance but argues that a systems approach is required to overcome the 'decision disconnect' between various components in the governance structure. Her main focus is 'the relationship between ethics, governance, the enterprise, and information and communications technology (ICT) organisations'. She argues that value systems are central in governance and provides a detailed account, using a systems-thinking approach, of how governance can be undertaken that minimises the risk of decision disconnect arising from the aforementioned components of ethics, governance, the enterprise and information and ICT.

References

Alexandra, A & Miller, S, *2010, Integrity systems for occupations*, Ashgate.

da Cruz, M, nd (a), 'No duty of care: the governance of ICT', Ramin Communications, <http://www.ramin.com.au/itgovernance/nodutyofcare-the-governance-of-ict.html>

da Cruz, M, nd (b), 'AS8015-2005 — Australian standard for corporate governance of information and communication technology (ICT)', Ramin Communications, <http://www.ramin.com.au/itgovernance/as8015.html>

The Department of Education and Children's Services, 'ICT governance in DECD', nd, <http://www.decs.sa.gov.au/it/pages/indecs/ICTgovernance>

---

1   We acknowledge the contribution of Dr Jeremy Moss to this discussion.

Miller, S, 2010, 'Integrity systems and professional reporting in police organisations', *Criminal Justice Ethics*, vol. 29, pp 241–57.

Weckert, J & Al-Saggaf, Y, 2008, 'Regulation and governance of the Internet' in HT Tavani and KE Himma (eds), *The handbook of information and computer ethics*, Wiley, Hoboken, NJ, pp 475–95.

Biographies

Dr Richard Lucas has over 30 years experience working and teaching in the ICT industry. He has worked professionally as an ICT project team leader, senior systems analyst, systems engineer, programmer, and instructor in the fields of agriculture, defence, engineering, health, health insurance, electoral systems, and taxation. He has worked in the public and private sectors as both an employee and independent consultant. For more than 15 years he has taught ethics and social responsibility to ICT students at the professional, vocational, and university levels in Europe and Australia. He has developed curricula and teaching support material for undergraduate and masters subjects in the fields of ethics, technology and professionalism. In 2009 he published *Machina ethica — a framework for computers as Kant moral persons*. Currently he is the head of discipline for information systems, in the Faculty of Information Sciences and Engineering, at the University of Canberra.

Dr Cecilia Ridgley is an executive consultant working with large, government greenfields programs. Cecilia's professional background began with the military as an army officer cadet at the Australian Defence Force Academy, moving through senior leadership positions in the Australian Government, before joining the private sector as an IT management consultant. Cecilia's leadership of high-profile projects and enterprise strategies for large whole-of-government and NGO programs over the last 20 years has provided insight into the ethical demands of IT in a public policy environment. Cecilia's background spans enterprise and information architecture, organisational modelling, program management, change leadership, risk management, budget control, business and IT analysis, transition and transformation leadership, security and assurance management, and people management. Cecilia brings these disciplines together to develop a strong, systemic response to values challenges in governance of IT in large businesses. Cecilia has consulted for international firms before managing her own practice — Redline Consulting, and has lectured at the University of New South Wales — Australian Defence Force Academy. Cecilia has published award-winning research in belief systems; learning; counter terrorism; and crucial decision-making and holds a PhD in computer science on a systems approach to ethical decision-making.

9. Virtuous IT governance: IT governors can't be virtuous!

Richard Lucas

University of Canberra

It is not that IT[1] cannot (or ought not) be governed,[2] but it is rather that, as a field of endeavour, IT governors have nothing against which to measure their virtue: nothing against which to measure their governance efforts.

Why should this be so? Well it turns out that the problem is not with IT governance as such, but in being able to measure governance efforts within a virtuous framework. To be virtuous there must be an ideal against which governance efforts can be measured. I will now shift my discussion from governance particularly to professionalism generally. I do this because the problem I address is not limited to simply governance: The problem is common to all aspects of being a professional. That is, if virtuous governance is not possible then, I claim, virtuous professionalism is not possible because they share a common malady. Conversely and more powerfully (but more straightforwardly), if virtuous professionalism is not possible then virtuous governance is not possible, if for no other reason than governance is a proper subset of being a professional. So, to make my claim that IT governors cannot be virtuous, all I need show is that IT professionals generally, cannot be virtuous.

But I get ahead of myself; before I get to far into this discussion I need to cover the basics of virtue and the basics of governance. That is, it seems best to look at what virtues are and what governance is before I say why virtuous IT governance is not possible. What are we talking about when we talk about virtues? What does it mean to engage in governance?

A brief outline of virtue

As most IT professionals are not versed in the details of any particular ethical theory especially virtue (they, mostly, just know what is right) I will now give a primer of what I mean by virtue.

---

1   I use the older but more accurate 'IT' over the more recent (in some countries) 'ICT' as the 'C' is a redundant reference. Technology that is used to contain information must capture it (a form of communication). To be useful, technology that contains information must have ways of extracting the information from the technology (also a form of communication).

2   The multiplicity of books addressing IT governance attests to that.

Firstly, a definition of virtue:

A virtue is a character trait that contributes to human flourishing.[3, 4]

Some examples of virtues are:

1. autonomy
2. beneficence
3. benevolence
4. caring
5. compassion
6. courage
7. empathy
8. fairness
9. forgiveness
10. friendship
11. integrity
12. judgement
13. justice
14. liberality
15. loyalty
16. knowledge
17. good temper
18. negligence
19. openness
20. perception
21. prudence
22. reflection

---

3   Aristotle defines virtuous character as: 'Excellence [of character], then, is a state concerned with choice, lying in a mean relative to us, this being determined by reason and in the way in which the man of practical wisdom would determine it. Now it is a mean between two vices, that which depends on excess and that which depends on defect.' Nicomachean Ethics II.7, 1106b36-1107a3.

4   I am aware that some accounts of human flourishing might mean the significant harm or even destruction of other species. This is not what I mean. By my account, human flourishing must include the consideration of other morally worthy entities (for example, in my view, trees).

23. respect

24. self-control

25. selflessness

26. proper shame

27. trustworthiness

28. temperance

Now, the reader might have a problem with believing that some in the above list are virtues. Try viewing a virtue as the mean sitting between two vices and relative to some situation or action. To understand what 'sitting between two vices' means I will give two examples: good temper and courage.

Firstly, good temper, (which, if you lose it, leads to anger). There are times when it is appropriate to be angry and times when it is not. Here the excess of (unjustified) anger makes for an irascible person and the deficiency of (justified) anger makes for an inirascible (that is, excessively peaceful) person. According to Aristotle to be either irascible or inirascible is to be morally deficient.

Courage is another example: courage is the mean between two fears; too much and too little. To have too much fear is to be a coward, to fear too little is to be reckless or foolhardy, a madman. To be courageous is to overcome unjustified fear; it is also courageous to heed justified fear.

It is obvious that not all virtues apply in all circumstances and not all virtues are equal across all professions/occupations. So, which virtues apply to a particular profession and in which circumstance? Before I can answer this question I need to explain a bit more about virtues.

What is it that virtues are relative to? In many professions, there is something against which (to be relative to) to measure the virtues and vices; there is an ideal. So, what are these ideals?

According to Justin Oakley and Dean Cocking:

> what counts as acting well in the context of a professional role is in our view importantly determined by how well that role functions in serving the goals of the profession, and by how those goals are connected with characteristic human activities. That is, good professional roles must be part of a good profession, and a good profession, ... is one which involves a commitment to a key human good, a good which plays a crucial role in enabling us to live a humanly flourishing life. (2001, 74)

To understand virtue in a professional context it is probably best to start with an easy example, to see how it works for something that all might agree is a profession. Take medicine, for example. For medicine the main goal is to serve the health of people. It is clear that health is central to human flourishing and so this qualifies medicine as a good profession.

The virtuous medical professional has:

1. autonomy
2. beneficence
3. compassion
4. courage
5. honesty
6. humility
7. integrity
8. non-maleficence
9. trustworthiness
10. truthfulness

Just so that you do not think that I have picked an exceptional example, consider architecture (Yoch, 1978). For architecture, the main goal is to design aesthetic and functional housing for people. It is clear that housing is central to human flourishing (ask the homeless about flourishing) and this qualifies architecture as a good profession.

The virtuous architect has:

1. beneficence
2. compassion
3. courage
4. honesty
5. humility
6. justice
7. trustworthiness
8. truthfulness

So, we have two examples, what difference does this make to my claim that IT workers cannot be virtuous? To answer this question I must first return to my claim that IT professionals cannot be virtuous because there is nothing against which to measure their virtue.

For each of my examples above, I started with a main goal, showed that this goal contributed to human flourishing, and then gave a list of virtues that I claimed were consistent with the main goal. To do this for IT, it is necessary to begin with some definitions.

A brief outline of governance

At its most general governance is: 'the exercise of power overall' (Távora, 2007,) or, more particularly,

> 'European governance' refers to the rules, processes and behaviour that affect the way in which powers are exercised at European level, particularly as regards openness, participation, accountability, effectiveness and coherence. These five `principles of good governance' reinforce those of subsidiarity and proportionality. (European Commission, nd)

As I used medicine to highlight virtue, I now do the same for governance; that is, I use medical governance as a way of examining IT governance. Again, for medicine the main goal is to serve the health of people. It is clear that the health of people is central to human flourishing: this qualifies medicine as a good profession that ought to be able to have virtuous medical governance.

Medical governance

So, good governance (for medicine) means ensuring that the management of medicine embraces openness, participation, accountability, effectiveness and coherence. Combining the values of good medical governance with the virtues of a medical professional provides a coherent notion of what is necessary for an IT professional to carry out good governance.

In general, IT governance means:

> specifying the decision rights and accountability framework to encourage desirable behavior in using IT. ... IT governance reflects broader corporate governance principles while focusing on the management and use of IT to achieve corporate performance goals. (Weill & Ross, 2004: 2)

and,

> The goal of an IT governance program is to establish chains of responsibility, authority, and communication to empower people in support of the overall enterprise goals and strategy. (Ambler & Kroll 2007)

Notice that there is no reference to anything that might resemble ethics, never mind virtue or human flourishing. Even organisations such as the IT Governance Institute[5] make no reference to the ends, moral or otherwise, of IT. Indeed, the closest they get is to mention some sort of IT–business alignment. It is not my purpose here to examine the merits of these definitions, rather to point out the omission of any mention of ethics in the standard literature on ICT governance. This is because, for IT, there is no main or direct goal that contributes to human flourishing! According to my claim, it is not surprising that the two definitions given above do not contain anything relating to virtuous governance: such a claim simply could not be included.[6] Now, this seems to leave no room to move, nothing more to say on the matter.

There may, however, be something in the indirect references to loyalty and human welfare in the code of ethics of Australia's IT professional group, the Australian Computer Society (ACS). In the ACS code of ethics (2012), section 4.1(b) refers to loyally serving the community; while, section 4.1(d) advises members to use their special knowledge and skill for the advancement of human welfare.

While there is nothing in the detail of the code that spells out what it means to advance human welfare, the notion does approach ideas of human flourishing and, perhaps, there is something here — in the general sentiment of advancing human welfare — that can be used to create a specific goal for IT. In practice, this would be a goal that promotes human flourishing and is, therefore, a goal against which virtues can be measured.

I propose that the following be the main goal of IT:

> Build and maintain information technology for the good of both society and individuals.

---

5   'The IT Governance Institute (ITGI) exists to assist enterprise leaders in their responsibility to ensure that IT is aligned with the business and delivers value, its performance is measured, its resources properly allocated and its risks mitigated.' (<http://www.itgi.org>). I could not find any spelling out of the the term 'value' or the phrase 'properly allocated' in terms of ethics or human flourishing. Presumably the ITGI leaves the justification up to the individual businesses.

6   Why should I make this claim though? After all, all I have done is to give a couple of examples and that does not constitute definitive proof. Read any literature on IT as a profession; it will be resplendent with admonitions to do good, to be professional, to be technically competent. It will not, however. contain any reference to what you are supposed to be technically competent about, what you are being good about. There is no reference to what it is that the IT professional is being good for. In this light, see the ACS (or any other IT professional society) code of ethics.

Unfortunately, there is nothing in the IT literature which states, promotes, or explains in detail how to carry out this main goal; that is, what is the good that I refer to? Recall that I established that both medicine and architecture met the criteria for being a profession that addressed human flourishing. With this in mind, we can now refocus the goal of IT to being in the service of human flourishing professions. So, now the main goal of IT (according to me) becomes:

> Build and maintain [insert name of qualifying profession, for example, 'medicine' or 'architecture'] information technology for the good of both society and individuals.

The process is now to match the lists of virtues for these professions (medicine and architecture) with these new ideals. For example, the virtuous medical IT professional has:

1. autonomy
2. beneficence
3. compassion
4. courage
5. honesty
6. humility
7. integrity
8. non-maleficence
9. trustworthiness
10. truthfulness

The virtuous architectural IT professional has:

1. beneficence
2. compassion
3. courage
4. honesty
5. humility
6. justice
7. trustworthiness
8. truthfulness

For IT, however, there is a problem of implementation. In a 2008 study of ethics in the IT profession in Australia, John Weckert and I asked IT professionals for whom they did their IT work. Their responses were revealing: 64.7 per cent of the respondents listed ICT, their own group, as the end purpose of their work. That is, they did not see their work as contributing to any other goal (Lucas & Weckert, 2008: appendix 69). Now, this cannot be right. As we have seen for medicine and architecture, to be a good profession, the goal of that profession ought to be some higher ideal: the health of people in medicine's case. In the raw data, the respondents described themselves as being affiliated with some 41 different industry groups. As some of them made too fine a distinction between industry spheres, some of the responses were combined. For example health, medical and hospital were combined in a single health group. This combining of spheres resulted in a reduced set of 19 identifiable industry groups and one unspecified group. Here are the combined results.

Table 1: Responses by industry group

Response	Number	% of Total Responses
Agriculture	1	0.3
Aviation	2	0.6
Commerce	7	2.0
Defence	1	0.3
Education	23	6.5
Entertainment	1	0.3
Finance	4	1.1
General	26	7.3
Government	16	1.5
Health	2	0.6
ICT	229	64.7
Insurance	3	0.8
Media	2	0.6
Religion	1	0.3
Science	5	1.4
Security	6	1.7
Service	16	4.5
Statistics	1	0.3
Unspecified	9	2.5

Source: Based on Weckert and Lucas, 2008

This clearly shows that the majority of IT professionals have no clear idea that they are in the service of another profession nor of the goal or purpose of their work in relation to that qualifying profession. In spite of this difficulty I will now give some concrete examples of how the IT professional can be measured against virtue criteria. The way to do this ought to be clear: pick an area of work that can clearly be seen to be contributing to human flourishing, a 'qualifying' profession.

The significant areas listed in the study by IT professionals who had an awareness of the end goal of their work listed identified were the following:

Table 2: Responses by service category

Response	Number	% of Total Responses
Education	23	6.5
Government	16	4.5
Health	2	0.6
Religion	1	0.3
Science	5	1.4
Security	6	1.7
Service?	16	4.5

Source: Based on Weckert and Lucas, 2008

While I have taken some liberties with this list (especially by including the service category), others will, no doubt, criticise my choices of those spheres that are either included or excluded from the list. A detailed analysis of all 41 distinct response groups would be necessary to arrive at a list of work areas that might count as contributing to human flourishing, but that is beyond the scope of this chapter. Also, it is not my purpose here to show how IT professionals might carry out their work so that it counts as ethical.[7]

My point here is to show that the set of work areas identified by IT professionals while small, is not empty. There is some work by IT professionals that can clearly count as, if done for the right purpose or goal and done in the right sort of way, contributing to human flourishing.

---

7   There is much work to be done to show how these virtues (and others) can be used to serve an ideal. For the details of the application of virtue specifically, and ethics generally, to IT you will need to wait for a forthcoming book by John Weckert and I, *An integrity system for ICT — incorporating ethics into industry practice*.

What is the relationship of this to governance?

It seems that I have gotten a long way away from governance in writing about IT professionals generally. It serves the purpose, however, of setting the virtuous context for IT governance. The next step is to restate my requirement of the virtuous IT professional:

Create and enact [insert name of qualifying profession] information technology governance that contributes to human flourishing.

Examples

Because respondents to the research listed health as a work area, I will now give some examples of how the IT professional working in that field uses virtues to measure IT governance. Recall that some of the virtues of the medical IT professional are:

1. autonomy

2. compassion

3. courage

4. honesty

5. integrity

6. justice

7. non-maleficence

8. trustworthiness

9. truthfulness

The four principles that are most commonly referred to in medical governance literature,[8] and how they might be modified to fit IT, are:

Autonomy, in medical IT governance, means asking the following simple questions —does the IT being proposed (or used — when already in place) enhance the autonomy of those charged with making medical governance decisions?; and, do the resulting IT governance procedures enhance the autonomy of those affected by subsequent governance decisions?

Beneficence, in medical IT governance, means asking the following simple questions: does the IT being proposed (or used, when already in place) enhance

---

8   See for example, Hussein (2010); Gillon (2003); Baines (2008).

the beneficence of those charged with making medical governance decisions?; and, do the resulting IT governance procedures allow for greater beneficence to be shown towards those affected by subsequent governance decisions?

Non-maleficence, in medical IT governance, means asking the following simple questions: does the IT being proposed (or used — when already in place) enhance the non-maleficence of those making medical governance decisions?; and, do the resulting IT governance procedures minimise harm towards those affected by subsequent governance decisions?

Justice, in medical IT governance, means asking the following simple questions: does the IT being proposed (or used — when already in place) ensure the equitable decision-making of those making medical governance decisions?; and, do the resulting IT governance procedures enhance justice towards all vulnerable groups affected by subsequent governance decisions?

Of course, getting clear and unambiguous answers to these questions is difficult. In the case of proposed systems developments it is imperative that they be answered before any work begin on said systems. This will in turn add to the economic cost and project timeline of these projects.

Conclusion

Surely, there is some work that IT governance professionals do that is morally worthy but does not fall into the category of virtuous? Surely, there are things that IT governance professionals do that are morally worthy but do not contribute to human flourishing? Well, one could be anthropometric about this and claim that all things humans do either adds to, or diminishes, human flourishing. Unfortunately, this stance does not get us to the heart of the problem that I am trying to address. Claiming anthropocentricism trivialises the issue. My claim here is not that all IT professional governance activity can be evaluated in terms of virtue ethics but rather that, for those that can, it is not IT itself that the activity ought to be measured against: it is the end goal of the IT work that ought to be the yardstick.

For the issues of IT governance and professionalism to reach a point at which they can be seriously considered, each IT system needs to be considered in light of the following two fundamental questions:

Does the IT system that was built contribute to human flourishing (in each particular field such as medicine and architecture)?

Do the governance activities that IT professionals engage in, in constructing IT systems that do contribute to human flourishing, themselves contribute to human flourishing?

References

Ambler, SW & Kroll, P, 2007, 'Best practices for lean development governance. Part I: principles and organization', *The Rational Edge*, 15 June, <http://www.ibm.com/developerworks/rational/library/jun07/kroll/index.html>

Australian Computer Society (ACS) 2012, 'Code of ethics', June, <https://www.acs.org.au/__data/assets/pdf_file/0005/7835/Code-of-Ethics_Final_12.6.12.pdf>

Baines, P, 2008, 'Medical ethics for children: applying the four principles to paediatrics', *Journal of Medical Ethics*, vol 34, pp 141–45.

European Commission, nd, 'Governance in the EU: a white paper', <http://ec.europa.eu/governance/index_en.htm>

Gillon, R, 2003, 'Ethics needs principles — four can encompass the rest — and respect for autonomy should be "first among equals"', *Journal of Medical Ethics*, vol 29, pp 307–12.

Hussein, G, 2010, 'When ethics survives where people do not', *Public Health Ethics*, vol 3 no 1, pp 72–77.

Lucas, R & Weckert, J, 2008, *Ethics and regulation in the ICT industry*, report for the Australian Computer Society, Centre for Applied Philosophy and Public Ethics, Charles Sturt University, Canberra.

Oakley, J & Cocking, D, 2001, *Virtue ethics and professional roles*, Cambridge University Press.

Távora, F, 2007, 'Globalisation and governance', *Pós-Graduado em International Economics*.

Weill, P & Ross, J, 2004, *IT governance*, Harvard Business School Press.

Yoch, JJ, 1978, 'Architecture as virtue: the luminous palace from Homeric dream to Stuart propaganda', *Studies in Philology*, vol 75, no 4, Autumn, pp 403–29, <http://www.jstor.org/stable/4173981>

10. The decision disconnect[1]

Cecilia Ridgley
Business Development PMO, ICT Shared Services,
ACT Government

Introduction

What is the relationship between ethics, governance, the enterprise, and information and communications technology (ICT) organisations? Why should an ICT practitioner seeking to create organisational value endeavour to understand these relationships? In considering these questions I employed a systems thinking approach to make explicit commonalities between these entities.

In this paper I will begin by establishing the need for understanding these relationships. I will then address how we come to (mis)understand these relationships and concepts through extant definitions, and the implications for design and practice of governance.

I present a top down analysis, examining first the societal and operational levels (state and corporate), and secondly, the functional level of governance — in application to ICT organisations. Finally, I present a systems-thinking framework, for practitioners and theorists alike, to explicitly plan and design for the dynamic interrelationship of values, people, process, and purpose in enterprise governance.

Throughout the paper I focus on the unambiguous establishment of the critical nature of value systems in the practice of governance in the enterprise, and particularly the ICT enterprise. I argue for the prevention of a decision disconnect that arises from a lack of awareness of the pervasiveness of complexity emerging from the interactions of values, governance, and ICT organisation systems or through designing and planning for a single instance of a system component, such as a single moral framework. The framework advocates designing for multiple and diverse instances of the systems components — to design for requisite diversity.

---

1 A version of this chapter was published in *The Australasian Journal of Information Systems*, 2009, vol 16, pp 185–207. It is published here with permission.

Understanding why values are critical

ICT practitioners seeking to create organisational value should endeavour to understand the relationship between ethics, governance and ICT because there are clear examples where ignorance of the nature of these relationships has caused organisational failure.

Media reports, since the early 1990s, of high-profile organisations coming to an untimely demise due to the perceived lack of ethical behaviour of those charged with their leadership, have increased public awareness of the concepts of governance and business ethics. There have been a large number of scholarly enquiries into the unfolding of events at organisation such as Barlings Bank, Mirror Group Newspapers, Enron, and even major sports corporations (Drennan, 2004; Sama and Shoaf, 2005; McNamee and Fleming, 2007; Zattoni, 2007; Cacioppe et al, 2007; Schnebel and Bienert, 2004).

The first attempts to codify governance, occurring as a reaction to these high-profile failures, was the Cadbury Report in 1992, followed by other reports including the Hampel Report of 1998 and the Turnbull Report in 1999 (Drennan, 2004). Ethical codes of conduct such as that of the Australian Computer Society (ACS, 2008) also attempt to provide clear guidance to individual practitioners on ethical behaviour. Providing a list of principles or values in the form of a code or standard does not, however, assist the practitioner in resolving conflicts in principles or amongst stakeholders. This point is supported by Graeme Pye and Matthew Warren who write:

> Although the code addresses a wide area of principles, the advice only serves as guidance to the IT professional from a personal behaviour aspect and does not seek to deliver a methodology for resolving ethical dilemmas between individuals or an individual and the business practice or goal. (2006: 202)

In the literature, the treatment of ethics as it relates to governance and organisational success is variably placed at the feet of the individual practitioner, the corporate governors, or corporate culture, and enshrined in codes and principles that seek to influence the membership of the entire organisation. This perspective is exemplified by the statement that 'a pattern of values expressed in an explicit code or in shared understanding will permeate the organisational environment' (Minkes et al, 1999: 328).

The difficulties in placing the ethical responsibility solely with individuals is best expressed in EE Arthur's relating of Harold Geneen's assessment of governance:

'Among the boards of directors of Fortune 500 companies, I estimate that 95% are not fully doing what they are legally, morally, and ethically supposed to do. And they couldn't, even if they wanted to.' (1987: 59)

The affect of moral agency is an important constraining factor that is not addressed in the extant standards and definitions for governance that are addressed in this paper. Alan Lovell (2002) examines this issue in more detail.

It is not my intention in this paper to dwell on an analysis of the public failures of corporations, but, rather, to highlight three observations stemming from a review of the literature:

1. There is a continuing and pressing need to examine the role of governance and ethics in organisations.

2. The current focus for remedy remains on the key behaviours/processes of the individuals; groups of individuals empowered with the leadership of the organisations, or, a vague conception of 'corporate culture' that is still, at its heart, deemed the responsibility of those named above.

3. The authors variably exhort the importance of regulation, personal and corporate cultural ethical values, proper risk management, and effective internal controls as the panacea against corruption.

Certainly there is no argument from this author that these external pressures and internal processes provide *necessary* mitigations for the risk of direct unethical behaviour leading to predictable circumstances of organisational failure, but I question whether they are *sufficient* mechanisms to prevent the emergence of systemic failure of an organisation through indirect and indeterminate ethical dynamics.

I propose that emergent-system effects, such as corporate corruption and cultural ignominy, may be best dealt with through a *system* of governance rather than a set of governing processes or tasks. Arthur's call for action is as relevant today in supporting a systems thinking approach:

In this time of concern with the ethics of corporate governance, corporate cultures, and the corporation as a moral agent, it is perhaps time to reconsider a theory which unites the three in looking at corporate governance. (1987: 70)

It seems to me that the factor that contributes most to a lay person's confusion over, and failure in application of, the role of ethics and governance in organisations stems from the very issue of confusing processes with systems, and subsequently discounting the exigent complexity that is characteristic of systems, but not of processes, in decision-making. The approach offered here is to deal with process failure through processes, and to deal with system failure via systems — to take a systems-thinking approach to the problem.

But what difference does this make? The key difference, and the basis for the definition and framework offered later in this paper, is the recognition and conscious 'design' for learning — single and double feedback loops — between the system actors and these interacting elements, and across functional hierarchical layers of the organisation. Chris Argyris (in Smith and Hitt, 2005) argues that 'human beings produce action by activating designs stored in their heads' and, furthermore, that 'at the core of acting effectively is learning'. He also proposes that 'organisations create designs for action that they teach individuals to produce skilfully in order to achieve the organisation's goals effectively' and, 'Double-loop learning occurs when, in order to correct an error, it is necessary to alter the governing values of the master programs' (262–63). Argyris' 'master programs' are the identifiers of the governing values of an organisation, and deviations from the intended outcomes of the master programs trigger double-loop learning to effect a change in the governing values. These master programs are essentially the 'designs' in the 'heads' of those leading an organisation (not just those who are tasked to do so). Argyris is establishing a conception of a system model for the interaction of values (ethics), governance and action in an organisation through its people, with the key relationship dynamic being learning.

The advantage of recognising learning as a key coupling dynamic between these system entities is to realise that learning itself is a system and what we are actually seeking to understand is the nexus of at least four co-evolving systems — governance, ethics through value systems, learning and, action through operational systems. If the learning system is consciously designed for purpose, it may affect prevention of emerging multi-systemic consequences through double feedback loops. This learning is targeted at making explicit and challenging the assumed value systems within the 'master program' that drive the co-evolving systems, rather than simple remediation of failing processes of the action system alone through single feedback loops such as quality control, audit, and risk management. The locus of attention is changed from remediation of operational governance issues to prevention of strategic governance issues as illustrated in Figure 1.

Figure 1 demonstrates two different foci for attention: the first operates internally in the governance system (at the 'operational level'), and the second at the nexus between the governance system and its co-evolving systems in organisation (at the 'strategic level'). The actor, process, values, and data presented in the top representation are instances of what might be focused on from an operational viewpoint, and are not intended to represent the entire focus of governance at the operational level. The intent of the diagram is to support the analysis of governance at the systems level — a step up from the 'process' level of operation.

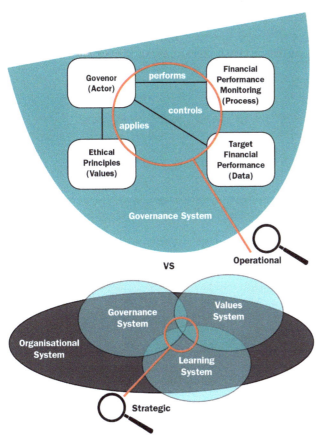

Figure 1: Focal hierarchy for remediation versus prevention of governance issues

Source: Author's research

Once we can establish the locus of attention to the level of the system and its co-evolution with other systems, we can more cogently define the concepts under examination, and identify shortfalls in existing definitions and practice that may inhibit value creation. As this makes clear:

> the difficulty in applying the skills demanded for effective and ethical organisational governance is more easily understood by recognising the complex interrelated issues involved in leading an effective [organisation]. (Karri et al, 2006: 75)

Definitions

A symptom of the tendency to confuse focus levels arises in the literature on governance when the entity that is known as *corporate governance* and the entity that can be described as *state governance* seem to be treated as entirely theoretically divorced concepts with their own language, rules, instruments and methods. Examples of this can be seen in the difference between Jan Kooiman's work *Governing as governance* (2000), in which the concepts examined are those of state governance, and the 'Australian Standards 8000–2003' series which examines corporate governance.

Kooiman's work focuses on the strategic interpretation of governance, but with little to inform the practitioner on how to operationalise the model, while AS 8000–2003 seeks to provide a guide to practice at the operational level, without reference to the interdependencies of the strategic.

Furthermore, from a practitioner's perspective, it is vexing to delineate between the act of governing, government as both noun and verb, and the label 'governance'. Governance is sometimes used by lay people to describe the act of governing rather than a system in its own right. This issue of the conflation of nouns and verbs in the ICT industry is common, for example: the use of the term *methodology* to describe *methods*, and *applications* to describe *programs*.

The systems approach proposed is an attempt to highlight a definition for governance that is relevant and insightful, regardless of the specialisation and application of its use, as well as sensible and coherent for the practitioner. It makes little sense to have a definition of governance that is not equally applicable to nation states and to corporate bodies or even household units. Gerry Stoker writes 'Governance is ultimately concerned with creating the conditions for ordered rule and collective action. The outputs of governance are not therefore different from those of government. It is rather a matter of a difference in processes' (1998: 17).

The definition should account for all levels of focal hierarchy. The application of a systems approach is an attempt to provide not so much a unified theory as a simple pattern for describing governance, and the interaction of value systems with it that may be modified for scale where required. That is, my definitions will describe the key objects, relationships, boundaries, and purpose of governance, ethics, organisations and ICT.

The purpose of defining in this way is to provide a 'template' or framework for the practitioner and researcher to make sense of the problems encountered with these concepts in application, and to effectively communicate solutions across specialisation boundaries. Stoker reinforces the validity of this approach when he states that:

> The value of the governance perspective rests in its capacity to provide a framework for understanding changing processes of governing. (1998: 18)

It is not my intention to create another taxonomy for each of these concepts, but, rather, to create an appropriate nexus from which specialised taxonomies for particular applications may conjoin.

In this section, I provide an example extant definition at each level of the systems hierarchy — from societal to functional — to illustrate why this definition is more cogent theoretically and practically.

In reviewing extant definitions of governance, I examined the Australian Standards 8000 series for corporate governance 2003, the International (2008) and Australian (2005) standards for corporate governance of information technology, and Kooiman's *Governing as governance* (2000).

These definitions are situated in the governance hierarchy illustrated in Figure 2.

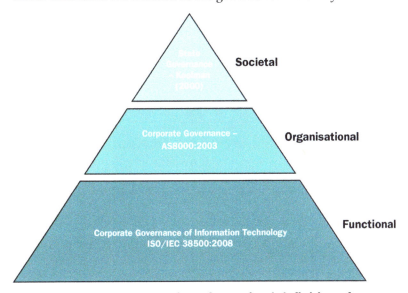

Figure 2: Hierarchical location of examined definitions for governance

Source: Based on International Organisation for Standardisation, 2008, Koolman, 2000 and Standards Australia, 2003

The key criticism behind each definition is the lack of recognition and design for requisite diversity in the many interrelating components of the systems that co-evolve with governance and, particularly, in the values systems. Where values systems are recognised, only one instance of the possible value systems actually at play within the organisation seem to permeate the design of the governance system. This paper maintains that governance systems, through their definitions, must recognise and design for diversity in all components of the system, and especially values systems.

The societal level

Kooiman describes governance through the act of governing:

> Governing can be considered as the totality of interactions, in which public as well as private actors participate, aimed at solving societal problems or creating societal opportunities; attending to the institutions as contexts for these governing interactions; and establishing a normative foundation for all those activities. (2000: 4)

In this definition, Kooiman provides us with an almost complete system's model for governance. His description identifies actors, purpose, environment and pressures. The following description defines Kooiman's processes and interacting systems for the model:

> The three elements of governing, image formation, choice of instruments and action, and the structural conditions they are nested in, respectively value systems, resources and social-political capital, can be supposed to not be equally distributed for governance purposes. (2000: 224).

But Kooiman's definition lacks two important theoretical and practical notions:

1. He does not specify the dynamics of the system — that is how the 'interactions' occur and how they influence each other nor how the interdependent systems or 'structural conditions' arise and facilitate or impede the achievement of the system purpose.

2. He does not specify an evaluative mechanism for assessing the performance of the system, or a means for distinguishing and treating undesirable properties within the system, but rather includes undesirable or unhelpful interactions, for example, as valid parts of the 'totality'.

Consequently, we are left with a snapshot in time of what may constitute governance through governing practice, but not any means by which to design for ourselves an effective and successful governance system. Without such an indication, although we possibly have the correct ingredients, we are missing the recipe.

Kooiman's seminal theoretical work provides a much richer development of the concepts of governance than what is represented here, however, I propose that a definition should capture not only the components of the system, but also their key dynamic relationships — as it is in definitions on which most practitioners build their understanding and expectations of the use of a notion such as governance. The risk of providing a static representation of the system in a definition is that practitioners and perhaps even theorists will apply and extend the practice without heed to the impact of the system dynamics, and fail to achieve the stated purpose. This risk is (at least partly) dealt with in the definition provided by this paper through the surfacing of the key coupling dynamic of learning.

One of the key messages from Kooiman's work in providing a strategic definition of governance, which is important for this discussion, is the encapsulation of the organisation and its processes within a wider context and environment. This is further exemplified in a definition of governance related to the university enterprise by Simon Marginson and Mark Considine:

> Governance is where the identity of each university as a distinctive social and cultural institution is shaped, within a 'global knowledge economy'.(2000: 8–9)

In this definition, the university's governance system is explicitly encapsulated within a wider system of the 'global knowledge economy' whereby, presumably, the creation of value by such a university contributes to the economy and feeds back to enhance the university's individual worth. Identity is another idea that is important for our systems-model surfaces in this definition. Marginson and Considine link the emergence of identity to the notion of governance. In the definition proposed by this paper, identity is a key outcome for the interaction of values, governance and organisation systems. This concept is an important addition to the model and is addressed again later in this paper as part of the offered framework.

The operational level

The 'Australian Standards 8000–2003' series defines corporate governance as 'a system by which entities are directed and controlled'.

This conception, while vague, is supported by a range of considerations and tasks for the practitioner, with values systems represented by a set of principles, thus:

> A clear statement of corporate governance principles helps boards identify their functions and roles [and that] a system of good corporate governance is recognised to include the following principles:

- Accountability: being answerable for decisions and having meaningful mechanisms to ensure adherence to all applicable standards.
- Transparency/openness: clear roles and responsibilities and clear procedures for decision-making and the exercise of power.
- Integrity: acting impartially, ethically and in the interests of the organisation, and not misusing information acquired through a position of trust.

- Stewardship: using every opportunity to enhance the value of the public assets and institutions that have been entrusted to care.
- Leadership: leadership from the top is critical to achieving organisation-wide commitment to good governance.
- Efficiency: the best use of resources to further the aims of the organisation, with a commitment to evidence-based strategies for improvement.

These principles provide individual guidance to the actors tasked with corporate governance of organisations and amount to statements of personal ethics.

The definition provided does recognise that governance is a system but makes no attempt to define that system. Furthermore, it proceeds to list a set of processes being:

- answer to decisions
- delineate roles and responsibilities
- act impartially
- act ethically
- act in the interest of the organisation
- do not abuse trust
- seek and act on opportunities to enhance organisational value
- show leadership
- use resources efficiently
- seek evidence for improvement

The following information elements are also identified:

- procedures for exercise of power and decision making
- mechanisms to adhere to standards

But we do not know why we are doing this — these principles do not provide a statement of purpose, or a clear set of boundaries. Furthermore, it is self-evident that achieving some of the processes can conflict with the achievement of others — which thereby creates an internal incoherency in the model. There is no recognition of external pressures and the encapsulation of the system in wider systems, nor are there any provisions for designing and measuring the performance of the governance system except against personal judgements and interpretations of what 'ethically' might mean, what 'clear' constitutes or, what 'best use' entitles a manager to do.

The definition provided by the standard is ambiguous, subject to personal interpretation in both practice and evaluation of practice, and assumes a range of existing conditions and organisational maturity that may or may not be present. This may be a matter of intentional design, taking the philosophy of minimal intervention in the judgement and activities of corporate governors, but this assumption reinforces the concept of the accountable individual rather than a systemic failure.

> A feature of such ... interest is the apparent desire to place the blame on an individual, or individuals, for a corporate failure, as has been seen in the case of Enron and its Chief Executive, Jeff Skilling ... This selective emphasis on the individual often disguises the culpability of the organisation as a whole, and the series of management and systems failures, which created the circumstances in which the abuse could take place. (Drennan, 2004)

Taking the definition alone, without the several volumes of supporting material and handbooks associated with it, this definition is not sufficiently directive to build an acceptable governance system design. Of course, the argument is to study the supporting material to find out how to do 'governance', or to use good business judgement — however, I again argue that the lay person will seek to understand and build their expectation of practice through definition and, if this is not well conceived, then assumptions of practice will inevitably flavour the interpretation of the rest of the supporting documentation.

The functional level

The next level of governance is the functional level. I have chosen the ICT governance function for examination here, although any other function of the organisation would be equally appropriate. Nada Korac-Kakabadse and Andrew Kakabadse (2001) support ICT governance as a hierarchical child of corporate governance: IS/IT and corporate governance are integrally inter-related thus making IS/IT governance a subset of corporate governance (2001: 9).

The following definitions are provided for examining ICT Governance.

> Governance is the description of the necessary performance of the operation of a system to ensure it achieves its goals, meets the wishes of all those concerned, or uses its assets efficiently. (Lewis, 2008)

> Governance is the strategic alignment of IT with the business such that maximum business value is achieved through the development and maintenance of effective IT control and accountability, performance management and risk management. (ISO/IEC 38500:2008)

> Governance is the system by which the current and future use of ICT is directed and controlled. It involves evaluating and directing the plans

for the use of ICT to support the organization and monitoring this use to achieve plans. It includes the strategy and policies for using ICT within an organization. (AS8015:2005)

IT Governance is the decision rights and accountability framework for encouraging desirable behaviours in the use of IT. (Weill and Ross, 2004)

ISO/IEC 38500:2008 takes the Australian Standards 8000 series definition further with a set of system-performance measures: effective, efficient, and acceptable use of IT within an organisation; and a purpose statement — use of IT within organisations. It also specifies a wide range of actors and includes internal and external views of the organisation. The definition also provides a boundary of scope: 'applies to the governance of management processes (and decisions) relating to the information and communication services used by an organization', and recognises that the locus of control for these processes does not necessarily reside within the organisational hierarchy boundaries. The ISO standard was developed from the Australian Standard AS 8015:2005, which combines principles, tasks, roles and responsibilities and purpose, and recognises an evolutionary and dynamic aspect of the governance system through the changing 'needs of people in the process.' Furthermore, 8015 recognises the encapsulated systems in the lower order of system hierarchy (below functional) such as portfolio and project governance as well as specifically requiring alignment with higher level (corporate/operational) systems.

An extract from 8015 is provided in Table 1.

This Standard combines the six principles of governance with the three primary tasks of Directors to form a list of tasks for the control of ICT. The six principles are:

1. Establish clearly understood responsibilities for ICT

   Ensure that individuals and groups within the organisation understand and accept their responsibilities for ICT.

2. Plan ICT to best support the needs of the organisation

   Ensure that ICT plans fit the current and ongoing needs of the organisation and that the ICT plans support the corporate plans.

3. Acquire ICT validly

   Ensure that ICT acquisitions are made for the right reasons in the right way; on the basis of appropriate and ongoing analysis. Ensure that there is appropriate balance between costs, risks, long term and short term benefits.

4. Ensure ICT performs well whenever required

Ensure that ICT is fit for its purpose in supporting the organisation, is kept responsive to changing business requirements, and provides support to the business at all times when required by the business.

5. Ensure ICT conforms

    Ensure that ICT conforms to all external regulations and complies with all internal policies and practices.

6. Ensure ICT use respects human factors.

    Ensure that ICT meets the current and evolving needs of all the 'people in the process'.

The three primary tasks are:

*   Evaluate the use of ICT.
*   Direct preparation and implementation of plans and policies.
*   Monitor conformance to policies and performance against the plans.

Informal supporting documentation in the form of checklists to assist in raising the awareness of governing actors of their responsibilities are available.

The standard embodies an ideal vision of governance as determined by the committee members that published the standard, and it stops short of defining the ethically charged value statements it espouses, such as appropriate, responsible, balanced, fully, proper, complete, safe, correct. This is also observed by Pye and Warren who state:

> From a philosophical prospective the ICT Governance Standard (2005) is based in the traditional ethics approach by focusing on the ideas behind what is deemed good governance from the practical reasoning aspect of what is right, dutiful and obligatory across the governing tasks of evaluating, directing, and monitoring as recommended in the Standard. (2006: 204)

It is these two last criticisms that I wish to examine in arguing for a more cogent definition, such as that offered earlier in the discussion.

In leaving the definition of the value statements to the actors involved, we find ourselves in exactly the same situation criticised in the media reports mentioned earlier — individuals are accountable for what is good and what is bad in governance and, as such, individuals — depending on whether they are personally and individually good or bad, or even as a group — will affect good or bad governance.

This does not account for the complex interdependency of value systems and its interaction with the learning and governance systems of an organisation. It does not account for system effects in governance, but, rather, implies that good governance is directly determined from the strength of the individual judgement and personal ethic of the governing actors. As before, I am not arguing that this is not true, but rather that it is insufficient in and of itself to describe and account for the emergent complexity of interactions in the modern enterprise and, particularly, in the modern ICT enterprise. A more cogent definition would look to the coupled and co-evolving systems and the dynamic feedback mechanisms (in addition to the control mechanisms) of the system.

Fundamentally, the error here is exemplified in the first criticism made of this standard, and that is that the ideal is a reflection of the values, assumptions, knowledge and experience of the committee commissioned to design the standard. The committee, an esteemed and recognised group of scholars and practitioners, in themselves are not the issue, but, rather, the whole system of using a committee to design a template for a system of governance. It is predictable that the committee would design in its own image. Ethically, this is a practice that is known to be a pervasive mode amongst technology minded practitioners. Peter Weill and Jeanne Ross's (2004) definition goes further by specifying the 'appropriate' ethical frameworks for use in IT governance — rights, duties and virtues. Of course, these are not the only frameworks for establishing ethical practice — they are the frameworks deemed acceptable by the authors based on their own value systems. It is interesting to note here that most of the extant governance standards take a consequentialist perspective, whilst Weill and Ross (2004) take a rights and duties perspective, but more on this point later.

So what is the problem with a committee of experts determining what constitutes good corporate governance of ICT? Andrew Feenberg's (in Doppelt, 2001) body of work on *essentialism* in technology design provides the answer.

Essentialism is the use of unified concepts that often neglect temporality and variation. My charge here is not that the standard itself is essentialist in nature (as I believe it does attempt to account for variation and temporality in people's needs, although not in values) but rather that the process to produce the standard is an example of essentialism at work and this is reflected and inherited in the lack of recognisable variation and temporality in the values concepts of the standard. For the lay person, Wikipedia.org (2008) states: 'In simple terms, essentialism is a generalisation stating that certain properties possessed by a group (eg, people, things, ideas) are universal, and not dependent on context, such as stating 'all human beings compete with each other for success'. My contention is not that a generalisation has been intentionally stated, but, rather, that due to the process of the conception of the design, certain values are assumed and embedded in this design as generally good and universally applicable and, ipso facto, that these values and characteristics of ideal governance are achievable and desirable regardless of context. As Jane Collier and Rafael Esteban write: 'The model of organization which underpins both "co-ordination" and "control-and-regulation" views of governance is hierarchical, authoritarian, and mechanistic' (1999: 175–76). It excludes, through primary order assumptions, alternative visions for governance. This is the hallmark of essentialism according to Feenberg.

> On [Feenberg's] analysis, [essentialism] masks the particularity, historicity, contingency, interest-ladenness, and politics of every specific technology that we confront in our built environment: buildings, hospitals, clinics, highways, cities, clinical trials, machines and devices of all sorts, factories, offices, etc. Each such technology embodies a design, and underlying that, a technical code embodying established experts' determination of what is and is not a relevant factor in designing this or that sort of thing. In turn, Feenberg shows that every such design and technical code embodies particular peoples' decisions/power over which among many possible considerations, interests, values, costs, functions and voices are to be included and which [are] excluded in that technology. (Doppelt, 2001)

Feenberg is examining design of physical technology. I am extending his analysis to include the design of systems generally, and the design of governance systems for ICT, specifically. For technology, read design of governance systems. Feenberg believed that technology was pervaded by a dominant ethical framework of essentialism whereby 'technology and the built environment are perceived as determined, or dictated, by necessary imperatives of efficiency and special bodies of expert professionals who enjoy a monopoly over knowledge of these imperatives' (Doppelt, 2001). The relationship to the process of developing the standard, and the consequences which can be observed in the standard, of this work become clear when Feenberg describes technology as emerging from 'the embodiment of a social process in which empowered groups of experts choose to express certain sets of specific interests and standards in specific technologies, which in turn are re-experienced, challenged, and redefined by their users.' Feenberg writes that the essentialists believe that their design embodies persons' interests. In the same way, we can see that the design of the 8015 standard seeks to embody persons' interests and recognises the variability of human needs, but does not extend this characteristic to other components of the system, notably values. Cristiano Antonelli, for example, recognises that knowledge (a resource component of the system) may be conceptualised in different ways given different stakeholders and contexts, and this impacts governance. He writes:

> Three different and rival concepts of technological knowledge can be identified: (a) knowledge as a public good, (b) knowledge as a proprietary good, (c) knowledge as a localized, collective and complex,

path dependent activity. Such a shift has relevant implications in terms of governance mechanisms, strategic attitude of agents and public policy making. (2005)

These different conceptions are not accounted for within the design of the standard and, furthermore, the values embodied in the standard are not justified in the standard (although it is noted that there are current attempts by members of the committee and others to provide this supporting justification). Thomas Ahrens states 'One problem ... is that rules of corporate governance exhort companies to follow those ethics without ever offering a clear justification. The ethics are simply adopted wholesale from economic rationalism' (2008: 149).

Further weight in support of this argument is provided by Gary Weaver with 'Pursuing shared ethical goals by means of culturally inappropriate management practices, in short, can undermine the effectiveness of ethics management efforts' and, 'Questions of culturally appropriate organizational structure remain even after questions of moral content are resolved' (2001: 3).

Designing for feminist ethics in addition to traditional rights and duties, or consequentialist-based frameworks and values systems is an example of an alternative approach for design. This approach is exemplified in Machold et al (2008).

> In seeing what a technology excludes, what it is not but could be, we gain a clearer and truer grasp of what it is — the bias of technology beneath the guise of efficiency and rationality. (Doppelt, 2001)

It is this final statement that summarises the major criticism with the extant definitions of governance from strategic through functional, and that is the inherent bias of the view that ethics and value systems are a personal and individual factor both directly within the design of the definitions and in the conception of those designs. The implication is that the inherent complexity and pervasiveness of value systems in co-evolution with governance and learning systems in organisations is neglected, leading to ignorance of factors contributing to the emergence of unintended and undesirable consequences, such as organisational failure.

Application to the ICT enterprise

To illustrate how unintended consequences may arise, I refer to the results from a recent survey conducted by the Centre for Applied Philosophy and Public Ethics (CAPPE) at The Australian National University and Charles Sturt University. In this survey by Richard Lucas and John Weckert:

empirical research was conducted to determine the current state of the following important two factors:

attitudes of ICT professionals towards ethics in their workplace,

and

ethics education of ICT professionals in the tertiary sector. (2008)

The key relevant survey finding for this paper is that most respondents use or prefer a duty- and rights-based moral framework, with consequences listed as the least favoured framework (Lucas and Weckert, 2008). Given that, according to most extant definitions, decision-making within governance aims at achieving outcomes — consequences — whilst to do what is right (duty) or to have what I rightfully am entitled to (rights) are notions seemingly abstracted by several steps from informing immediate decisions of practice, it seems that there is a moral disjunction between the value systems that lay ICT professionals believe are appropriate and the value systems informing the design of governance systems.

Furthermore, given that governance systems are the progenitors of organisational identity — if that organisational identity is embedded in an entirely different moral framework than that of the ideal identity of its membership, a serious inhibitor to collective ownership, engagement and subsequent organisational effectiveness is likely to result — a decision disconnect.

A moral conflict may even arise whereby the tools of governance that are used to resolve scarcity of resources towards consequences would be viewed as direct moral challenges to concepts of a right and duty to produce quality outputs. In simple words, a programmer would be offended by a manager's decision to trade-off quality for time, and would see this act as unethical — 'cutting corners' — and insulting to his/her professional expertise. In my experience, and it seems in the experience of those who undertook the survey, this is the 'ethical' issue that is most pervasive of the ICT organisation and forms the tension between business managers and technology specialists. This is an example of an internally generated risk as described by Richard Barber, and as such would not be reported through the normal mechanisms of, in this case, project governance through the risk register and therefore not be recognised as a governance issue. This demonstrates the point that there is a big difference between complying with governance standards (or a risk management standard in this instance) and actually governing (or managing risk).

I am not sure whether this finding is an artefact of the survey — respondents immediately associating 'ethics' with 'doing the right thing' and thus providing answers from a particular frame of reference — or a valid indication of practice amongst IT people. The key ethical problem identified by the respondents — compromising quality — seems to me to indicate that we are dealing with the latter rather than the former.

The IT project manager, or developer, or analyst, is in a situation where their envisaged identity is compromised because the framework they are working within is in conflict with the ideal structure and policies of the enterprise. Furthermore, their ideal values dynamics arising from a 'hardened' framework contrast with the 'moderating' processes ideally employed in governance. The result of such deterioration in ideal identity is a group of disenchanted and disempowered individuals, or a group bent on consolidating and fighting the good fight for what is right, what is good, and what they rightfully deserve, against what is perceived as an unethical management regime. 'Management', however, are ideally required to manage to outcomes and need to balance and control resources where time and money constrain function and effect. Considering consequences for outcomes is the means for prioritising how function and effect may be constrained either through scope or quality in order to achieve vision.

The extant definitions of governance, when applied to this situation, clearly represent a teleological action perspective as Habermas (Reynolds and Yuthas, 2008) describes it. A design that accounts for both a teleologic action view and Habermas' theory of communicative action view in engaging stakeholders in the governance process would possibly counteract the decision disconnect described.

This example is used to illustrate the pervasiveness of the values systems at play in an organisation, and how this may in turn impede or accelerate organisational learning and governance. Understanding the competing forces in the value systems at play in an enterprise can surface the opportunities and impediments to achieving the ideal integral enterprise state. Making explicit these opportunities and impediments is a process of the governance system. These may not emerge solely from processes within 'discrete' organisational systems, such as governance processes, financial or human resource management processes, or ethical decision-making processes, but may emerge from the nexus of the systems of governance, learning and value systems in action within organisations.

In this section I have shown how extant definitions of governance may not be sufficient in dealing with the complexity and variation of the dynamic relationship between governance, value systems, learning and organisation and have highlighted the importance of value systems in governance through examining how problems from value systems at play in an organisation can impede good governance. I have offered a more cogent definition of governance to account for the gaps in its existing understanding and practice and, in the next section, I address how to apply this definition through a framework for planning and designing governance systems.

The design of governance systems

Governance is a system for:

1. imagining
2. directing
3. controlling
4. influencing the:

   a. integration (disintegration)
   b. standardisation (pluralisation)
   c. transformation (stasis) of:

      i. organisation practice (values, learning, leadership, language)
      ii. structure (hierarchy, network, flat, cellular)
      iii. resources (technical, human, financial, knowledge)
      iv. process (unify, diversify, coordinate, replicate), towards:

         1. achieving an ideal organisational state —

            a. as envisaged by internal/external governing stakeholders
            b. constrained or nurtured by external pressures/co-evolving systems and internal drivers/subordinate systems
            c. resulting in beneficial (costly) outcomes
            d. measured in positive, negative, or absent value.

Ethics, in the way we are discussing it here, is the application of value systems within an enterprise.

Governance and ethics, therefore, intersect through enterprise practice. Practice, in this model is the application of value, learning, leadership and language systems in an organisation. Ethical enterprise practice is the desirable alignment of the actual with the envisaged ideal enterprise state.

This definition addresses the criticisms of the definitions examined in the previous section of this paper through: providing all system elements; recognising dynamics and encapsulation; and, not requiring a specific values framework for implementation.

If the *practice* of ethics (through value systems) amongst the members of an enterprise and its *ideal identity* are in conflict with those of the governing membership of an organisation in:

- structure
- assumptions
- dynamic or
- learning

then we have a decision disconnect that will constrain the effectiveness of that organisation.

This issue must be dealt with at the nexus of the governance, learning, values, and organisational systems of an enterprise by those charged with leading that enterprise — the governors. Taking a systems approach to understanding values in the enterprise the dynamic, complex nature of its relationship to identity and value creation will help bridge the divide of the decision disconnect.

The first step to resolving this issue is to articulate how values systems interact and transform organisation through the 'practice life cycle'.

If ICT people think management is unethical because it uses a consequence framework for decision-making, then what they are really saying is that there is no place for their framework in the governing model of business and ICT.

Governors can then reflect on how the organisation can integrate and support the interactions of value frameworks between the different aspects of organisations through practice — through learning and through language.

To articulate clearly how values step through the practice cycle, governors must firstly recognise:

- why and how values are important
- that the problem lies in viewing governance as a process, not as a system
- that values interact with governance as dynamic systems that frame enterprise practice
- that this interaction can be modelled and influenced through intelligent design
- that learning acts as a feedback mechanism to harness and create requisite diversity.

The 'practice life cycle' is a representation of the dynamics of the organisation and is illustrated in Figure 3.

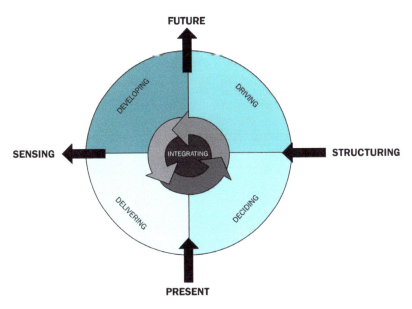

Figure 3: The practice life cycle

Source: Author's research

This diagram illustrates the different characteristics of organisation whereby integration practices form the driving force for value creation. These integration practices emerge from the nexus of values systems, governance systems, and organisation systems and, in the framework proposed, are values, language, learning, and leadership. These practices, coupled with structure, resources and processes, form the inter-system components that impede or improve organisational effectiveness through governance. This is illustrated in Figure 4.

Values systems, therefore, are critical and central to the integrating and transforming practices of organisation that takes the strategy off the shelf and into every product and service. The framework offered captures not only the values information — the principles statements, but captures and understands how values are used in an organisation. In mature organisations, the practitioner uses the framework to surface the assumptions behind the practice and determine whether these assumptions are valid in the envisioned future organisation.

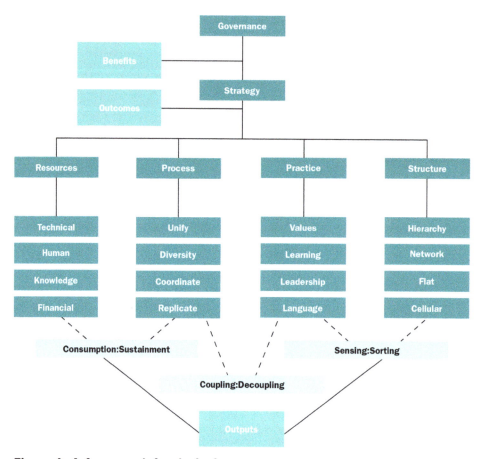

Figure 4: A framework for designing more cogent governance systems

Source: Author's research

With this framework we have a more cogent mechanism for guiding, encouraging, and constraining behaviour in the enterprise through intelligent design. Governors can understand the gaps in how employees, technical staff and business staff act to achieve or impede the vision, and more comprehensively design the organisation to harness this diversity and complexity; to capture and use the creative energy that emerges from difference rather than turn this energy onto itself to create churn and resistance.

To do this, governors, and governance theorists, must move away from an essentialist perspective in definitions, and recognise and design for different kinds of value systems, not only in the values they are informed by, but also in the way they are

- structured
- used
- evolved.

In other words, governors must recognise, plan, and design for requisite diversity in all elements co-evolving with the governance system.

An example of how alternate views of values systems can be accounted for in design is provided in Figure 5.

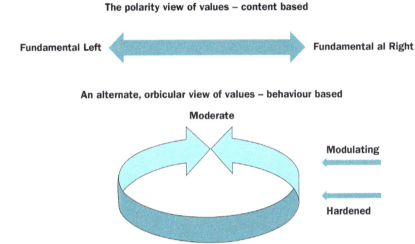

The polarity view of values – content based

Fundamental Left ⟷ Fundamental al Right

An alternate, orbicular view of values – behaviour based

Moderate

Modulating

Hardened

L   R
Fundamental

Figure 5: An example of viewing value systems differently for design

Source: Author's research

In the above illustration we have two differing values systems — fundamentalist and moderate. In the top view, these value systems are seen as two ends of a spectrum — polarities. Using this view, when these systems come into contact with each other in enterprise situations, conflict and impediments to transformation — resistance to change — may ensue. This is particularly the case when 'hardened' or 'root' value systems are at play or in conflict.

There are different kinds of value systems, not only in the values they are informed by, but also in the way they are structured, used and evolved. Hardened systems are a kind of set-form value system and root value systems are conditioned systems that are numb to transformational events. There is a clear and present need to review how these differing value systems interact:

> in an environment of global activity, of multinational companies, and of collaborative ventures across national boundaries, there are likely to be moral dilemmas if on the one hand, firms seek to apply their own value systems to business decisions in cultures other than their own. (Minkes et al, 1999: 329)

In the second view, these values systems are not seen as polarities but rather as having some shared characteristics. In taking this view, which allows for the existence of both systems side by side, resolutions to conflict may be proposed that harness shared conceptions of behaviour or passion, for example. By changing the view of how the value systems interact with each other towards a richer, more diverse design, impediments to organisational effectiveness may be overcome.

When we examine governance, concerned with policy and structure, and enterprise architecture, through which governance is assisted by processes of integration and alignment, we may visualise these components as an enterprise model, shown in Figure 6.

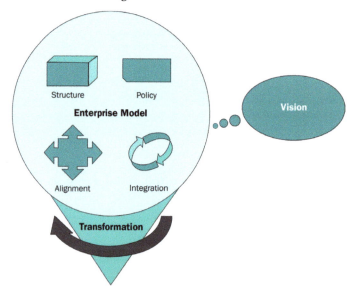

Figure 6: Enterprise model

Source: Author's research

Value systems, like enterprise systems, have a systems model with governing assumptions as policy, structure either as a sense form or a set form, and conditioned or inquiring dynamics (derived from structure) of alignment and integration, and finally identity as vision. The transforming property of value systems is learning, and so the interdependence between these elements of practice is reinforced. This is illustrated in Figure 7.

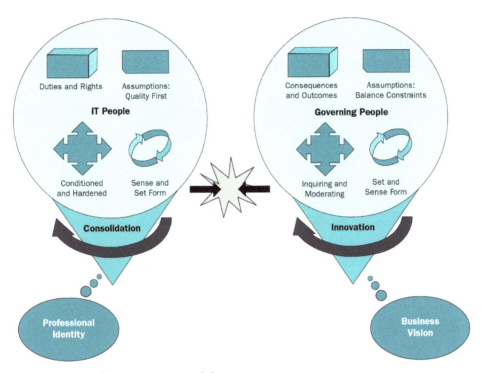

Figure 7: A value systems model

Source: Author's research

In the example given from the Lucas and Weckert (2008) survey, the values systems between 'management' and 'IT' are instantiated differently in assumptions, structure and dynamic. This fundamentally puts business vision, in its ideal form, and IT identity, in its ideal form, in conflict. Hence, the continuing experience of distrust between IT and 'management', the resistance to change, and the result of this misalignment is the inability to transform. This is illustrated in Figure 8.

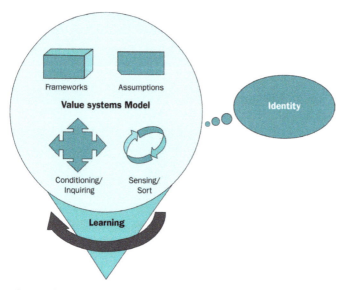

Figure 8: Misalignment of values systems — IT and management

Source: Author's research

Transformation is important in evolving systems; driving and moderating change; adaptation; renovation and renewal. Transformation is achieved through practice, as it is defined in the previous discussion. It is the application of value systems through ethics that informs practice so, along with learning, leadership and language, ethics creates organisational transformation.

Transformation, in turn, creates value.

Summary

The relationship between ethics, governance, the enterprise, and ICT organisations is important because it powers, through double-feedback learning, transformation of organisations, which creates value. An ICT practitioner seeking to create organisational value should endeavour to understand these relationships because, at the functional level of governance, these relationships are impeding or growing organisational value. I have employed a systems-thinking approach to surface commonalities between these entities towards a system purpose.

In this paper I have established the need for understanding these relationships. I addressed how practitioners and theorists come to (mis)understand these relationships and concepts through extant definitions, and the implications for design and practice of governance.

I have presented a top-down analysis, examining first the societal and operational levels (state and corporate) and, secondly, the functional level of governance — in application to ICT organisations and concluded with a systems-thinking framework, for practitioners and theorists alike, to explicitly plan and design for the dynamic interrelationship of values, people, process, and purpose in enterprise governance, using the principle of requisite diversity.

The paper has unambiguously established the centrality of value systems in the practice of governance in the enterprise, and particularly the ICT enterprise. Through a systems-thinking approach, governors can act to prevent a decision disconnect arising from the pervasiveness of complexity of the interactions between values, governance, and ICT organisation systems. Through the proposed framework, a practitioner or theorist may design and plan for multiple instances of system components, such as alternative moral frameworks, within the governance and co-evolving systems. Through planning, designing, and acting for requisite diversity in all aspects of governance, governors help drive organisational transformation towards the desired vision.

References

Ahrens, T, 2008, 'The hidden ethics of corporate governance and the practical uses of corporate governance codes: a commentary on Bhimani', *Journal of Management and Governance*, vol 12, pp 149–52.

Antonelli, C, 2005, 'Models of knowledge and systems of governance', *Journal of Institutional Economics*, vol 1, no 1, pp 51–73.

Argyris, C, 2005, 'Double loop learning in organizations: a theory of action perspective', in KG Smith & MA Hitt (eds) *Great minds in management*, Oxford University Press.

Arthur, EE, 1987, 'The ethics of corporate governance', *Journal of Business Ethics*, vol 6, pp 59–70.

Australian Computer Society (ACS), 2008, 'Australian Computer Society code of ethics', viewed September 2008, <http://www.acs.org.au/about_acs/acs131.htm>

Cacioppe, R, Forster, N & Fox, M, 2007, 'A survey of manager's perceptions of corporate ethics and social responsibility and actions that may affect companies' success', *Journal of Business Ethics*. Online Oct 2007, cited Sep 2008.

Collier, J, & Esteban, R, 1999, 'Governance in the participative organisation: freedom, creativity, and ethics', *Journal of Business Ethics*, vol 21, pp 173–88.

De Vore, PW, 'Do no harm: technology, ethics and responsibility', *Journal of Technology Transfer*, vol 22, no 1, pp 63–74.

Doppelt, G, 2001, 'What sort of ethics does technology require?', *The Journal of Ethics*, vol 5, pp 155–75.

Drennan, L, 2004, 'Ethics, governance and risk management: lessons from Mirror Group Newspapers and Barings Bank', *Journal of Business Ethics*, vol 52, no 3, July, pp 257–66.

Gregor, S, 2002, 'Design theory in information systems', *Australian Journal of Information Systems*, special issue, pp 14–22.

International Organization for Standardization (ISO), 2008, *Corporate governance of information technology*.

Karri, R, Caldwell, C, Antonacopoulou, EP & Naegle, D, 2006, 'Building trust in business schools through ethical governance', *Journal of Academic Ethics*, vol 3, pp 159–82.

Kooiman, J, 2003, *Governing as governance*, Sage Publishing.

Korac-Kakabadse, N & Kakabadse, A, 2001, 'IS/IT governance: need for an integrated model', *Journal of Corporate Governance*, vol 1, no 4, pp 9–11.

Lewis, EJE (cited 2008), available from <http://www.itee.adfa.edu.au/~ejl/Portal/Systems%20planning/SP%20pages/LJ_ICT_Governance_1_Starting_Out.htm>

Lovell, A, 2002, 'Ethics as a dependent variable in individual and organisational decision making', *Journal of Business Ethics*, vol 37, pp 145–63.

Lucas, R & Weckert, J, 2008, 'ETGovICT briefing paper for presenters', Centre for Applied Philosophy and Public Ethics, Australian National University, Canberra.

Machold, S, Ahmed, PK, & Farquhar, SS, 2008, 'Corporate governance and ethics: a feminist perspective', *Journal of Business Ethics*, vol 81, pp 665–78.

Mcnamee, MJ & Fleming, S, 2007, 'Ethics audits and corporate governance: the case of public sector sports organisations', *Journal of Business Ethics*, vol 73, pp 425–37.

Marginson, S & Considine, M, 2000, *The enterprise university*, Cambridge University Press.

Minkes, AL, Small, MW & Chatterjee, SR, 1999, *Journal of Business Ethics*, vol 20, pp 327–35.

Pye, G & Warren, MJ, 2006, 'Striking a balance between ethics and ICT governance', *Australasian Journal of Information Systems*, vol 13, no 2, May.

Reynolds, M & Yuthas, K, 2008, 'Moral discourse and corporate social responsibility reporting', *Journal of Business Ethics*, vol 78, pp 47–64.

Rogerson, S, 1997, 'Ethics and information technology', *Australasian Journal of Information Systems*, vol 4, no 2, pp 125–26.

Sama, LM & Shoaf, V, 2005, 'Reconciling rules and principles: an ethics-based approach to corporate governance', *Journal of Business Ethics*, vol 58, pp 177–85.

Schnebel, E & Bienert, MA, 2004, 'Implementing ethics in business organisations', *Journal of Business Ethics*, vol 53, pp 203–11.

Smith, KG, & Hitt, MA (eds), 2005, *Great minds in management*, Oxford University Press.

Standards Australia, 2003, *Good governance principles*, Australian standard series AS8000–2003.

Standards Australia, 2005, *Corporate governance of information and communication technology*, Australian standard AS8015.

Stoker, G, 1998, 'Governance as theory: five propositions', *International Social Science Journal*, vol 50, no 155, pp 17–28.

Tavani, R, 2003, 'Recent books on or related to ICT ethics', *Journal of Ethics and Information Technology*, vol 5, pp 177–80.

Weaver, GR, 2001, 'Ethics programs in global businesses: culture's role in managing ethics', *Journal of Business Ethics*, vol 30, pp 3–15.

Weill, P & Ross, J, 2004, *IT governance*, Harvard Business School, Boston, MA.

Zattoni, A, 2007, 'Morten Huse: *Boards, governance and value creation: the human side of corporate governance*', review article, *Journal of Management and Governance*, vol 11, pp 439–44.

Section V

Ethics education

The place of ethics in ICT courses

Professionalism, as we have noted previously, has strong links with ethics. A professional is someone who, amongst other things, behaves ethically with respect to his or her occupation. Education is also an important aspect of professionalism. A professional is an expert relative to the general population and this expertise is usually partly a result of being educated in a particular of body of knowledge. It is not surprising, then, that a component of ethics education is commonly considered to be an important element of a professional's education. This is the case in information and communications technology (ICT) and, for a course (by which we mean a specific set of subjects/units leading to a qualification, typically an undergraduate degree) to be accredited by the Australian Computer Society (ACS), it must contain some study of ethical and social issues. Both of the chapters in this section relate ethics education closely to professionalism, and neither discusses courses that are part of standard university degrees in ICT. Before returning to these two discussions, we will consider some unsettling evidence for anyone who thinks that ethics education is important.

ICT ethics education, in its current forms, does not seem to have as much effect as is desirable. A recent survey indicated this, and this view of the efficacy of ethics education is supported, if somewhat ambiguously, in the literature (Lucas and Weckert, 2008). We found only one paper that looked directly at the ICT situation (Sharma and Burmeister, 2004), and it found no significant indication of efficacy. A number of studies of business ethics courses also suggest that their effect is not strong in the business community (Buchko & Buchko, 2009; Hun-Joon Park, 1998) and the same appears to be true of medical ethics courses in the medical area (Campbell, 2007).

Before this rather bleak assessment of ethics education results in it being removed from ICT-course curricula, three considerations should be kept in mind. First, there is evidence that, at least in the medical field, ethics training is important in making a contribution to policy. There is no obvious reason why this should not also be true for the ICT industry or organisations that employ ICT workers. Second, there is evidence that company ethics training programs are effective in the business field. Again this could hold for ICT companies. Third, there is some evidence that what is taught in typical ICT ethics courses in Australian universities might not be what is most needed.

Further consideration of this third point could suggest ways of making ethics education more effective. The survey mentioned previously indicated a difference in the content of typical university ethics courses and the main ethical concerns of ICT professionals. As could be expected, privacy and intellectual property issues were prominent in the minds of ICT professionals, but what could generally be described as professional issues, that is, issues where professionals must make work-related decisions, dominated. High on the list were compromising quality to meet deadlines, unprofessional behaviour, making false promises and conflicts of interest. Also significant were compromising functionality and requirements to meet deadlines. Unprofessional behaviour also figured large in the extra comments section of the survey. Typical concerns included blaming others for one's own mistakes, poor team contributions, awarding contracts without due process, overpricing and under quoting of time and overstating of skills. The issue of professionalism was also of major concern in the interviews and most of the same worries surfaced. A number of related, new concerns were also mentioned, of particular significance were responsibility and informed consent.

A survey of the material covered in courses suggested that the common topics covered were the standard ones of privacy, security, cyber crime, intellectual property, regulating commerce and free speech and a few others, with professional ethics being one topic but frequently left unspecified. The most common text mentioned was Michael J Quinn's, *Ethics for the information age* (2005), so a reasonable assumption is that the topics covered in the text also form the basis of the courses in which it is used (and most other texts cover substantially the same material). The one chapter of Quinn's volume that deals with professional ethics covers the issue of whether software engineering is a profession, the software engineering code of ethics and codes of ethics in general, and whistleblowing. Notably, many of the issues raised in the survey and interviews are not explicitly mentioned either in Quinn's text or in the course descriptions. It does not follow from this that these topics are not covered, but at least it raises the question of whether the focus of ethics courses is on the most important issues as seen by the practitioners.

While the ACS does require the ethical component to be covered in order for a course to be accredited, it does not require that there be a discrete subject and it is clear that the most popular option for universities is for ethics to be part of other subjects. This reflects a debate regarding the best way of teaching ethics to ICT students. One school of thought is that there must be a discrete subject with rigorous assessment, otherwise students do not take the issues seriously. The other view is that discussion of ethics should not be conducted separately because then students do not see it as an integral part of the profession. A leading advocate of this view is Don Gotterbarn (1992). He argues that ethics education is much more effective if it is incorporated throughout the curriculum, rather than being taught as a discrete subject or course and, second, that ethics is not necessarily best taught by philosophers or theologians. His concern is not to denigrate a philosophical approach to computer ethics, but it is, rather, to understand how best to ethically educate computing professionals. The related debate mentioned by Gotterbarn concerns the teaching of ethics; by what sort of person should it be taught? On the one hand it is argued that it should be taught by experts in ethics and, on the other hand, there are arguments for it being taught by ICT professionals because they know what the real issues are and they have more credibility in the eyes of the students. In both of these debates there are good arguments on both sides.

Three issues, then, have emerged from the survey and the literature: there appears to be a mismatch between what is taught and what professionals see as the most important ethical issues; there might be a problem with the placement of ethics in ICT curricula and it may not be taught by the most appropriate people.

The first chapter of this section, by David Lindley et al, all of whom are or have been ICT practitioners, begins with a discussion of what the ACS takes an ICT professional to be and an outline of the knowledge and skills that the ACS requires of such professionals. This provides the context within which ethics education is considered. The ethics courses focused on here are not part of university degrees, but, rather, constitute components of the ACS professional development program. The content of the two courses mentioned is interesting in the light of the previous discussion of what appears to be a mismatch between practitioners' primary concerns and the common content of university ethics courses. Lindley et al describe content that is heavily weighted towards everyday professional concerns. Interestingly, too, student involvement is central and the tutors are all experienced ICT professionals.

The chapter by Stoodley et al builds on the notion of student involvement in ethics education and develops it in detail and methodologically. It first contrasts the approach being presented, called *prompting new experiences*, with three prevailing approaches to ICT ethics education: exercising behavioural control, developing decision-making skills, and developing moral sensitivity. The aim of prompting new experiences is to change the student's experiences of ethics. Four teaching methods are suggested: moral argumentation and dilemma discussion; role models and heroes; values clarification; and, logic, role taking and concept construction. A 'model of ethical IT' is proposed which illustrates ethics awareness expansion from 'my world' through a number of stages to 'the wider world'. The chapter concludes with an explanation of how the model can be used as a conceptual tool and as a learning tool.

References

Buchko, AA & Buchko, KJ, 2009, 'So we teach business ethics — do they learn?', *Journal of Business Ethics Education*, vol 6, pp 119–46.

Campbell, AV, Chin, J & Voo, T, 2007, 'How can we know that ethics education produces ethical doctors?', *Medical Teacher*, vol 29, pp 431–36.

Gotterbarn, D, 1992, 'The use and abuse of computer ethics', *Journal of Systems Software*, vol 17, pp 75–80.

Hun-Joon, P, 1998, 'Can business ethics be taught? A new model of business ethics education', *Journal of Business Ethics*, vol 17, pp 965–77.

Lucas, R & Weckert, J, 2008, *Ethics and regulation in the ICT industry*, report for the Australian Computer Society, Centre for Applied Philosophy and Public Ethics, Charles Sturt University, Canberra, unpublished.

Quinn, Michael J, 2005, *Ethics for the information age*, Addison-Wesley Longman, Boston.

Sharma, A, & Burmeister, OK, 2004, 'Professionalism in ICT: meeting the challenge of ethical dilemmas in the workplace', *AiCE 2004 Conference Proceedings*, 2004.

Biographies

Brenda Aynsley has over 30 years experience as an ICT practitioner, including 11 years as an active member of state and national bodies of the ACS. She was most recently the vice president membership boards (2010–11) and is currently the chair of the International Federation for Information Processing (IFIP)'s international professional practice partnership (IP3) whose mission it is to be a global partnership that will define international standards of professionalism in ICT; create an infrastructure that will encourage and support the development of both ICT practitioners and employer organisations; and, give recognition to those who meet and maintain the required standards for knowledge, experience, competence and integrity. Brenda is a fellow of the ACS and an honorary life member.

Prof Christine Bruce is professor in the Science and Engineering Faculty, Queensland University of Technology (QUT). She researches peoples' experiences of information and learning in academic, workplace and community contexts. Over the years she has had particular focus on information literacy, information technology learning, graduate capabilities, and research study and supervision.

Christine is presently director, Higher Degrees Research in the QUT Information Systems School, and convenor of the QUT Higher Education Research Network. In 2008 Christine was appointed a fellow of the Australian Learning and Teaching Council. In 2010 she received a State Library Board of Queensland Award for contribution to information literacy, information literacy education and research.

Michael Driver has over 40 years experience in technology with the last 25 in senior ICT management. He has worked throughout Australia with top-tier law firms and government entities and also spent 12 years in Papua New Guinea in the early 1980s, introducing the PC into that country and mentoring local users. Mike has been a tutor and mentor for the ACS Computer Professional Education (CPe) Program since 2007 specialising in enterprise architecture. Mike is a senior member of the ACS (Certified Professional).

Prof Sylvia Edwards is an adjunct professor with the Faculty of Science & Engineering at (QUT). She was the recipient of the prestigious Australian Award for University Teaching, 2006. Sylvia researches higher education leadership, information searching behaviour and information literacy. She specialises in applying her research findings to practical situations, and her work has significantly contributed to curriculum renewal at QUT

Robert Godfrey has had well over 50 years experience in the IT industry, starting in the United Kingdom as a software developer, then as a business systems analyst, and for the last 40 years as an academic. He has recently retired from the School of Computing and Information Systems at the University of Tasmania where he has worked since 1988. Since 2005 he has made several visits to China each year teaching in a number of Chinese universities. His early specialisation was in database design and development, which then broadened into human-computer interaction, in particular user interface design, and the ethics and practices of user/analyst interaction. Bob is a long-time member of the ACS and has been actively involved in their professional education programme for the last 25 years.

Bob Hart has over 45 years of experience in IT as a practitioner, user and academic. He was, until recently, responsible for standards and development at ACS and is now a principle advisor to ACS on standards. He was a founding director of the IFIP IP3 and was involved in the formation of the Seoul Accord. He currently provides advisory services to educational institutions, other IT professional associations and governments on professional standards.

Glen Heinrich worked for over 40 years as an IT practitioner, primarily in health but also in education and software development. Following his retirement he was involved in tutoring online courses in IT. Glen is a fellow and honorary life member of the ACS.

David Lindley has over 10 years experience as an IT practitioner and a further 17 as an IT educator and academic. He has worked in Australia, United Kingdom and Malaysia and, as a visiting lecturer, in Sri Lanka, India, Pakistan, Bangladesh and China. As an educator, his specialisation is learning by working adults using collaborative online teaching methods. David is a fellow of the ACS and an honorary fellow of the Australian Catholic University.

Ian Stoodley holds a research position at QUT. He has conducted qualitative research into undergraduates, postgraduates, academics and industry practitioners in the areas of public health, higher education teaching and learning, meta-research, and professional ethics. He is joint editor and author of the book *Transforming IT education* (2006). His principal academic interest is in the ethical formation and support of technology professionals and he was awarded a QUT Faculty of Science and Technology Deans Award for Academic Excellence for his doctoral work on IT professionals' experience of ethics and its implications for IT education.

Bhuvan Unhelkar has more than two decades of strategic as well as hands-on professional experience in the ICT industry. As a founder and consultant at MethodScience he has developed mastery in business analysis and requirements modelling, software engineering, agile processes, mobile business and Green IT. He has authored 16 books and numerous papers and executive reports (mostly for Cutter, Boston, United States). He is winner of the Computerworld Object Developer Award (1995), Consensus IT professional award (2006) and IT writer award (2010). His doctorate is in the area of 'object orientation' from the University of Technology, Sydney, in 1997. He is an adjunct academic at the University of Western Sydney and a tutor at the ACS CPe Program. Bhuvan is an engaging and sought-after speaker, a fellow of the ACS, life member of the Computer Society of India.

Kim Wilkinson has worked in the IT industry for over 40 years, almost entirely in commercial application development and support for large companies and government agencies. During this time he has worked as a programmer, analyst, project manager and data administrator. He holds two postgraduate qualifications. Kim is a member of the ACS and has been a mentor in its CPe Program course for the last four years.

11. Educating for professionalism in ICT: Is learning ethics professional development?

David Lindley,[1] Brenda Aynsley,[2] Michael Driver,
Robert Godfrey,[3] Robert Hart,[4] Glen Heinrich,[5]
Bhuvan Unhelkar[6] and Kim Wilkinson

Introduction

This paper considers professionalism as the product of a process; a status that can be achieved and justified by completing a series of activities. It does not attempt to explain what professionalism is, rather, it explains what the Australian Computer Society (ACS) deems professionals should know and be capable of doing.

In this paper, we aim to convey the following messages:

- Professionals require education beyond that offered in a typical university degree.

- The ACS Computer Professional Education (CPe) Program is constructed on the established Skills Framework for the Information Age (SFIA, 2008a) plus an additional skill set labelled Professionalism.

- To recognise information and communications technology (ICT) practitioners as ICT professionals, the ACS requires that they have a practical knowledge of ethics.

- Formal research is necessary to ascertain the effectiveness of the ACS approach to ethics education.

And, to justify these messages, we address the following questions:

- What is an ICT professional?

- Are there grades of ICT professionalism?

- What is professional development for ICT professionals?

1   Australian Catholic University
2   IFIP International Professional Practice Partnership (IP3)
3   University of Tasmania
4   Independent Consultant
5   Australian Computer Society
6   Method Science Pty Ltd

- Where, on the scale of academically oriented education versus competency-based training, does ICT professional development lie?
- How does the ACS achieve its learning objectives in teaching ethics?
- Are the techniques that the ACS uses good enough?

What is an ICT professional?

The ACS Professional Standards Board (2007) deems an ICT professional as someone eligible for its professional level of membership. That individual should possess the following knowledge, skills and capabilities:

- Factual and theoretical knowledge of ICT in broad contexts.
- Advanced, coherent body of knowledge in a discipline/field involving critical understanding of theories and principles.
- Advanced skills, demonstrating mastery and innovation required to solve complex and unpredictable problems in a discipline/field of ICT.
- Exercise management and supervision [skills] in contexts of work activities where there is unpredictable change.
- [Competent to] Take responsibility for complex technical and professional activities or projects.
- [Competent to] Review and develop performance of self and others.

In this context, the descriptors the ACS uses to define levels of membership can be stated as:

- Knowledge: facts, information and skills acquired through experience and education.
- Skills: the ability to perform a task.
- Capability: a standard necessary to perform a specific job.

Underpinning this choice of descriptors is the skills framework for the information age (SFIA), which is described as providing 'a common reference model for the identification of the skills needed to develop effective information systems (IS) making use of information technologies (IT)'.

SFIA is a two-dimensional table that represents skill sets on one dimension and levels of responsibility, or capability, on the other. A tabular view of SFIA subset is provided in Figure 1 below.

The ACS considers that graduates from most, but not all, Australian ICT degree programs can, after 18 months relevant industry experience, perform at SFIA Level of Responsibility 4 (Enable). With further experience, and studies within the ACS CPe program, its graduates are expected to have achieved SFIA Level of Responsibility 5 (Ensure, advise) and be eligible for professional level membership.

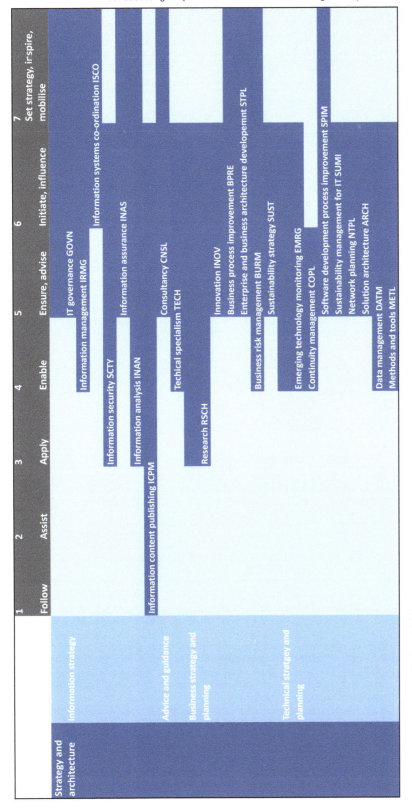

Figure 1: A subset of the SFIA

Source: Extracted from SFIA, 2008c

SFIA Levels of Responsibility 4 and 5 are defined Table 1a and 1b below.

Table 1a: Definition of SFIA Level of Responsibility 4

SFIA Level of Responsibility 4 (Enable)	
Autonomy	Works under general direction within a clear framework of accountability. Substantial personal responsibility and autonomy. Plans own work, to meet given objectives and processes.
Influence	Influences team, and specialist peers internally. Influences customers at account level and suppliers. Some responsibility for work of others and allocation of resources. Participates in external activities related to specialisation. Decisions influence success of projects and team objectives.
Complexity	Broad range of complex technical or professional work activities in a variety of contexts.
Business	Selects appropriately from applicable standards, methods, tools and applications and use. Demonstrates analytical and systematic approach to problem solving. Communicates fluently orally and in writing and can present complex technical information to both technical and non-technical audiences. Is able to plan, schedule and monitor work activities in order to meet time and quality targets and in accordance with health and safety procedures. Is able to absorb rapidly new technical information and apply it effectively. Good appreciation of wider field of information systems, its use in relevant employment areas and how it relates to the business activities of the employer or client. Maintains awareness of developing technologies and their application and takes some responsibility for personal development.

Source: Extracted from SFIA, 2008c

Table 1b: Definition of SFIA Level of Responsibility 5

SFIA Level of Responsibility 5 (Ensure, advise)	
Autonomy	Works under broad direction. Full accountability for own technical work or project/supervisory responsibilities. Receives assignments in the form of objectives. Establishes own milestones, team objectives and delegates assignments. Work is often self-initiated.
Influence	Challenging range and variety of complex technical or professional work activities. Work requires application of fundamental principles in a wide and often unpredictable range of contexts. Understands relationship between specialism and wider customer/ organisational requirements.
Complexity	Broad range of complex technical or professional work activities, in a variety of contexts.
Business	Advises on the available standards, methods, tools and applications in own area of specialisation and can make correct choices from alternatives. Can analyse, diagnose, design, plan, execute and evaluate work to time, cost and quality targets. Communicates effectively, formally and informally, with colleagues, subordinates and customers. Demonstrates leadership. Clear understanding of the relationship between own area of responsibility/ specialisation to the employing organisation and takes customer requirements into account when making proposals. Takes initiative to keep skills up to date. Maintains awareness of developments in the industry. Can analyse user requirements and advise users on scope and options for operational improvement. Demonstrates creativity and innovation in applying solutions for the benefit of the user.

Source: Extracted from SFIA, 2008c

While comprehensive in the range of skills it encompasses, SFIA has no skills category relevant to professionalism, or professional behaviour. To cover this area, the ACS has expanded on SFIA with an additional skill set that, for university and CPeP graduates, are defined in Table 2 below.

Table 2: Definitions of ACS professionalism skills

SFIA Level of Responsibility 4 (Professionalism skills of university graduates)	Develops a basic risk management plan for simple projects including the impact on social, business and ecological environments.
	Identifies legal requirements and constraints imposed on the work/project and contributes to compliance.
	Commits to a code of ethics, standards and practice and can apply these in basic projects.
	SFIA Level of Responsibility 5 (Professionalism skills of CPeP graduates).
	Develops a risk management plan for projects including the impact on social, business and ecological environments and ensures compliance.
	Ensures compliance with all legal/regulatory requirements.
	Ensures compliance with appropriate professional codes of ethics, standards and practice.

Source: Extracted from SFIA, 2008c

An ICT professional, therefore, is someone who has full accountability for their own technical work and responsibilities; whose decisions can impact on the success of projects; who develops business relationships with customers; who must apply fundamental principles in a wide and often unpredictable range of contexts; and, who can analyse, diagnose, design, plan, execute and evaluate work to time, cost and quality targets. In addition, they can communicate effectively, demonstrate leadership, and keep their skills up to date. They are creative, innovative, and aware of their impact on social, business and ecological environments. Their knowledge and actions are able to influence direction within the organisation, their peers and industry.

Are there grades of ICT professionalism?

An ICT professional, in the view of the ACS, is someone eligible for its professional level of membership. This level is not easily achieved and thus, in the view of the ACS, ICT professionals are a subset, perhaps a small subset, of the generality of ICT practitioners.

But the ease or difficulty of gaining ACS membership at the professional level is not the consideration here. It is whether membership at this level means something about the person who gains the professional level of membership and, therefore, also suggests something about other practitioners who are not members at the professional level.

The ACS professional level of membership aims to be a differentiator between ICT practitioners, who the ACS verifies as reliable and competent at SFIA level of responsibility 5, and others, who might be less than reliable at that level. Those with the ACS professional level of membership can use their membership as evidence for prospective employers and clients of their professional abilities. Those who are not ACS professional level members will require additional evidence, and additional corroboration, to justify similar claims.

Note that the ACS is not aiming to be elitist or exclusive with its professional level of membership. There are other levels of membership with less onerous prerequisites. The associate level, for instance, is mapped to the SFIA level of responsibility 3 (Apply). But it is the objective of the ACS for its members at these other levels to raise their knowledge, skills and capabilities to the level of the professional. The ACS aims to be both an inclusive organisation, and an organisation that encourages continuing professional development amongst its members.

Indicative of the ACS view of professionalism is that the majority of assessments in its CPe program do not have a grade for exceptional achievement. Instead, most items of assessment are graded 0, 1, or 2 — where 2 is measured as at or exceeds expectations. The ACS does not view professionalism in multiple grades. Rather, a person is either an ICT professional, or they are not. They either meet the criteria for membership at the professional level, or they do not. They can either take on professional responsibilities at SFIA level 5, or they cannot.

What is professional development for ICT professionals?

The ACS specifies its professional level of membership in terms of knowledge, skills and capability. Professions Australia defines a profession in similar terms; the possession of 'special knowledge and skills in a widely recognised body of learning derived from research, education and training at a high level' (Professions Australia, 1997).

Considering, again, the definitions of an ICT professional provided above, it seems reasonable now to define professional development as the acquisition of, and the continuing possession and use of, facts, information, and skills necessary to perform a task.

It is important to distinguish between the 'acquisition of ... facts, information and skills', and the 'continuing possession of ... facts, information and skills'. The first is initial professional development (IPD), and the second is continuing professional development (CPD).

The UK Initial Professional Development Forum defines IPD as 'a period of development during which an individual acquires a level of capability necessary in order to operate as an autonomous professional'. It goes on to clarify the concept with the statement 'Professional bodies may recognise the successful completion of IPD by the award of chartered or similar status' (nd).

Engineers Australia states 'Continuing Professional Development (CPD) helps you maintain up-to-date technical skills and knowledge of processes, technology and legislation. It also enables you to attain and maintain your Chartered Status' (2009).

The ACS now follows a similar strategy. To achieve its professional level of membership, an applicant is required to have a minimum of four years relevant experience and, in addition, have completed a course of study encompassing the ACS core body of knowledge. This entitles an applicant to use the postnominal MACS (Member of the ACS).

But, to use the post-nominal MACS CP, indicating an ACS-recognised certified professional, a MACS must complete the ACS CPe program. Initial professional development, therefore, is a course of study, typically a university degree in some aspect of ICT, plus experience, plus postgraduate studies in professionalism (that is, the ACS CPe program).

Then, to maintain their CP status, a member must complete, annually, at least 30 hours of professional development; which is to say, continuing professional development. Unlike IPD, the ACS view of CPD is that it should be self-directed. Except for its quantity, and that it must be relevant to a practicing computer professional at SFIA level 5 or above, the ACS does not prescribe what the professional development must comprise.

Where, on the scale of academically oriented education versus competency-based training, does ICT professional development lie?

To achieve the ACS professional level of membership, an applicant requires a minimum of four years relevant experience plus the completion of a course of study encompassing the ACS core body of knowledge. The ACS, therefore, deems professionalism to be acquired only through a combination of education and experience.

It seems obvious that, if professionalism is defined in terms of knowledge, skills and capability, then a professional requires education beyond that offered in a typical diploma or degree. They require more than just knowledge, and more than just technical skills.

The professional requires contextual awareness; that is, how ICT, and the numerous ICT roles in industry, fit within and influence the world of business, society, and the environment. They also need a reasoned and objective assessment of themselves; their own knowledge, skills and capabilities. A person seeking a job on the basis of an innocent but incorrect assessment of their own ability is as unprofessional as a person who deliberately falsifies their résumé. And these two requirements lead to the skill set the ACS has added to those of SFIA, namely, the skills concerned with risk management.

So, where does ICT professional development lie on the academically oriented education versus competency-based training scale? It lies across the entire scale with, perhaps, IPD centred more towards the academically oriented end, and CPD more towards the competency-based end.

What is achieved by teaching ethics?

The ACS describes itself as a professional association. This suggests that the practice of ICT and computing-related activities is, or should be, a profession. An ICT practitioner in Australia, however, can claim to be an ICT professional even though, unlike practitioners in disciplines such as law and medicine, they do not need any form of registration or belong to a professional association.

Professions Australia describes a profession as 'a disciplined group of individuals who adhere to ethical standards ...' (1997). While it should not be assumed from this that ICT practitioners in Australia who do not belong to a professional association will necessarily behave unethically, it can be reasonably argued that the possibility of unethical behaviour is less likely if a person is educated and trained in the interpretation and application of ethical standards. Further, it can be argued that a person is less likely to behave unethically if they are subject to disciplinary processes by their peers; which is to say, their professional association.

ACS rules and regulations (nd) include a code of ethics, extracts of which are provided in Table 3 below.

Table 3: Extract of ACS code of ethics

4.1 To uphold and advance the honour, dignity and effectiveness of the profession of information technology and in keeping with high standards of competence and ethical conduct, a member must:	(a) be honest, forthright and impartial, and
	(b) loyally serve the community, and
	(c) strive to increase the competence and prestige of the profession, and
	(d) use special knowledge and skill for the advancement of human welfare.
4.3 Values and Ideals	I must act with professional responsibility and integrity in my dealings with the community and clients, employers, employees and students. I acknowledge:
4.3.1 Priorities	I must place the interests of the community above those of personal or sectional interests.
4.3.2 Competence	I must work competently and diligently for my clients and employers.
4.3.3 Honesty	I must be honest in my representation of skills, knowledge, services and products.
4.3.4 Social Implications	I must strive to enhance the quality of life of those affected by my work.
4.3.5 Professional Development	I must enhance my own professional development, and that of my colleagues, employees and students.
4.3.6 Information Technology Profession	I must enhance the integrity of the information technology profession and the respect of its members for each other.

Source: Taken from ACS rules and regulations (ACS, nd)

While the code of ethics is readily accessible to ACS members and people affected by their work, it comprises general and non-specific statements and often requires guidance in interpretation and application. This point is supported by the ACS standards of conduct, which state that 'A member is expected to take into account the spirit of the Code of Ethics in order to resolve ambiguous or contentious issues concerning ethical conduct'. In addition, the ACS has a code of professional practice and professional conduct, which is designed 'to provide members with authoritative guidance on acceptable standards of professional conduct and … is not intended to include a multitude of detailed rules'. It goes on to say that the code should not be 'narrowly interpreted' (ACS, nd).

While the ACS offers formal education programs in professional ethics, a specialist intermediary is often required to apply the ACS code of ethics to professional practice in the real world.

Together with its code of ethics and supporting education activities, the ACS has implemented disciplinary procedures, see Table 4 below, which can be applied in the event that a member behaves in a manner inconsistent with the Code.

Table 4: Nature of complaints and disciplinary action

7.1. Nature of Complaints and Disciplinary Action	
7.1.1. A complaint may be made against any member who acts or fails to act in the course of his or her professional activities in such a way as to justify the taking of disciplinary action, including;	(a) failing to observe, whether intentionally or unintentionally, the Rules or the National Regulations,
	(b) failing to comply with, whether intentionally or unintentionally, any ethical, professional or technical standards published by the Society,
	(c) acting or failing to act so that, as a consequence, whether intended or not, the Society is or is likely to be brought into disrepute or suffer loss or damage,
	(d) any combination of (a), (b) and (c).
7.1.2. If the complaint is substantiated, the member may be disciplined by:	(a) expulsion from the Society, or
	(b) suspension from the rights of membership for a period of 3 years or less or until imposed conditions are met, or
	(c) being required to comply with conditions imposed as to the carrying out of the member's occupation, or
	(d) being required to complete specified courses of training or instruction, or
	(e) caution or reprimand, or
	(f) any combination of 2 or more of (a), (b), (c), (d) and (e).

Source: ACS rules and regulations (ACS, nd)

The ACS encourages its members, and their employers, to view ethical behaviour as a risk-management strategy. Philip Argy, while president of the ACS, wrote 'the standards set for ACS membership and the ethics of professionalism provide a guaranteed credential for employers and board directors wanting to minimise their risk', and 'membership of the ACS at the professional level immediately conveys to the world that you are committed to the highest standards and maintain a level of knowledge, expertise and mental acuity that ensures you are more able to deliver requirements on time, on budget, and with less risk' (2006).

This notion of ethical behaviour as a risk management strategy is formalised such that an ICT practitioner, to graduate from the ACS's initial professional development program, must demonstrate capabilities listed in Table 2 (above). Additionally, the practitioner must demonstrate that he or she;

> Carries out risk assessment within a defined functional or technical area of business. Uses consistent processes for identifying potential risk events, quantifying and documenting the probability of occurrence and the impact on the business. Refers to domain experts for guidance on specialised areas of risk, such as architecture and environment. Co-ordinates the development of countermeasures and contingency plans.

This second skills definition is taken from SFIA, specifically the business risk-management skill at the SFIA level 5 of responsibility titled 'Ensure, advise' (SFIA Foundation, 2008c).

How does the ACS achieve its learning objectives in teaching ethics?

The ACS delivers initial and continuing professional development programs using its in-house educational group called ACSEducation. The initial professional development programs are all delivered online, while the majority of continuing professional development programs are delivered on-ground.

ACSEducation has three full-time staff, none of whom are routinely involved in active teaching. The teaching staff of ACSEducation are engaged on a sessional or voluntary basis and all are senior professional members of the ACS. Typically they have no formal training as educators and most work outside of ACSEducation as ICT practitioners. ACSEducation is supported by an advisory panel comprising senior ICT professionals from Australian public- and private-sector IT organisations and, from 2010, it has been governed by an independent academic board comprising senior university academics.

The ACSEducation Learning Management (LM) system is an open-source product that its developers, Moodle Pty Ltd (2008), claim to have designed on sound pedagogical principles. The product is used by many organisations, such as The Open University, United Kingdom (2005), both for managed learning and collaboration.

Professionalism and ethics are addressed in three ACS programs; at the postgraduate level through the ACS Professional Year (PYear) and CPe programs; and at the undergraduate and vocational levels through the ACS Diploma of Information Technology. In the diploma, professionalism and ethics topics are only assessed, with teaching provided through colleges independent of the ACS.

Professional ethics at the postgraduate level

Similar processes and practices are used for teaching professional ethics in the PYear and CPe Programs.

The PYear is a 12-month, job-readiness program comprising three formal subjects and a 12-week internship with a host company. Participants, typically, are recent graduates of university courses accredited by the ACS and most are international students seeking permanent residence in Australia. The PYear participants are student members of the ACS and thus bound by the ACS rules, especially as pertaining to the ACS code of ethics.

The CPe program comprises four subjects that are completed on a part-time basis, plus a period of mentoring where a participant works one-to-one with a senior member of the ACS. Participants in the CPe program, normally, are graduates of an ACS-accredited university degree with at least 18-months experience who are employed in an ICT-related role. Most are professional grade members of the ACS who, like their PYear colleagues, are bound by the ACS rules pertaining to ethics and professional conduct. Participants in the CPe program, typically, will start with the professional ethics subject titled Business, Legal and Ethical Issues.

Like their colleagues in the PYear, participants studying professional ethics in the CPe program use the ACSEducation LM system to access learning materials, submit assessment items, and contribute to weekly discussion forums. The following discussion provides a general overview of the ACS postgraduate professional ethics subject.

Students in the first week of the professional ethics subject are presented with the following statement:

> In preparing this subject it has been assumed that you and your fellow students are over-achievers; self-motivated, disciplined, and determined to succeed. You have extensive prior knowledge and experience relevant to your study; you are open-minded about sharing your work and educational experiences; and you accept critical thinking as part of the learning process.

In this statement, the ACS is recognising that postgraduate study — specifically, postgraduate study in professional ethics — differentiates its participants from the majority of the population, including the majority of ICT practitioners. It implies that professional ethics is a specialised pursuit critical to the success of someone seeking to be, and to be recognised as being, an ICT professional.

All subjects in the CPe program require 8 to 10 hours of study per week for 13 weeks. The content of the professional ethics subject is structured as in Table 5 below.

Table 5: Content of Business Legal & Ethical Issues subject

Module 1 (weeks 1–3) The role of IT Professionals in Business.	What is a professional?
	The client/professional relationship.
	Corporate culture and personal values.
	Frameworks to resolve ethical issues in the workplace.
Module 2 (weeks 4–6) Risk Management Frameworks.	Risk management principles and concepts:
	IT Risk Management.
Module 3 (weeks 7–9) IT Governance.	IT Governance Regulatory Frameworks.
	IT Governance's role in ISO/IEC38500:2006 IT GOVERNANCE standard.
Module 4 (weeks 10–13) Managing Risk in the Business.	Risk management issues, challenges and compliance with respect to social, business and ecological environments.

Source: ACSEducation

The ACS takes a more practical than normative approach to teaching professional ethics. Nowhere in the required readings, for example, is there mention of deontology or utilitarianism, though participants may encounter such concepts in their ancillary readings. Most effort is given to examining practical situations of ethical significance and discussing with colleagues the applicability to those situations of the ACS code of ethics, different risk-management frameworks, and standards of IT governance.

A teaching week in the ACS professional ethics subject has two sessions; Sunday to Wednesday and Thursday to Saturday. In the first session, participants work individually, reading and, based on their personal and professional experiences, answering tutor-supplied questions. In the second session they work collaboratively in cohorts of up to 20 discussing and debating the questions previously answered individually.

The role of the tutor in the professional ethics subject is to set questions and monitor discussions. Tutors must redirect dialogue that strays from the required theme, encourage less assertive participants to enter into an exchange of ideas, gently restrain dominating participants, and reprove participants who fail to contribute.

For their contribution to weekly discussions, participants are graded, as noted previously, with no differentiation between work at, and work above, an expected standard. It is assumed that work above an expected standard will necessitate a participant not fulfilling their responsibilities in another aspect of their life; maybe professional or family. In a similar vein, participants who contribute below an expected standard are not only awarded low marks, but they are told that their behaviour is unprofessional in that they are not assisting with the learning of their colleagues.

Assessment of discussion forums accounts for 20 per cent of the marks in the subject.

Weekly discussions rely on each participant sharing their knowledge, skills and experience with their colleagues with the aim to address, as expressed here by the International Federation of Accountants, the reality that;

> In the case of complex ethical situations it is unlikely that there will be only one 'right' answer. While analysis may not give a single 'right' answer to a problem or dilemma, it may lead to one or more answers that are more consistent with the fundamental principles … (2007)

The approach of using weekly discussions separates the CPe program from many others courses that teach professional ethics and, in the view of ACSEducation, this process increases its educational effectiveness. Participants are obliged to reflect upon professional ethics every day in their workplaces and debate points of view with colleagues whose workplaces may be different from their own.

In Week 6 of their 13 weeks in the professional ethics subject, participants submit an assignment addressing ethical issues as described in three case studies. Again, quoting the International Federation of Accountants, the aim is;

> By learning to analyze case studies and examples of ethical threats, individuals realize that problems and ethical dilemmas do have solutions. (2007)

For each case, the participants are expected to deliberate on the facts and provide an analysis of 750 to 1000 words. Most participants have difficulty in curtailing their responses to the word limit, and many complain that they need more space in which to express their arguments. A typical tutor's response to such complaints is something like: 'a professional must be able to express themselves clearly and succinctly so that their opinions and actions can be understood by those outside their profession'.

In addressing the case studies, the participants are expected to use an ethical framework, examples of which are provided in their readings, to guide them through the deliberation process. They must also compile a personal ethics matrix detailing the values they hold highly in the professional setting. For example;

> Honesty, that quality of being honest in dealings with clients, colleagues and myself. Clauses might include a direct inclusion from the ACS Code of Ethics eg.

> 4.7.1 I must not knowingly mislead a client or potential client as to the suitability of a product or service.

> 4.7.2 I must not misrepresent my skills or knowledge.

> 4.7.3 I must give opinions which are as far as possible unbiased and objective.

> 4.7.4 I must give realistic estimates for projects under my control.

> 4.7.5 I must qualify professional opinions which I know are based on limited knowledge or experience.

> 4.7.6 I must give credit for work done by others where credit is due.

In dealing with this part of the assignment, participants are also asked to disclose their sources of inspiration which, by way of example, might be as diverse as the Bible, parents, partners and grandparents and even a popular television program. All such sources are acceptable, so long as the reason for their inclusion is argued satisfactorily.

The first assessment accounts for 30 per cent of the marks in the subject.

At Week 13, students must submit a second assignment. This 1500-word paper reviews the values of either the organisation for whom the participant works, or another with which they are familiar. The participants must consider the relationship between values and ethics in the workplace and propose a plan to develop a new code of ethics for the organisation. They are not required to write an actual code, only to justify its need and prepare a plan for its implementation.

The second assignment accounts for 50 per cent of the marks in the subject.

The marking guide for the first assignment, which is provided to both tutors and participants, explains that marks are awarded according to criteria, such as:

- How extensive were your efforts to link the relevant theory, its challenges and the issues to the practical business situation?

And, for the second assignment:

- How comprehensive was the stakeholder perspective you developed?
- How relevant and workable are your recommendations?
- How well did you analyse the link between values and ethics?

Professional ethics at the undergraduate level

At the undergraduate level, in the ACS Diploma of Information Technology, ACSEducation offers a subject called Professional Practice. The undergraduate subject is not a cutdown version of the postgraduate subjects, but, rather, a unit of professional development refined over many years for technician-level membership of the ACS. Learning outcomes of the Professional Practice subject include:

- Understanding of the role of professional societies and the implications for professionals of their codes of ethics and practice, in particular the ACS code of ethics and code of professional conduct and professional practice.
- Understanding of the ethics and social responsibilities associated with being an information technology professional.
- Investigation of current issues associated with the use and abuse of information technology.
- Development of skills in analysing the social and ethical impact of information technologies.

In addition, the Professional Practice subject includes a unit of competency from the Australian Quality Training Framework (AQTF) (ACS, 2005). Called 'Ensuring privacy of users', the unit requires that students are able to:

1. Review privacy policy in relation to legislation.

2. Determine policy shortfalls.

3. Update and review policies.

Assessing a unit of competency involves observing the demonstration of a skill; in this case, reviewing a privacy policy, determining its shortfalls against relevant legislation, and making updates.

To assess against both the learning outcomes and the unit of competency, ACSEducation provides candidates with workplace scenarios such as, that of a company that manufactures office furniture, a company that sells insurance, and an employment agency. In a three-hour exam, students are asked to answer questions such as those in Table 6 below.

Table 6: Typical questions in an ACS Professional Practice examination

Question: Your project manager has asked you to organise a meeting to investigate the impact of the proposed upgrade of the company's Customer Interaction software. Provide recommendations to your project manager about:	who should attend the meeting;
	how the meeting should be held, bearing in mind that attendees may be based at any of the organisation's offices; and
	what documents should be circulated to attendees prior to the meeting.
	Prepare a draft agenda for the meeting using an agenda template you are familiar with. The agenda should include proposed location(s) for the meeting, and a list of agenda items that may address the project manager's brief.
Question: Prior to attending the meeting your project manager has asked you to undertake an ethical analysis of the Interactions as described in the question above, to identify any issues or concerns that should be addressed as part of the new development process. To assist you she has provided the following list of questions for you to respond to:	Who are the stakeholders in this scenario?
	Which facts raise ethical concerns? Why?
	Identify a major ethical dilemma which must be resolved.
	What are the rights and duties of each of the stakeholders?
	What guidance does the Australian Computer Society's Codes of Ethics provide in this situation?
	How would you resolve the dilemma identified in point (c) above? Justify your decision.
	How are the stakeholders affected by your decision?
Question: You decide to investigate existing privacy legislation to prepare for the meeting. You also investigate existing policies and procedures that may relate to the legislation but find none that are relevant.	What type of legislation would you expect to find in your search?
	What type of policies and procedures would you recommend to be implemented?
	What actions would need to be taken to implement the proposed policies successfully?

Source: Extracted from SFIA, 2008c

Techniques the ACS uses, are they good enough?

The selection of ACSEducation teaching and assessment strategies, including those for teaching professional ethics, is influenced by the characteristics of its students, ACS certification requirements, and external accreditation frameworks.

Student characteristics

Gilly Salmon (2004), in her 'Model of online learning', provides the basic elements of the ACSEducation teaching strategy. An explanation of this model, and its associated principles, is provided in Lindley (2007).

While the appropriateness of the Salmon model for ACSEducation students — that is, adult learners who, typically, are highly motivated, IT literate and geographically dispersed — might seem self-evident; no thorough evaluation of its effectiveness has been undertaken. To overcome this deficiency, Lindley (2008) proposes a research methodology to answer the questions:

- Is the ACS CPe program effective and efficient?
- How can the ACS CPe program be improved to achieve its objectives better?

In essence, the methodology involves working with university researchers to compare ACSEducation techniques with those employed by organisations such as:

- Overseas ICT professional associations, for instance, the British Computer Society.
- Specialised ICT associations, for instance, the Information Systems Audit and Control Association (ISACA).
- Professional associations in disciplines other than ICT, such as the Institute of Chartered Accountants and Engineers Australia.

Focusing more broadly than solely on education in professional ethics, the aim is to investigate the areas listed in Table 7 below.

Table 7: Points of comparison with other professional associations proposed for investigation by Lindley (2008)

Association	History and purpose.
	Membership categories and entry requirements.
	Number of members and recent membership trends.
	Policies concerning professional development
Professional Development Activities	Objectives for professional development.
	Experience and history.
	Processes for recognition of prior learning, content development, course delivery, assessment and quality assurance.
	Staff and resources.
	Costs to students, subsidies, and scholarships.
	Articulation arrangements for further education.
	Marketing and promotion procedures.
	Development plans.
Specific Professional Development Program	Rationale and desired outcomes.
	Initial or continuing professional development.
	Recognition under the Australian Qualifications Framework (or equivalent).
	Expected student commitment.
	Student numbers and number of offerings.
	Formal methods and models employed.

Source: Lindley 2008

ACS certification requirements

Successful completion of the ACS CPe and diploma programs can be used as evidence in support of an application for ACS membership. The learning outcomes of the postgraduate Business Legal & Ethical Issues subject, for instance, are designed to be consistent with the professional ethics requirement for ACS professional level membership, and successful completion of the subject certifies that an applicant possesses the required ethics knowledge and skills.

A potential difficulty lies in assessment against desired learning outcomes. As noted previously, most teaching staff are not trained educators and the challenge is to ensure assessment accuracy and consistency. An obvious formative technique is for the small number of teaching staff who are professional educators to mentor their colleagues, but such activities cannot by themselves guarantee assessment accuracy. In the assessment-only undergraduate subject, Professional Practice, a simple comparison of pass rates with other subjects and other sessions provides a reasonable measure of consistency. In the postgraduate subject, however, because pass rates are consistently high, a more sophisticated approach is required. A useful measure, considered in combination with others, might be comparative 'did not complete' and 'deferred' rates across offerings.

External accreditation requirements

External education frameworks used by ACSEducation include the SFIA and the AQTF, defined by training.com.au as a 'set of standards which assures nationally consistent, high-quality training and assessment services' (2007).

External education frameworks offer two major advantages for ACSEducation programs. First, they impose discipline on course design. For example, the AQTF requires that a diploma-level course incorporate a minimum number of units of competency, each of which adheres to a prescribed level of difficulty. The second advantage is that they enable the community at large, beyond the ACS, to locate ACSEducation programs in the context of national and international systems of education.

The price, however, is that external frameworks change over time, and not always in a timely manner that is consistent with changes desired by ACSEducation programs. A recent change in the SFIA framework, for instance, from version 3 to version 4, has left some ACSEducation postgraduate subjects requiring changes to their learning outcomes, and perhaps also their content.

References

Argy, P, 2006, 'Professionalism is the best form of risk management', viewed 30 December 2008, <http://web.archive.org/web/20060919023218/http://www.acs.org.au/news/oz010106.htm>

Australian Computer Society (ACS), nd, Rules & Regulations, <http://acs.org.au/about-the-acs/governance/rules-and-regulations>

——, 2005, 'Professional practice syllabus', as at 1 July 2005, viewed 29 September 2009, <http://www.acs.org.au/dit/index.cfm?action=show&conID=ditprofessionalp> <http://www.acs.org.au/__data/assets/pdf_file/0017/10178/Diploma-of-IT-ICA50105-Subjects-Syllabuses.pdf>

Australian Computer Society Professional Standards Board, 2007, 'Redefining and building the ICT profession: core body of knowledge review discussion paper', viewed 25 September 2009, <http://www.acs.org.au/__data/assets/pdf_file/0007/7792/The-ICT-Profession-Body-of-Knowledge.pdf>

Australian Research Council, Linkage Projects, nd, viewed 1 July 2008, <http://www.arc.gov.au/ncgp/lp/lp_default.htm>

Engineers Australia, 2009, viewed 25 September 2009, <http://www.engineersaustralia.org.au/education/continuing-professional-development/continuing-professional-development.cfm> <http://www.engineersaustralia.org.au/professional-development/continuing-professional-development>

International Federation of Accountants, 2007, 'IEPS 1 — approaches to developing and maintaining professional values, ethics, and attitudes', viewed 30 September 2009, <http://www.ifac.org/Members/Pubs-Details.tmpl?PubID=119151330213823&Category=Education> <http://www.ifac.org/sites/default/files/publications/files/ieps-1-approaches-to-deve.pdf>

Lindley D, 2007, 'Computer professional education using mentored and collaborative online learning', SEARCC 2007, Proceedings of the South East Asia Regional Computer Conference, 18–19 November, Bangkok, Thailand, viewed 30 December 2008 <http://www.ijcim.th.org/v15nSP4/P09SEARCC_ComputerProfessionalEducation.pdf> <http://www.ijcim.th.org/SpecialEditions/v15nSP4/P09SEARCC_ComputerProfessionalEducation.pdf>

———, 2008, 'A research proposal to assess the efficacy of initial professional development offered by professional associations, in particular, the Computer Professional Education Program offered by the Australian Computer Society, SOCIAL INCLUSION — engaging the disengaged in life wide learning', Proceedings of the Adult Learning Australia 48th Annual National Conference, November, Fremantle Western Australia.

Moodle Pty Ltd., 2008, 'Philosophy', viewed 30 December 2008, <http://docs.moodle.org/en/Philosophy>

The Open University, 2005, 'The Open University builds student online environment with Moodle and more', viewed 30 September 2009, <http://www3.open.ac.uk/media/fullstory.aspx?id=7354>

Professions Australia, 1997, 'Definition of a profession, viewed 30 December 2008, <http://www.professions.com.au/defineprofession.html>

Salmon, G., 2004, *E-moderating: The key to teaching and learning online*, 2nd edn, Taylor & Francis, London.

SFIA Foundation (2008a), Getting it right with SFIA, viewed 25 September 2009, <http://www.sfia.org.uk/cgi-bin/docs.pl/IntroLeafletSFIAv4.pdf>

———, 2008b, Skills framework for the information age, viewed 25 September 2009, <http://www.sfia.org.uk/cgi-bin/docs.pl/ChartFramework_SFIAv4.pdf>

——, 2008c, Framework reference SFIA version 4: Skill definitions in categories, subcategories and skills, viewed 25 September 2009, <http://www.sfia.org.uk/cgi-bin/docs.pl/ReferenceGuideSFIAv4.pdf>

training.com.au, 2007, AQTF 2007 — A better system for everyone, viewed 30 December 2008, <http://www.nssc.natese.gov.au/nqc_archive/aqtf>

UK Initial Professional Development Forum, undated, viewed May 2008, <http://www.ukipdforum.co.uk>

12. Experiential ethics education for IT professionals

Ian Stoodley, Christine Bruce and Sylvia Edwards
Queensland University of Technology

This chapter presents a new approach to IT ethics education that may be used by teachers in academic institutions, employees responsible for promoting ethics in organisations and individuals wanting to pursue their own professional development. Experiential ethics education emphasises deep learning that prompts a changed experience of ethics. We first consider how this approach complements other ways of engaging in ethics education. We then explore what it means to strive for experiential change and offer a model which may be useful in pursuing IT professional ethics education in this way.

Different ways of approaching IT ethics education

Since education may be thought of in different ways, we need to be clear that the term 'education' is used here in a broad sense. Education is thought of here as any space where learning or personal development takes place. It may thus involve formal training, informal assistance or personal retreat. It may occur in pre-service formation or in-service courses, in organisational and private contexts, or within the walls of educational institutions.

The prevailing approaches to IT ethics education are characterised in this chapter as:

1. exercising behavioural control

2. developing decision-making skills

3. developing moral sensitivity.

Our additional approach, experiential ethics education, may be characterised as:

4. prompting new experiences.

Ethics education seen as exercising behavioural control

Hypothetical 1	Carol Anderson, CEO of Ventures in IS, wants to influence her employees towards deeper ethical action. In consultation with the management team, she adopts the Australian Computer Society (ACS) code of ethics, requires employees to sign a statement agreeing to abide by this code and runs periodic training sessions that highlight aspects of the code.

When ethics education is considered to be *exercising behavioural control*, the emphasis is on instilling authorised standards. For this reason, codes of conduct figure prominently in this approach and the predominant technique utilised is the presentation and enforcement of such codes. Codes have been found to be useful instruments to a) enable action to be taken in response to a complex ethical issue within a reasonably short period of time, b) help introduce some objectivity into the decision-making process, and c) serve as a guide for people who are not trusted to make a valid choice independently (Colnerud, 2006). The promotion of codes of conduct is considered to exert an influence on attitude, which has been identified as influencing intention, which in turn influences behaviour. Codes, therefore, are a valid element of ethics education whose aim is to ensure conformity to institutional standards (Cronan & Douglas, 2006; Leonard, Cronan & Kreie, 2004).

Ethics education seen as developing decision-making skills

Hypothetical 2	Dan Glover, national sales manager for an international company providing IS services to businesses, is concerned that employees are able to respond ethically in a rapidly changing environment. He engages a training institute to develop workshops based on Leonard Kohlberg's stages of moral development, to help employees in their on-the-spot decision-making when interacting with clients.

When ethics education is considered to be *developing decision-making skills*, the emphasis is on cultivating individuals who are able to take independent decisions about moral issues. Kohlberg's stages of moral development are typically referred to in this context. Kohlberg proposed six developmental stages:

1. avoidance of punishment

2. seeking reward or benefit

3. avoidance of others' disapproval

4. avoidance of dishonour

5. respect for people

6. desire to not violate personal principles.

(Kohlberg, 1981: 121–22)

These stages emphasise rational thinking and judgement, and tend to esteem an intellectual and critical approach to norms. They promote independence, logic and universal principles.

This approach to ethics education focuses on logical ethical processes and is synonymous with training in decision-making (Maner, 2004).

Ethics education seen as developing moral sensitivity

Hypothetical 3	Andrew Davis, academic in the IT Faculty of the University of Northern Queensland, wants to sensitise students to IT ethical issues. He prepares a number of case studies, which are integrated into his course. The students discuss the ethical implications of these in small groups, present the issues and their conclusions to the larger class, and the class responds.

When ethics education is considered to be *developing moral sensitivity*, the emphasis is on imbuing an ethical outlook in individuals.

Moral sensitivity is reflected in Carol Gilligan's care approach to moral development, which focuses on the relational, contextual and enacted aspects of ethics (Tronto, 1993). Gilligan proposed the developmental stages of:

1. exclusive self-concern

2. questioning of self-concern as a sole criterion

3. balanced self-and-other concern.

(Gilligan, 1982)

These stages are understood by some to be accommodated by James Rest in a 'Defining issues test' which has been employed in a number of IT ethics studies (for example, Paradice & Dejoie, 1991; Staehr & Byrne, 2003; Woodward, Davis & Hodis, 2007). This test identifies four aspects of ethical behaviour, with a view to offering a comprehensive representation of the requirements for moral behaviour:

1. Sensitivity — ethical awareness of the existence of an issue requiring a moral response.

2. Judgement — Kohlberg's stages of development.

3. Motivation — reconciliation of conflicts between moral demands and personal goals.

4. Character — a willingness to pursue moral behaviour, despite difficulties.

(Rest, Narvaez, Bebeau, and Thoma, 1999)

According to this approach to ethical education, students of ethics are more than functional skill-bearers and must be approached as whole human beings who need to be influenced at the level of their values and ideals. Only by attending holistically to the professional, will ethics become integrated in a coherent way into their lives. In organisational settings many personal traits associated with ethical conduct (for example, compassion, generosity and idealism) may be suppressed rather than encouraged (Goodpaster, 1996). Professionals who wish to be ethical in this context need to be able to discern what is ethical, independent of their environment.

The development of a core disposition of responsibility is thus recognised as integral to professional ethics formation. This is expressed as ethical sensitivity (Grodzinsky, 1999), a readiness to engage in moral deliberation (Tavani, 2004), having a moral horizon (Grodzinsky, 2000) and a commitment to moral action (Coady & Bloch, 1996). Thus, 'good ethical judgement' forms the basis for ethical analysis and constitutes the core need of IT professionals (Bynum, 2004). This is ultimately gained through experience, however, heuristic methods of analysis (based on pattern recognition more than formal logic) are suggested aids to the process. This ability may be developed through case analysis.

We suggest that these three approaches to IT ethics education are increasingly powerful. *Exercising behavioural control* focuses on the standards for ethical practice and aims at producing compliant professionals who understand the expectations of external authorities to whom they are accountable. *Developing decision-making skills* focuses on the method of ethical deliberation and aims at producing logical professionals who know techniques of reasoning about ethical dilemmas. *Developing moral sensitivity* focuses on the necessary attitude of ethical responsibility and aims at producing responsive professionals who are discerning of ethical issues in the course of their practice.

Each of these builds on the former and enhances it. Compliant professionals require clear guidance concerning how to act in the specific circumstances that they may face and tend to approach ethics as the avoidance of harm rather than the pursuit of good. Logical professionals can apply standards to new situations and generate their own understanding of the good in situations, however their readiness to engage in reflection depends on their propensity to be ethical in the first place. Responsive professionals are attuned to ethical possibilities and are thus alert to apply reflective skills in accordance with ethical standards.

These approaches emphasise respectively content, process and motivation, and thus focus on different aspects of professional ethical practice. Rather than suggesting that any one of these should be abandoned, it would seem that a comprehensive educational approach would encompass all of them. We now turn to another approach to ethics education, which has not yet been explored extensively in the IT ethics literature but which may also be usefully applied. This approach focuses on the need to bring about a new kind of learning, which is based on variation of experience. Learning ethics in this approach is defined as becoming aware of ethics in a different way.

Ethics education seen as prompting new experiences

Hypothetical 4	Carol Simons, personnel director of a national software development company, calls a series of roundtable meetings of employees, management, clients and other stakeholders. At these meetings each group is invited to contribute their understanding of ethics as it applies to the company. Together they produce a joint expression of the organisation's ethics, in diagrammatic and prose form.

When ethics education is approached as *prompting a new experience of ethics*, the emphasis is on stimulating individuals to change their conception of ethics. Their future actions will then flow from these new conceptions. Therefore, from this perspective, educational experiences should be designed to provide a lasting impact on the learner at a personal level, engaging learners in fundamental ways. We suggest that this approach provides a foundation upon which the approaches outlined earlier may build. The strength of this approach lies in its promotion of reflective, deep and authentic learning.

Promoting reflective learning

Such an approach is called a relational, in contrast to a cognitive or behavioural, approach to learning. That is, to stimulate learning we show the interrelationship of the different facets of ethical experience and prompt interaction with these facets. The central role of reflection and communication needs to be recognised and incorporated into the learning experience.

Relational learning may be summarised by the following key features (Ramsden, 1988: 26–28), together with their suggested application to ethics formation:

1. Promoting conceptual change. Learning is about changes in conceptions — a fundamental change of engagement with ethics is more important than knowing philosophical schema, decision-making techniques or arguments surrounding a specific issue.

2. Learning always has content as well as a process — engagement with actual ethics scenarios is more productive educationally than addressing process questions without also giving attention to the ethical content.

3. Improving learning is about relations between learners and subject matter, not teaching methods and student characteristics — the personal engagement

of learners with ethics is more important than how the content is conveyed or what learning styles are represented.

4. Improving learning is about understanding the student's perspective — the learner's existing engagement with ethics is the starting point from which instruction may begin.

This approach, succinctly summarised by Ramsden 20 years ago, was originally developed from Swedish research and has been communicated in detail to the university sector through two important titles, *Learning and awareness* (Marton & Booth, 1997) and *The university of learning* (Bowden & Marton, 1998).

The approach highlights the difference between knowing something intellectually and effective learning. The goal of teaching thus changes from the transmission of facts to the stimulation of personal engagement with the phenomenon. It recognises the IT professional as an integrated human being, rather than simply a physical, mental or emotional entity. It privileges the professional's viewpoint as the foundation for ethical instruction.

Promoting deep learning

Attention is drawn in the IT literature to questions concerning the quality of learning that is achieved in the ethics curriculum. The difference between *deep* and *surface* learning is highlighted by these questions and the key to this difference lies in the extent to which students engage personally with the course content. Shallow learning may be evidenced by the students distancing themselves from personal ethical engagement and adopting an uncontested self-defining approach to ethics.

> One danger for students who adopt a 'shallow' approach to ethics education is that they risk disowning the content by externalising it and, for example, equating it with law ... a 'shallow' approach to learning might equally result in a reduction of the content to a matter of opinion, a state of solipsism in which ethical dilemmas are completely addressed by internal dialogue. (Greening, Kay & Kummerfeld, 2004: 93)

Consideration, therefore, needs to be given to 'valued learning' in the context of ethics education. In a relational approach, learning will focus on those aspects of the object of learning (in this case, professional ethics) which are qualitatively and critically different, in order to bring about wider awareness.

Promoting authentic learning

Seen this way, instruction needs to be more than the teaching of rules, the presentation of simple scenarios and reactions to dilemmas. Each of these may result in a particular (and unhelpful) way of viewing IT ethics, including externalising ethics, equating being legal with being ethical, oversimplifying solutions and failing to take a proactive stance towards ethics. In order for professionals to take ethics seriously and be equipped to practice ethics in more than superficial ways, they need to be engaged at a personal level. A greater emphasis, therefore, needs to be placed in IT ethics education on the authenticity of learning tasks and lifelong learning (Greening et al., 2004).

Such an approach to learning does not allow the learners to avoid the question of having to change their fundamental understanding (Ramsden, 1988). This engagement with the course content cannot be achieved through simple instruction — the learner has to interact with the content on an experiential level.

> … teaching methods that enable students to work on discrepancies in a supportive environment — one that permits incorrect thoughts to be retraced and remedied — are likely to be appropriate. Simply telling students the 'right' conception cannot work, because change involves an active working upon and interaction between the old way of thinking and the new; there is a real sense in which new conceptions grow from older ones. Change in conceptions requires teachers to arrange situations where students must confront the discrepancies between their present way of thinking about the subject matter and the new way desired by the teacher, and where students can come to realize the personal value of the new way. This realization may take a very long time to mature. Time for contemplation, reflection, working things out, and discussion with others learning the same subject matter is thus not a luxury, but a necessity. (Ramsden, 1988)

Elaboration of an experiential approach to ethics education

The primary goal of the educational approach we are advocating, then, is to promote a change of experience. Promoting a change of experience is understood to be more than introducing the learner to new knowledge or drilling them in new behaviour. It is instead about establishing a new relationship between the learner and the object of their learning. In this relational approach the learner addresses the qualitatively significant aspects of the phenomenon and incorporates them into their world. New understanding and comportment result from such change.

Various words or phrases are used here to express the idea of 'experience'. Conception, understanding, view and way of seeing are all synonyms. These indicate the relationship between a subject (here IT professionals) and the object of that subject's attention (here ethics).

The idea of *experience* as used here can be understood by contrasting two alternative approaches to knowledge. According to one approach, reality is separate from the individual and knowledge is gained by the individual being aligned with this external reality. This objectivist approach is dualistic, whereby the individual and reality are distinct from each other. According to the other approach, reality is intimately connected to the individual and knowledge is gained by the individual's construction of meaning. This is a constructivist approach, whereby the individual determines the meaning of reality (Svensson, 1997).

The relational approach understands that the individual and reality are distinct but intimately related. Even though there is an external reality, humans always interpret that reality through their experience. The subject (the one experiencing) stands in relation to the object (the phenomenon which they are experiencing), with each (both subject and object) contributing to the experience. This is a *constitutionalist* (or interpretive, or relational) approach, whereby meaning is constituted by the interaction of subject and object (Bruce, 1997; Ramsden, 2003; Trigwell, 2000). Knowledge is the developing relation between the individual and reality.

Building on a constitutionalist view of knowledge, learning is defined as a change of experience, rather than being simply cognitive or behavioural (Ramsden, 1988). Education is therefore about stimulating a change in the way a person (in this case, an IT professional) relates to a phenomenon (in this case, ethics). The learner's personal engagement with and understanding of the relevance of the phenomenon are integral to this approach. The retention of facts or mastery of processes may be expected to result from, but are peripheral to, the aims of such an education.

The nature of learning is presented here as constitutionalist rather than cognitivist. A change from a cognitivist to a constitutionalist approach requires a re-evaluation of the nature of knowledge and the goals of education. Learners, according to a cognitivist approach, 'are supposed to learn the truth and reproduce it in tests', so that, when they begin a non-cognitivist course, they have to learn again what it means to learn and to 'construct their own answers and be critical' (Stahl, Wood & Howley, 2004: 13). Instead of just learning to reproduce vocabulary and arguments from 'authoritative' sources, learners are expected to engage with the material and develop their own understanding.

Professional bodies and employers rely heavily on authoritative sources, such as codes of ethics, to provide a standard of conduct for their members and employees. A reliance on unmediated codes and regulation may indicate that a cognitivist approach is being adopted. Such an approach assumes that:

1. Reality is independent and separable from humans.

2. Reality is objective, and its qualities and meaning are inherent to itself.

3. Language corresponds directly to objective reality.

4. Causal relationships explain human behaviour.

5. Human behaviour has inherent characteristics and is independent of particular situations.

6. Humans are bound by their circumstances to act in a particular way.

In contrast, a constitutionalist perspective believes that:

1. Reality and humans are inseparably related through the individual's experience. People behave under the influence of their experience of the world, not outside of it.

2. People's understanding of reality is constructed through their own experience and through their interactions with others.

3. People's actions are controlled by their interpretation and understanding of a situation. This includes their ideas about: the 'facts' and their alternatives; other people's feelings, intentions, expectations and reactions; and what is right and wrong, and appropriate.

(Sandberg & Targama, 2007: 22–30)

In the light of these observations, a constitutionalist stance does not assume that statements of standards in codes will be understood in the same way as they were authored, or that regulation will be interpreted in the same way as it was intended. Knowledge formation is understood to be highly communicative and dependent on specific circumstances.

Therefore, communication about ethics is most effectively conducted in the professional's particular circumstances and, most usefully, pursues those issues that are experienced as valuable by the professional in those circumstances. In the light of the difficulties faced by professionals in the organisational setting (for example, standing up to those who control the provision of their basic needs) such conversations about ethics are most realistically initiated and maintained by the organisational leadership, to achieve the 'legitimation of moral discourse as a voice in the professional and managerial conversation' (Goodpaster, 1996: 445).

We are not suggesting here that standards as expressed in codes are no longer needed. Since professionals interpret these guidelines based on their experience, however, complementary means are required to make the codes effective as guides for conduct. Conversations about codes would start from the professional's perspective and seek to see ethics from their point of view. They then would introduce the professional to new ways of seeing ethics (in conformity with the code), with a view to widening their experience.

Another approach employed in IT ethics education has been to emphasise decision-making. This aims at producing independent thinkers. A reliance on the internal processes of the individual, and emphasis on critical thinking and problem-solving, may indicate that a constructivist approach is being adopted. This presumes that individuals have the capacity in themselves to discern what is ethical.

The constitutionalist approach, in contrast to cognitivist or constructivist approaches, emphasises not just the environment external to the learner or just the internal functioning of the learner, but the relation between these. Both of the external and internal aspects are seen as contributors, exerting an influence on each other.

A constitutionalist understanding of knowledge formation, for example, deduces that ethical support will not be effective if approached in a top-down managerial style (Sandberg, 1994). Management following a constitutionalist method would start from an understanding of ethics as seen through the eyes of employees, then seek to build a shared meaning of ethics in their specific organisational context. Working collaboratively with employees, they would arrive at an understanding of ethics constituted from both their perspective and the employees' perspectives. This requires a longer time and energy commitment than a directive approach, however, it is suggested it would be more effectual in achieving an ethical outcome. It also proposes a moral means of promoting ethical practice by valuing employees and their perspectives, and could provide a platform to respond to the concerns of IT professionals (for example, about profit-making in business) (Stoodley, 2009b).

The following ideas serve to indicate the kinds of support that may be offered from a constitutionalist perspective.

1. Since IT professionals see ethics in terms of relationships, one such means may be to clarify the connection of a code with real human interactions. This could be achieved by presenting exemplary scenarios in conjunction with the code's articles. It is of note that this is offered by some professional bodies, for example the ACS (ACS, 2004; Ferguson, Salmond, Al-Saggaf, Bowern & Weckert, 2005).

2. Since a code presumes a prior ethical orientation, another such means may be to promote other-centred attitudes on the part of the employees. This could be achieved by including ethical formation, based on the citizenships revealed in this study, in employees' ongoing professional development.

3. Since ethical guidelines are open to the kind of interpretation that relies on professionals' previous experience, another such means may be to examine the extent to which the code aligns with professionals' perspectives. This could be achieved by a study similar to this project or through tools developed out of this project. This would provide a means of understanding how professionals differ from the expected standard and provide a basis from which to develop learning experiences.

4. Since understanding is constituted rather than simply received, another means may be to embark on a collaborative process of ethical meaning-making. This may result in a rewriting of organisational guidelines, to reflect the broader perspective. Such collaboration could involve stakeholders at all levels of the organisation's operations.

Integral to these suggestions is the goal of changing a professional's relationship with ethics, rather than simply striving for a change in memorised knowledge or drilled behaviour. A professional's previous experience is not considered 'wrong', rather, it is viewed as incomplete. The purpose of formation and support activities is to expand on, rather than replace, their experience. In this way, a professional's experience of ethics is stimulated to become more comprehensive.

A comparison of the approaches to IT professional ethics education

What we propose here, then, is an approach to IT professional ethics education that focuses on the relationship of the practitioner to ethics, bringing about experiential change. When a practitioner experiences themselves as being an ethical professional, as portrayed below, they will see themselves and their world differently and, we propose, that this will affect their practice. This is a relational (or constitutional) approach to education that draws on existing educational research. Its originality lies in the fact that it has not previously been pursued extensively in IT ethics education.

The way this approach differs from the alternatives reviewed earlier is summarised in Table 1. When ethics education is approached as *Exercising behavioural control*, the educational focus is on standards (for example, codes of ethics), the educational intention is to produce conformity (for example, according to organisational expectations) and the educational act is presentation (for example, external standards which must be known).

Table 1: Comparison of approaches to IT professional ethics education

Educational approach	Focus	Intention	Act
Exercising behavioural control	Standards	Conformity	Presentation
Developing decision-making skills	Problem-solving	Reflective thought	Enquiry
Developing moral sensitivity	Attitude	Engagement	Heuristics
Pursuing a new experience of ethics	Experience of ethics	Changed experience	Challenging experience

Source: Author's research

When ethics education is approached as *Developing decision-making skills*, the educational focus is on problem-solving (for example, methods of decision-making), the educational intention is to produce reflective thought (for example, critical thinking) and the educational act is enquiry (for example, posing questions rather than answering them).

When ethics education is approached as *Developing moral sensitivity*, the educational focus is on attitude (for example, from a psychological perspective), the educational intention is to produce engagement (for example, personal moral sensitivity) and the educational act is heuristics (for example, exposure to many cases in order to recognise patterns).

In contrast, when ethics education is approached as *Pursuing a new experience of ethics*:

1. *The Focus* on the *experience of ethics* recognises the need for the practitioner to grow in their relationship with ethics before they will engage with others ethically. For example, they may need to change from an experience of professionalism bounded by a traditional view of IT that admits little room for consideration of the information user (Stoodley, 2009b).

2. The *Intention* of a *changed experience* recognises the need for education to focus on experience as the basic unit of its engagement. For example, practitioners may be more effectively influenced by prompting them to reflect on their approach to the information user rather than requiring them to learn lists of regulations.

3. The *Act* of *challenging experience* recognises the need to promote a change in the practitioner's existing experience. For example, stimulating practitioners to extend their experience beyond that which they currently take for granted.

This alternative approach proposes that the most important goal of ethics education is a change of relationship with ethics, out of which ethical behaviour may flow.

A rationale for the experiential learning of ethics has also been based on the concept of *stage disparity* (identified by Kohlberg (1981) and Rest (1986)), which makes use of contrast to prompt moral growth (Vartiainen, 2005b). Confrontation with this experience is prompted by solving dilemmas through discussion with others. This provides fertile ground for differences of opinion and exposes the participants to alternative viewpoints. An academia- and industry-based collaborative-learning environment provides the range of experience required to provoke this kind of interaction. Self-criticism and evaluation are used to encourage ethical growth.

> The framework broadens the traditional idea of learning professional ethics in an educational institute with hypothetical moral conflicts to a form of experiential learning in which IT professionals together with students critically deliberate about moral conflicts they have confronted in real life ... clients, students, and instructors may be perceived as co-learners of professional ethics. (Vartiainen, 2005b: 10)

This approach looks beyond the individual to a community of practice, engaging the IT community in the process of ethical development. It also promotes experiential learning in real contexts.

Suggested teaching methods, reflecting the four aspects of Rest's model, include: moral argumentation and dilemma discussion (for moral sensitivity and judgement); role models and heroes (for moral motivation and character); values clarification (to avoid indoctrination); and logic, role-taking and concept construction (for principled moral reasoning). These methods induce change by presenting individuals with the opportunity to discover inconsistencies in their own thinking, with a view to prompting them to make up their own minds and not accept uncritically what is being presented to them (Vartiainen, 2005a). Thus, a project course is an example of an environment which satisfies the requirements of those who wish to strive to avoid indoctrination, with reliance on experienced practitioners to offer a perspective which aligns with professional standards (Vartiainen, 2005b).

In pursuing a deep-learning outcome, the main aim of our approach is that the student be prompted to have a changed experience of ethics. This, then, is an experience-based approach to learning, which understands learning as a change of relationship, rather than just a change of cognition or behaviour. This moves the focus away from receiving external truth, towards constituting personal understanding. Such a change is significant because it shifts professional ethical formation away from an instructor's perspective and moves it to a learner's perspective.

We now present a model which may be a useful tool in approaching IT ethical formation from an experiential perspective.

A model of ethical IT

A model of IT professionals' experience of ethics is represented diagrammatically in Figure 1. The model combines the experiences of ethics that is found in a group of IT professionals with a representation of an evolving definition of the field of IT. These two aspects are understood to be intimately related. They are also supported in the IT literature and in ethical theory (Stoodley, 2009b).

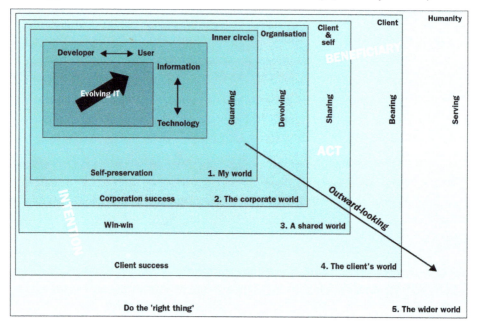

Figure 1: A model of ethical IT

Source: Author's research

Evolving IT is represented in the model by the innermost rectangles. Acceptance of the evolving nature of the field of IT, towards an increasingly information and user-centred (rather than a technology developer-centred) perspective, is understood as an important aspect of opening IT professionals to ethical considerations. It is significant because it moves the focus of IT away from its products and towards its purposes, orienting professionals away from themselves and towards those who are affected by their work.

The 'Model of Ethical IT' illustrates IT professionals' expanding awareness of what it means to be ethical. Five facets of a professional's experience, called 'citizenships', are represented in expanding rectangles which are labelled with the citizenship name in abbreviated form in their rectangle's bottom right-hand corner; for example, the second citizenship is called 'Citizenship of the corporate world' and is labelled in the diagram as '2. The corporate world'. Each citizenship, moving outwards from 1 to 5, progressively encompasses the previous citizenships and exerts an influence over them.

The object of ethics is identified as the 'beneficiary' and is included in the model in the top right-hand corner of each citizenship rectangle, for example the organisation being the beneficiary in 'Citizenship of the corporate world'. The activity associated with ethics is identified as the 'act' and is included on the right-hand border of each citizenship rectangle; for example devolving being the act in 'Citizenship of the corporate world'. The underlying objective of ethics is identified as the 'intention' and is included on the lower border of each citizenship rectangle; for example corporation success being the intention in 'Citizenship of the corporate world'. This level of detail serves to show the expanding nature of the experiences as they progress outwards.

The model offers a perspective on how the categories progressively expand outwardly, from a practitioner-centred outlook to a more other-centred outlook. The movement towards greater other-centredness is evident in the change in perspective concerning the beneficiary, act and intention. The beneficiary expands from the professional's inner circle, through the organisation and client, to humanity. The act expands from guarding, through devolving, sharing and bearing, to serving. The intention expands from self-preservation, through corporation success, win-win and client success, to doing the 'right thing'. These increasingly embrace others. The model, then, offers a means of comprehensively conceiving IT professionals' experience of ethics.

The IT professional's experience of ethics may thus be understood to lie along a continuum, with the professional progressively broadening their scope of engagement with others. As a result, they accept that their rights diminish as others' rights are accorded increasing priority, and their responsibility expands as it is defined in terms of others. As the professional's experience of ethics expands, they undergo a dual change of experience, with rights increasingly accorded to others and responsibilities increasingly defined by others' needs.

Each citizenship in the continuum represents a qualitatively different way of experiencing ethics. Movement into a citizenship that lies further along the continuum signifies a definite experiential change, which crosses a perceptual boundary. The experience of ethics is thus represented in five distinct citizenship categories, along an increasingly other-centred continuum.

The usefulness of the representation of experience in discrete units is its ability to reveal critical differences in experience, which may become objects of our attention, either as reflective professionals or as supporters of professionals. These aspects of ethical experience could otherwise be overlooked and left undeveloped.

The representation here of ethical experience along an other-centred continuum is consistent with the measure of other-centredness presented as an ethical ideal in Stoodley (2009b). Other-centredness recognises the primacy of other people when defining the nature of ethics, in general, and when defining the ground of professional ethics in particular. According to such a view, an ethical perspective places others at the centre of our world, instead of ourselves.

Following this reasoning, the key to understanding the critical differences between the citizenships is the extent to which IT professionals willingly accept that their rights and responsibilities are defined in terms of others. For example, an IT professional may choose to work late in order to complete the testing on a job due to be delivered the next day, rather than simply sign off on the project at the end of the afternoon. The professional thus allows the client's right to receive a properly tested product to limit the professional's own right to a full night's rest. The professional also allows the client's needs to define the scope of their responsibility. Rather than just working the hours required by their employer, the professional accepts the extra hours and allows responsibility for the client to fall on their shoulders. Naturally, we preserve our own welfare, preferring to define our own rights (where we benefit most) and our own responsibilities (where we bear the least responsibility) — taking an ethical perspective challenges both of these attitudes and requires a significant change of perspective.

The progression from 'Citizenship of my world' to 'Citizenship of the wider world', then, represents a dual attitude change, with respect to rights and responsibilities. This change moves from a relatively closed circle of interest and engagement to a more open one. The professionals' awareness of others in their world expands as they move through the citizenships. This is not to imply that 'Citizenship of my world' is unethical, however, increasing recognition of and response to others' needs is understood here to indicate an increasingly comprehensive understanding of what ethics is about. From this point of view, an ethically mature IT professional is one who has in their portfolio the full range of citizenships characterised in this account.

It needs to be noted that the citizenships do not represent stages of development but an increasingly sophisticated understanding of ethics. Citizenship 1 is not left behind when a professional experiences Citizenship 2, but their experience of Citizenship 1 is modified by their experience of Citizenship 2. The citizenship experiences of ethics are thus understood here to be cumulative, so a professional who experiences ethics as 'Citizenship of the wider world' will also experience ethics as all the other citizenships. Each citizenship, however, sheds light on the citizenship(s) preceding it. When a professional experiences ethics as 'Citizenship of the wider world', they will experience ethics as 'Citizenship of my world' differently to a professional who is not experiencing ethics as 'Citizenship of the wider world'. In other words, each subsequent experience illuminates the previous experiences in a new way. For example, when a professional experiences ethics as 'Citizenship of my world', they may view altruistic acts as a means of salving their own conscience — 'you've got to be able to live with yourself, I guess, afterwards. It's all very selfish, really' (Participant 26)[1] — whereas, when a professional experiences ethics as 'Citizenship of the wider world' they may view altruistic acts as fulfilling their responsibility towards other people — 'it's about what I want to achieve and what I think that I should be doing to contribute to society and mankind and whatever and I would rather have my skills used in an area which I think is … better' (Participant 13). Therefore, the citizenships represent pervasive changes of perspective on the part of the professional.

The 'Model of Ethical IT', then, indicates a developing relationship between IT professionals and the world in which they work. The outward-looking axis represents ethical conduct at all points. As argued earlier, however, a professional who has a comprehensive experience of ethics will evidence in their practice all of the facets which lie along this axis.

The outer layers in this model influence professional practice at the inner layers. For example, professionals experiencing ethics as 'Citizenship of the wider world' are aware of their own needs, however, they intentionally set them aside for the sake of others; and, professionals who experience ethics as 'Citizenship of the client's world' are aware of the organisation's needs, however, they risk censure by challenging the corporation to accept additional expense for the client's sake. Thus, the wider world experience influences a professional's answer to the question: 'What if a client asks you to exploit other people on their

---

1   Quotes are of participants in research detailed in Stoodley 2009b.

behalf?' The wider world experience also results in the professional including activities in their portfolio that offer no reimbursement, but which help meet the needs of the underprivileged.

A conceptual tool

The 'Model of Ethical IT' may be used as a tool to support education which strives to prompt an experiential change of ethics amongst IT professionals.

Throughout this chapter, we have presented a constitutionalist understanding of how knowledge is gained about the world. From this viewpoint, learners contribute to the experience of learning as much as the object of learning itself. The past experience of the learner and a circumstantial perception of relevance determine how they will understand their encounter with the phenomenon they are facing. This includes the IT professional's encounters with ethics.

From the point of view of a constitutionalist approach to knowledge, significant influence over people is exerted, not at a behavioural level, but at an experiential level. In other words, our influence over others is most effectively exerted in the realm of their relation to their circumstances.

> Human behaviour is not controlled by structures, systems and prescriptions as such. Instead, it is how people understand those structures, systems and prescriptions that determine their behaviour. (Sandberg and Targama, 2007: 175)

Influence over an IT professional's practice of ethics is thus best instigated through influencing their conceptions of ethics. This changes an educating leader's role from one of controlling, to that of guiding and stimulating understanding.

For example, as previously noted, the enforcement of codes has been observed to exert an influence over employee behaviour. What, however, has such enforcement taught practitioners? The lessons learned may have been that the corporate environment is a punitive one; that the letter of the law is key and that if behaviour seems to conform to the law then that is sufficient. The enforcement of codes does not need to convey such lessons, however. To ensure these perceptions are not imparted, enforcement needs to be accompanied by meaningful communication concerning its significance.

The 'Model of Ethical IT' serves as a conceptual tool, to represent the relationships among people and artefacts in the IT space. It illustrates these relationships in a way that is readily grasped and easily remembered. A diagram can be more effective at conveying such conceptual information than written prose. The model, then, offers a means of interacting with IT professionals on an experiential level.

The model may be used, in a strategic sense, to orient activities towards certain goals or outcomes, or it may be used, in an operational sense, to evaluate specific actions. It may be usefully employed to:

- reconceptualise the IT professional space
- plan future directions
- provide guidance of ethical formation and support
- define the scope of the IT profession's ethical responsibility
- provide inspiration for individual and organisational guidelines
- set standards of conduct and aspiration
- aid communication between stakeholders.

This conceptual tool could be used on an individual, group, organisational, professional or discipline level. At the individual level, IT professionals may refer to it in order to conduct a self-assessment of their practice. At the group level, a learning facilitator may present it to a gathering of professionals as a prompt for discussion. At the organisational level, a corporation may use it as a yardstick against which to measure their goals. At the professional level, a professional body may consider it as they formulate visionary statements for their members. At the discipline level, disciplinary leaders may adopt it as a means of envisioning new conceptual connections and future directions.

Our conceptualisations of IT play a key role in the deployment of IT, determining our expectations of IT and the way we interact with IT. This indicates the potential power of such a model.

A learning tool

The model may be employed as a learning tool, to stimulate professionals to make the conceptual shifts that it represented. It would thus prompt professionals to enlarge their experience of ethics.

The insights into learning offered by variation theory indicate how powerful conceptual shifts may be deployed for the effective formation of professionals. Variation theory is a theory of learning derived from the approach taken in this research (phenomenography). It identifies the need to confront new ways of experiencing a phenomenon in order for effective learning to take place. That is, in order to understand the complexity of a phenomenon such as ethics, you need to experience the breadth of its complexity. If you are not offered the opportunity to understand the levels of complexity in the experience, then you cannot learn about it (Edwards, 2007). Therefore, the model presented here may be used as a curriculum guide, to ensure practitioners in formation have the opportunity to encounter all of these different ways of experiencing professional ethics.

Activities which apply the insights of variation theory, using the model to engender the expansion of the learner's experience, could include:

- presentation and discussion of the range of citizenships
- self-assessment of the learner's practice against the citizenships
- examination of case studies of IT professionals representing the range of citizenships
- involvement in practical projects which expose learners to the range of citizenships
- journaling of personal engagement with the citizenships.

The key differences between the citizenships lie in the practitioner's dual experience of rights and responsibilities. These aspects of the citizenships identify critically different perspectives out of which different behaviours proceed. They specify key educational turning points for ethical training and key elements of variation, which need to be intentionally brought into awareness.

Thus, three sources of opinion on IT ethics agree. IT professionals, Emmanuel Levinas (1998) and Daryl Koehn (1994) all demonstrate that the most significant variations in ethics are the nuances of relationship between the professional and others (Stoodley, 2009b). An emphasis on the diversity of possible relationships (as indicated in the citizenships) and personal reflection in the light of that diversity are central to the learning process.

Similar approaches to IT ethics education have been recommended by others. One suggestion has been the collaborative creation of a code of ethics that encourages a reflective and critical learning experience. It is signalled that this constructivist approach to education contrasts with the cognitivist educational approach which students typically encounter, so it may cause them some confusion. Thus, an early orientation of the learners to constructivism is necessary for them to gain the most for the experience (Stahl et al., 2004). This would also be true of a constitutionalist to bring differing points of view together in an environment where different stakeholders are forced to communicate and collaborate. This is proposed out of a moral development perspective forwarded by Rest (1986) in which exposure to complexity fosters development. Participating experienced

IT professionals are expected to represent the accepted standard of ethical conduct against which students are able to measure their own ideas (Vartiainen, 2005a).

The central message of variation theory is the need to intentionally focus on variation of experience. The role of the instructor in variation theory is to ensure that the learner's experience is being expanded in ways that are potentially educationally meaningful. It is incumbent on the facilitator to determine what those educationally meaningful experiences are, a task which is aided by phenomenographic investigation which seeks to describe qualitatively different ways of experiencing phenomena (Stoodley, 2009b). This chapter provides a framework of the appropriate differences of experience for representation in a program of study of IT professional ethics. In other words, it is important that the learners be exposed to each of the citizenships. This does not necessarily need to occur in a progressive way, the main concern is that the learners have the possibility of meaningful contact with each citizenship. Nevertheless, systematic exposure to the citizenships could help learners grasp the expanding nature of the experiences represented by the citizenships.

The citizenship progression may also be used as the basis of a pre-test and post-test. This would draw the attention of learners to the citizenships, to the level of their identification with them and to the change in their relationship with them. A pre-test could serve the purpose of orienting learners to the core concepts of the citizenships and to their relationship with them. A post-test could serve the purpose of reminding learners of these aspects of ethics, of directing their attention to their relationship with them and of providing a means of self-assessing their change over the course of instruction. A subsequent question, in the light of these tests, such as 'How has your experience of ethics changed and does this appear to be a good thing?' may serve to prompt learners to reflect on any changes in their experience and whether those changes are desirable.

A short survey instrument could serve as a pre-test and post-test tool (for example, Figure 2). A strength of this instrument is that it could be rapidly administered and it would allow learners to situate their views incrementally in relation to the citizenships. A weakness is that it may lead their thoughts, for example, by framing their experience of ethics too strongly in terms of responsibility (and rights, which are implied in the statements) and by suggesting the 'correct' answers too obviously. In the light of this, it is probably useful to view this survey instrument at the pre-test stage not only as a means of eliciting the learner's views but also as a means of introducing them to the concepts that are central to the educational program on which they are embarking.

Your experience of IT professional ethics	
Mark how you experience IT professional ethics on the scales provided. This is a self-assessment tool, it is not for anyone else's eyes.	
In my professional practice my primary responsibility is to meet my needs, and the needs of my family and friends.	Strongly disagree - Strongly agree
In my professional practice my primary responsibility is to meet my employer's goals.	Strongly disagree - Strongly agree
In my professional practice my primary responsibility is to meet both my own needs and my client's needs.	Strongly disagree - Strongly agree
In my professional practice my primary responsibility is to meet my client's needs.	Strongly disagree - Strongly agree
In my professional practice my primary responsibility is to meet the needs of the wider society.	Strongly disagree - Strongly agree
What thoughts about IT professional ethics has this survey promted in you? (Note on the back)	

Figure 2: A survey of learner's experiences of ethics

Source: Author's research

The 'Model of Ethical IT', then, may serve as an overview of the citizenships. Such a model provides a means of encapsulating the main concepts succinctly and representing how they relate. The learners could be asked to devise their own model of ethical IT before viewing this model, then asked to compare and contrast the two models, and to provide a rationale for their preferences. This would serve the purpose of drawing their attention to the phenomenon of professional ethics on a global level and prompt their thoughts on the structural aspect of the experience of ethics. It would also provide a means of contact with learners who relate to diagrammatic representations of knowledge.

Evidence provided by phenomenography suggests that we all start at a global level and progressively situate details in accordance with our global understanding of the phenomenon of our attention (Marton & Booth, 1997). This is the way we make sense of new experiences as we are confronted with them. For example, we will not learn to read and write if we do not first understand somewhat what it means to be literate. Similarly, an IT professional will not be able to make sense of the detail of ethics if they do not first have at least a partial understanding of what it means to be ethical.

A course of instruction based on the citizenships could, after an initial overview, concentrate on the citizenships one at a time. This would provide the opportunity to explore the meaning of each citizenship, the relevance of each to professional practice, how each would be expressed in daily work and how each is related to the others.

A gambit of activities could serve to draw out the citizenship meanings — case studies, guest speakers, field visits, peer presentations and debates. In addition, in order to engage the learners on an experiential level, a practical project could be chosen by the learners, which would provide a stimulus for group discussion. For such a project to offer the range of citizenship experiences it could be conducted in collaboration with a local employer and be offered as a community service. Such a project could meet a community IT need. The effectiveness of reflection in professional development (Moon, 1999, 2006) indicates the appropriateness of the use of journaling in the learning process, through which the learners could consider their relationship with the citizenship continuum in the light of the demands of their practical project.

Conclusion

IT ethics education may be approached from various educational standpoints, resulting in different outcomes. We present here an additional approach to those predominantly espoused by IT ethicists to date. Educating for experiential change offers a means by which deep learning may occur, which will evidence itself in professional practice, to the benefit of the computing profession and the world which it impacts.

Acknowledgements

Various parts of this chapter are adapted from the report of a doctoral study on the experience of IT ethics (Stoodley, 2009a, 2009b).

References

ACS, 2012, ACS Code of ethics Case Studies & related clauses of the code of conduct, viewed 6 February 2013 <http://www.acs.org.au/__data/assets/pdf_file/0020/12179/ACS-Code-of-Ethics-Case-Studies.pdf>

Bowden, J & Marton, F, 1998, *The university of learning*, Kogan Page, London.

Bruce, C, 1997, *The seven faces of information literacy*, Auslib Press, Adelaide.

Bynum, TW, 2004, 'Ethical decision-making and case analysis in computer ethics', in TW Bynum & S Rogerson (eds), *Computer ethics and professional responsibility*, Blackwell, Malden, pp 60–86.

Coady, M & Bloch, S (eds), 1996, *Codes of ethics and the professions*, Melbourne University Press, Carlton South.

Colnerud, G, 2006, 'Teacher ethics as a research problem: syntheses achieved and new issues', *Teachers & Teaching*, vol 12, no 3, pp 365–85.

Cronan, TP & Douglas, DE, 2006, 'Toward a comprehensive ethical behavior model for information technology', *Journal of Organizational and End User Computing*, vol 18, no 1, pp i–xi.

Edwards, SL, 2007, 'Phenomenography: "Follow the yellow brick road!"', in S Lipu, K Williamson & A Lloyd (eds), *Exploring methods in information literacy research*, Centre for Information Studies, Charles Sturt University, Wagga Wagga, pp 87–109.

Ferguson, S, Salmond, R, Al-Saggaf, Y, Bowern, M & Weckert, J, 2005, 'Case studies and codes of ethics: the relevance of the ACS experience to ALIA', *The Australian Library Journal*, August, pp 299–308.

Gilligan, C, 1982, *In a different voice: Psychological theory and women's development*, Harvard University Press, Cambridge.

Goodpaster, KE, 1996, 'Praxiology and the moral agenda for professional education', *Praxiology: The International Annual of Practical Philosophy & Methodology*, pp 437–53.

Greening, T, Kay, J & Kummerfeld, B, 2004, 'Integrating ethical content into computing curricula', paper presented at the 6th Australasian Computing Education Conference, Dunedin, New Zealand.

Grodzinsky, FS, 1999, 'The practitioner from within: revisiting the virtues', *Computers and Society*, vol 29, no 1, pp 9–15.

——, 2000, 'The development of the "ethical" ICT professional: and the vision of an ethical on-line society: how far have we come and where are we going?', *Computers and Society*, vol 30, no 1, pp 3–7.

Koehn, D, 1994, *The ground of professional ethics*, Routledge, London.

Kohlberg, L, 1981, *Essays on moral development, Volume 1: the philosophy of moral development*, Harper & Row, San Francisco.

Leonard, LNK, Cronan, TP & Kreie, J, 2004, 'What influences IT ethical behavior intentions — planned behavior, reasoned action, perceived importance, or individual characteristics?', *Information & Management*, vol 42, pp 143–58.

Levinas, E, 1998, *Entre nous: on thinking-of-the-other*, MB Smith & B Harshav, trans, Athlone Press, London.

Maner, W, 2004, 'Unique ethical problems in information technology', in TW Bynum & S Rogerson (eds), *Computer ethics and professional responsibility*, Blackwell, Malden, pp 39–59.

Marton, F & Booth, S, 1997, *Learning and awareness*, Lawrence Erlbaum Associates, Mahwah, NJ.

Moon, JA, 1999, *Reflection in learning & professional development*, Kogan Page, London.

——, 2006, *Learning journals: a handbook for reflective practice and professional development*, Routledge, London.

Paradice, DB & Dejoie, RM, 1991, 'The ethical decision-making processes of information systems workers', *Journal of Business Ethics*, vol 10, pp 1–21.

Ramsden, P, 1988, 'Studying learning: improving teaching', in P Ramsden (ed), *Improving learning: new perspectives*, Kogan Page, London, pp 13–31.

——, 2003, *Learning to teach in higher education*, RoutledgeFalmer, London.

Rest, JR, 1986, *Moral development: advances in research and theory*, Praeger, New York.

——, Narvaez, D, Bebeau, M & Thoma, SJ, 1999, *Postconventional moral thinking: a neo-Kohlbergian approach*, Lawrence Erlbaum Associates, Mahwah.

Sandberg, J, 1994, *Human competence at work: an interpretative approach*, BAS, Goteborg, Sweden.

—— & Targama, A, 2007, *Managing understanding in organisations*, Sage, London.

Staehr, LJ & Byrne, GJ, 2003, 'Using the Defining Issues Test for evaluating computer ethics teaching', *IEEE Transactions on Education*, vol 46, no 2, pp 229–34.

Stahl, BC, Wood, C & Howley, R, 2004, 'Teaching professional issues in computing through the development of a student code of conduct', *Ethicomp 2*, viewed 6 February 2013, <http://www.cse.dmu.ac.uk/~bstahl/publications/2004_professional_issues_ethicomp.pdf>

Stoodley, I, 2009a, 'IT professionals' experience of ethics and its implications for IT education', PhD thesis, Queensland University of Technology.

——, 2009b, *Professional ethics: The IT experience*, VDM Verlag Dr Muller.

Svensson, L, 1997, 'Theoretical foundations of phenomenography', *Higher Education Research & Development*, vol 16, no 2, pp 159–71.

Tavani, HT, 2004, *Ethics and technology: ethical issues in an age of information and communication technology*, John Wiley & Sons, Hoboken.

Trigwell, K, 2000, 'A phenomenographic interview on phenomenography', in J Bowden & E Walsh (eds), *Phenomenography*, RMIT Press, Melbourne, pp 62–82.

Tronto, J, 1993, 'Beyond gender difference to a theory of care', in MJ Larrabee (ed), *An ethic of care: feminist and interdisciplinary perspectives*, Routledge, New York, pp 240–57.

Vartiainen, T, 2005a, 'Moral conflicts in a project course in information systems education', PhD thesis, University of Jyväskylä, Finland.

——, 2005b, 'Morally successful collaboration between academia and industry — a case of a project course', paper presented at the 14th International Conference on Information Systems Development, Karlstad, Sweden.

Woodward, B, Davis, D & Hodis, F, 2007, 'The relationship between ethical decision making and ethical reasoning in information technology students', *Journal of Information Systems Education*, vol 18, no 2, pp 193–202.

Section VI

Codes of ethics

Are codes of ethics useful?

Codes of ethics or codes of conduct (or both) are often seen as one of the defining parts of professions and of professional organisations. Members of a profession or a professional body are expected to abide by the code. The Australian Computer Society (ACS) is no exception and, in addition to its own code, has endorsed the 'Software engineering code of ethics and professional practice' (ACM).

Despite the fact that codes of ethics are commonplace, controversy surrounds their usefulness and this is evident in the two contributions of this section. Both argue for the value of codes, but such arguments would not be necessary were it not for the existence of many doubters. In this introduction we will set out some of the more important arguments on both sides.

Various benefits of codes of ethics for information and communications technology (ICT) professionals have been put forward. Codes of ethics motivate members of an association to behave ethically and inspire them because they 'provide a positive stimulus for ethical conduct'. Codes of ethics also guide members through the ethical problems that they face in their working life and educate them about what is acceptable and unacceptable in relation to their interaction with others. Codes of ethics can also be used to discipline members (if the code has teeth) by, for example, causing a member to be sacked from his/her job for violating the code of ethics (Tavani, 2007: 101). In addition to these, George Reynolds (2003) lists four more benefits of codes of ethics. According to him, codes of ethics improve ethical decision-making since adherence to them means that professionals will use a common set of core values and ideals to serve as guidelines for ethical decision-making. Codes of ethics promote high standards of practice and ethical behaviour because adherence to them reminds the members of their ethical responsibilities and duties when they are tempted to compromise or go against the code amidst competing demands from self, employer, clients, and community. Codes of ethics also enhance the trust and respect from the general public because the trust of the general public is built on the expectation that a professional will behave ethically and adhere to a code of ethics. Finally, codes of ethics provide an evaluation benchmark because professionals can use them to self assess their behaviour at work (Reynolds, 2003).

On the other hand, there are several criticisms of codes of ethics that are adopted by ICT professionals. In his discussion of the criticisms of the codes of ethics for ICT professionals, Herman Tavani (2007) notes that these codes of ethics, unlike the ones in the field of law and medicine, have no 'teeth' in the sense that a violation of the code does not necessarily result in the termination of the employment, or indeed any punishment at all. Furthermore, when limited to only four traditional areas of concern: privacy, accuracy, property and accessibility, codes of ethics can ultimately be vague, self-serving, inconsistent, unrealistic, unnecessary and incomplete (Tavani, 2007). Next, codes of ethics do not tell professionals what to do when two or more principles in the code conflict with each other. Finally, codes of ethics can give professionals the mistaken notion that all they need to do is to locate a directive and then follow it blindly, when they should have examined, explored, discussed, deliberated, and argued for or against a particular course of action (Tavani, 2007). In addition to the above, there are two more points of criticism against codes of ethics for ICT professionals. The first is that they can be too detailed, static and inflexible for this area: an area that is dynamic and, as such, needs a code that can easily be adapted. The second is that most of the codes are the product of the technological thinking in developed countries and so, those who develop these codes, tend to neglect the differences in cultural and social values (Berleur and Brunnstein, 1996). The discussion above does make some useful points about codes, but it does not tell us if they in fact make any difference to behaviour. In a review of 24 studies of codes, Karen Mather (2007) found that, while the results were indecisive, 15 of those studies supported codes while nine were negative, equivocal or said nothing of their efficacy. Here we mention just a few of the studies that focus on empirical studies of the efficacy of codes. It must be noted that most of the studies involved codes of corporations and not of professional bodies. These studies are, however, still relevant and we comment on their relevance at the end of the section.

Margaret Pierce and John Henry (1996 and 2000) are positive about the efficacy of codes. In their first study they reported that 78 per cent of respondents said that they used professional codes to held guide their ethical decision-making and, in their second, they say that 'the results suggest that the organization is perceived as more disapproving of questionable actions by those computer professionals who work for organisations which have CT [computer technology] ethics codes' (2000: 319). They make the obvious, but easily overlooked, point that 'in order for the codes to be effective, the organization must communicate the codes to the members and make the membership aware of the philosophy embedded in the codes' (1996: 427). Similar findings are reported by Adams, et al (2001): 'This study found the existence of a corporate code of ethics affected both employee ethical behaviour and perceptions of ethics in several ways. Respondents who worked for companies having a code of ethics judged subordinates, co-workers, themselves and especially supervisors and top managers to be more ethical than respondents employed in organizations not having a formal code of ethics'. Furthermore they say, 'Our findings suggest that the mere presence of a code is more important than the content of the code per se'. This latter comment is interesting in the light of criticisms that say that codes are in general too vague to be useful.

Not all research, however, has led to positive conclusions. Chris Cowton and Paul Thompson claim that, while there were some differences in ethical behaviour between the banks that they studied, none were statistically significant different: 'the findings fail to provide firm support for those who would claim that codes have a substantial impact on business practice' (2000: 173). Another report on a study in Hong Kong found that company codes of ethics have a positive impact on the conduct of employees and on the image of the company but not much on relationships between staff or between staff and suppliers and clients (Snell et al, 1999). In another study, which was reported in the same paper, no differences were found and only weak and qualified support was found by Sheila Harrington (1996). Generic codes, as opposed to specific information systems IS codes did have an impact on those who normally have a strong tendency to deny responsibility, but IS codes have less impact and neither had much impact on those more willing to accept responsibility for their actions. Finally, Pearson et al reported two findings of some concern. First, 'The results … provided very little support for the idea that IS managers are guided by the ethical codes of conduct put forward by the IS professional associations' (1997: 95) and, 'The IS profession should also be concerned by the apparent lack of commitment to respondents proclaim for professional associations … This raises an interesting question about the role and/or effectiveness of these organizations in the development of ethical standards for the IS profession'(1997: 99).

As mentioned already, the two papers following both argue that codes of ethics can be effective. In the first, Don Gotterbarn, a computer ethicist, after a general discussion of ICT governance that provides a context for codes of ethics, presents a case that even if they are, in a sense, 'toothless tigers' that are not backed up by disciplinary mechanisms, such codes still have an important role to play in professions. According to him, a code of ethics

> is a statement to members about the ethical stand of an organisation and profession, a conscience of the profession, an announcement to non-members what the profession standards for …, it imposes functions on an ethics committee to educate the membership, and it imposes responsibilities on the professional organisation itself.

The last point on responsibility is a particularly important one and reflects Gotterbarn's stance on stakeholders as all those who are impacted by the technology. This is highlighted, too, by his discussion of what he calls the 'micro-macro ethics confusion'. Codes are, according to him, too often understood as speaking just to individuals and not to organisations. Consequently their importance in the bigger picture of ICT's contribution to the quality of life is frequently overlooked.

Michael Bowern, a retired long-time practitioner, and Oliver Burmeister, an ICT academic, provide a different sort of defence of codes of ethics, appealing to, at least implicitly, self-interest and reputation as they were outlined in the introduction to the book. They provide two analogies to support their argument that a strong business case can be made for codes of ethics. First, they note that being environmentally friendly provides a commercial advantage, although this was not so in the recent past. If being environmentally friendly is commercially valuable, promoting professionalism based on a code of ethics amongst their ICT staff could have the same effect. They outline some strategies as to how the ACS could help promote this. Their second analogy is the Australian pharmaceutical industry that, they argue, benefits from compliance with its code of conduct, particularly in the areas of risk management and mitigation and consumer confidence. The final part of their paper discusses the importance of keeping a code up to date, particularly with respect to the industry case for the code's importance. This is based on their experience in revising the ACS code of ethics.

References

Adams, J, Tashchian, AS, & Shore, TH, 2001, 'Codes of ethics as signals for ethical behaviour', *Journal of Business Ethics*, vol 29, no 3, pp 199–211.

Association for Computing Machinery, nd, 'Software engineering code of ethics and professional practice', version 5.2, <http://www.acm.org/about/se-code>

Berleur, J & Brunnstein, K (eds), 1996, *Ethics of computing: codes, spaces for discussion and law*, Chapman and Hall, London.

Cowton, CJ & Thompson, P, 2000, 'Do ethical codes make a difference? The case of bank lending and the environment', *Journal of Business Ethics*, vol 24, no 2, pp 165–78.

Harrington, SJ, 1996, 'The effect of codes of ethics and personal denial of responsibility on computer abuse judgments and intentions', *MIS Quarterly*, vol 20, no 3, pp 257–78.

Mather, K, 2007, 'Ethics in the infosphere: extending the work of Effy Oz on ethical standards in ICT', *CEPE 2007 Conference Proceedings*.

Pearson, JM, Crosby, L & Shim, JP, 1997, 'Measuring the importance of ethical behaviour criteria', *Communications of the ACM*, vol 40, no 9, pp 94–100.

Pierce, MA & Henry, JW, 1996, 'Computer ethics: the role of personal, informal and formal codes', *Journal of Business Ethics*, vol 15, no 4, pp 425–37.

——, 2000, 'Judgments about computer ethics: do individuals, co-workers and company judgments differ?', *Journal of Business Ethics*, vol 28, no 4, pp 307–22.

Reynolds, G, 2003, *Ethics in information technology*, Thomson, Toronto.

Snell, RS, Chak, AM-K & Chu, JW-H, 1999, 'Codes of ethics in Hong Kong: their adoption and impact in the run up to the 1997 transition to sovereignty to China', *Journal of Business Ethics*, vol 22, no 4, pp 281–309.

Tavani, H, 2007, *Ethics and technology: ethical issues in an age of information and communication technology*, 2nd ed, John Wiley & Sons, Hoboken, NJ.

Biographies

Dr Michael Bowern has over 45 years experience as a practitioner, manager, consultant and researcher in the public and private sectors of the ICT industry and academia. His research interests include professionalism and the ethical development of ICT systems, particularly for electronic democracy. He is a senior member of the ACS, and previously has been a member of the ACS committee on computer ethics (CCE). In this position he played a major role in the review of the ACS code of ethics, and in writing the new version.

Dr Oliver Burmeister is a senior lecturer in the School of Computing and Mathematics, at Charles Sturt University. His research is focused on informing the design of mental healthcare systems, particularly those involving seniors. Oliver has a keen interest in ICT ethics and is currently pursuing projects involving privacy considerations for mental healthcare providers, and confidentiality considerations for staff in aged care facilities. Oliver is also the chair of the ACS CCE.

Prof Don Gotterbarn, the Director of the Software Engineering Ethics Research Institute at East Tennessee State University, is also a visiting professor at the Centre for Computing and Social Responsibility in England. He worked as a computer consultant on international software projects for the US navy, the Saudi Arabian navy, vote counting machines and missile defence systems. He has actively researched and promoted professional computer ethics for over 25 years. His awards for this work include: the 'Making a Difference' award (2002) from the Association for Computing Machinery (ACM) group on computing and society, the 'ACM Outstanding Contribution' award (2005) for promoting ethical behaviour of professionals and organisations, and the International Society for Ethics and Information Technology Joseph Weizenbaum award (2010) for his contributions to the field of information and computer ethics.

13. ICT governance and what to do about the toothless tiger(s): Professional organisations and codes of ethics[1]

Don Gotterbarn
East Tennessee State University

Introduction

Information and communications technology (ICT) is infamous for unfortunate incidents in planning, development, and delivery. A typical response to these incidents is to both complain about the toothless tiger of technical and professional standards that are not enforced, or enforceable, and to also advocate the development and implementation of strong government regulations — licensing and legislation. These regulations constitute one form of what has been called 'ICT governance'. Unfortunately, there are significant limitations to both approaches to ICT governance.

The purpose of this paper is to define strategies, which professional organisations can use to meet their responsibilities to the ICT profession and the ICT professional; strategies that move toward regulation without curtailing ICT's potential with ineffective sanctions. Professional organisations also need strategies for reducing negative incidents and for improving professional responsibility without the introduction of sanctions that apply only to practitioners who happen to be members of that organisation. There are ways in which the toothless tiger(s) can have a significant positive influence.

Outline

ICT has been with us for many years and, in the past 10 years, there has been a growing interest in ICT governance as a means of reducing information system disasters. National organisations have been formed, professional organisations have organised subcommittees to address ICT governance, and ICT has even been

---

1   A version of this chapter was published in *The Australasian Journal of Information Systems*, 2009, vol 16, pp 165–84. Published here with permission.

called a 'discipline'. The unwrapping of the concept of ICT governance results in inconsistent interpretations and ineffective implementations in industry. In some cases, the concept has been modified to meet a particular sectors needs. This broadening of concepts to fit individual needs is not new, and is sometimes useful. There is, however, a fundamental mistake in the narrowness of most interpretations of ICT governance, which make it less likely that it will achieve its ultimate goals. I believe this mistake can and should be addressed by professional computing organisations. In what follows, I will examine the various approaches to ICT governance, the difficulty it tries to address and I will argue for what I consider its critical limitations. I will then show how professional organisations can address the weakness of ICT governance using tools that they already have at hand.

The problem from the software perspective

In the early days of computing, the 1960s, people worried about 'the software crisis' — a term coined in 1968 — or the failure of software systems. Software workers addressed this possibility by developing models for building well-engineered software. The focus of the computing community was primarily internal; focusing on how to develop and test a program. This was the period during which mathematical modelling of software development was published in books on The Elements of Software Science (Halstead, 1977) and The Discipline of Programming (Dijkstra, 1976). The focus was on making computing a reliable engineering-like discipline and the impacts and concerns addressed were local to the particular system being developed.

The response to the perceived 'software crisis' generated and continues to generate many single-mode solutions that suggest undertaking one particular process will solve all of the perceived problemshe particular single mode solutions trie shifted from emulating an engineering approach to the development of software, then to a structured approach to program design, to a formal proof of software system requirements, then to object-oriented software development, a focus on individual programmers counting the number of errors they make ('Personal Software Process') and, now, an agile or extreme programming approach to software development. These single-mode methods have been regularly interspersed with approaches that emphasise measuring software's size, reliability, and space and time efficiency. The indication of the lack of success of these approaches is the rise of ICT governance to address the negative impacts of the software crisis on industry.

I believe there were three major difficulties with the approaches adopted by the software community. First, the software crisis is a complex problem and single-mode approaches overemphasise one piece of the problem and tend to ignore other pieces of the problem. Second, computing technology and applications are constantly changing, and changing at a rapid rate. This means the domain of the software crisis is also constantly changing; developments, such as computerised robotic surgery, were not issues of concern in the 1970s. E-commerce had no meaning 15 years ago. The software crisis is still a problem with software development, but the software being developed has expanded applications and the relevant stakeholder communities have increased correspondingly. The third problem is that the solution to the problem of software interaction with business and society has been addressed only from the software side of the problem. These single-mode approaches are focused on what software developers do. Since 1968, the answer has been the same, worded differently — and the mistake is the same. If I am an honest person and work hard, following a good process, then the problems will go away. Unfortunately this has consistently not worked. There are still significant systems failures, which lead to major corporate failures. It took many years, for example, for the world economy to recover from the negative effect of programmed trading on the stock market in October 1987.

There are numerous standards organisations, such as the American National Standards Institute (ANSI), the Institute of Electrical and Electronics Engineers Inc. (IEEE), the International Organization for Standardization (ISO), and Standards Australia, which have worked on developing rules and standards to control and monitor software developments. The diverse and developing nature of ICT, however, makes the application of these rules to software development difficult.

- Since there is no required common education program for ICT workers, many have not been taught the standards/rules.
- Since there are few sanctions and minimal oversight, and no threat of losing an ICT license, obeying the rules is voluntary and it is not always clear what is to be gained by following them.
- Since the ICT environment is changing so rapidly, rules either no longer apply by the time they are written, or they are inconsistent because they address different sides of an issue. Who, for example, enforces fraud legislation on international internet sales?

The solution to solving the software crisis has been seen as internal and focused on the technology and on how to be a good technician; the software developers were going to fix it. Another manifestation of these single–mode, internally focused approaches was a method of software development called 'over the wall development'. At its worst, such development had the following scenario. The skilled software technician would gather requirements for a business software project and then develop a complete system without any further consultation with the client. The system developed (as the software person misunderstood it) would be presented to the customer (thrown over the wall). The assumption being that the technician had, somehow, understood all of the important business issues and addressed them in the software. This systemic disconnection of developer and customer led to the delivery of working systems that had nothing to do with the needs of the customer. Since corporate executives did not fully understand how ICT worked, when a corporate manager complained about the style or function of the delivered system, ICT personnel could simply declare that what they delivered was the only way the computer could do the requested task.

This problem is not unique to software development, but occurs anywhere there are discrete project elements, and no communication between those working on different parts of a project. There is a disconnection between the world of the computer-astute developer and the customer who is skilled in their own domain. The two parties speak different languages. The programmer, who understands the syntax of a programming language, is overwhelmed when requested to write a system to model genetic mutations that calculates the sequential effects of genetic drift, inbreeding, selection, gene flow, and mutation upon the proportion of a population's gene pool comprised of the second of two alleles. To think that the difficulty of developing this project could be resolved by focusing on software technology is a mistake.

History repeats itself

In the development of ICT governance, I think history is repeating itself. A similar set of mistakes is being made in the way in which the industry is attempting to address the issues of ICT governance as those that it made in facing the problems that confronted ICT systems. Many areas in ICT governance are taking a single-mode approach, which focuses primarily on the industry's side of the problem, on the business element; thereby minimising the ICT elements and giving no consideration to the rapid change of the domains of ICT and business.

ICT governance

In general, ICT governance emerged as an attempt by business to deal with the impact of major software system failures on business. Although it was primarily a part of corporate governance, which focused on ICT, many ICT professional organisations contributed to the process of developing standards. These organisations advocated that their members adhere to these standards.

In 2003 Australia published several standards for corporate governance including: 'Good governance principles' (AS8000), 'Fraud and corruption control' (AS8001), 'Organisational codes of conduct' (AS8002), 'Corporate social responsibility' (AS8003), 'Whistle-blower protection programs' (AS8004), thus setting the pattern for a document from the corporate side addressing ICT governance. Standards, such as AS8015 (2005), were developed to deal with issues of in-house development and the fact that outsourcing to benefit the vendor's interests was not always consistent with those of a user organisation. This is outsourcing problem is similar to one of the problems of over-the-wall software development.

The interest in ICT governance is international and quality development from the corporate side is defined in the Information Technology Infrastructure Library <http://www.itil.co.uk/>; a detailed framework with hands-on information on how to achieve a successful governance of IT. As in the attempt to deal with the software crisis described above, there is an ISO standard to deal with corporate governance. 'Because inadequate information technology (IT) systems can hinder the performance and competitiveness of organisations or expose them to the risk of not complying with legislation, the new ISO/IEC 38500 standard provides broad guidance on the role of top management in relation to the corporate governance of IT.' <http://www.38500.org/>

As the multitude of standards illustrates, there is no single standard of ICT governance, and nor is there a consistent approach to the appropriate stage at which to implement ICT governance principles. The material on ICT governance is not consistent in describing the level where ICT governance comes into play. One should not be misled to think that there is a single standard of ICT governance. On the other hand, Peter Weill and Jeanne Ross p.14 (2004), in a study of 300 enterprises around the world, 'did not identify a single best formula for governing IT'. They say that 'IT governance specifies accountabilities for IT-related business governance and helps companies align their IT investments with their business priorities.' IT governance performance for them involves the corporation's 'Cost effective use of IT, IT for growth, IT for asset utilisation and, IT for business flexibility.' They argue that 'IT governance is the decision rights and accountability framework for encouraging desirable behavior in the use of IT.' This is very different from viewing IT governance as managing the interface between ICT development and corporate management.

Sense of 'control'

But, if we follow the Australian definition of ICT governance, it uses evaluation and control but it takes many forms, both in what is controlled — ICT systems development — and in the current and future corporate use of ICT. This same diversity exists in the standards being developed by the ISO. There is also a difference in what is meant by 'control'. Control is characterised both as rigorous highly delineated control, and general directional guidance principles to help the professional make judgements.

Single mode

As in software development, ICT governance takes a single-mode-solution approach, and there are a variety of single modes. There are also 'structuralists' who think everything is handled by structure and that the primary goals for ICT governance are to ensure that ICT generates business 'value' and reduces the risks that are associated with ICT by implementing a carefully defined organisational structure. This is sometimes connected with who is in charge or has ownership of the system; again, an over-the-wall problem.

Over the wall — it is mine!

The definition of ICT governance is tied to corporate governance and relates the business focus of an organisation to ICT management. It mandates that ICT decisions are owned by the corporate board, rather than by ICT managers. This results in the same problems as over-the-wall development. There are indeed limitations to what ICT can do and the ways in which it does things will have different effects. Balanced ICT governance needs the ICT side in their systems guided by ICT workers.

Limited view of stakeholders

Several of the problems associated with software development are recurring in ICT Governance. There is an additional problem which is common to the each of the issues of software development, namely, there is a limited view of who constitutes the stakeholders in a project. Because the view of who constitutes the stakeholders in a project is limited to the developer and the customer, the effect of an IT system on a business or on extended stakeholders is not considered.

The current concept of ICT governance is modelled on the traditional concepts of business ethics regarding who needs to be considered as the relevant stakeholders, namely those who have some financial interest in the business (Agle et al, 1999). The current concept of ICT governance stakeholders is 'IT governance implies a system in which all stakeholders, including the board, internal customers, and in particular departments such as finance, have the necessary input into the decision making process'. This view of the system context and the stakeholders in that context is also supported by ISO standard 38500 "ICT Governance which characterises ICT governance as the management system used by directors. 'IT governance is about the stewardship of IT resources on behalf of the stakeholders who expect a return from their investment.'

The briefing paper on a recent survey of ethical issues conducted by the Centre for Applied Philosophy and Public Ethics (CAPPE) (Lucas, 2008) the limited understanding of ICT workers about stakeholders is clearly indicated when most of them say they do not consider their work to be related to larger segments of society.

The concept of ICT governance is closely tied to the concept of control. The use of words like 'control' and 'govern' imply enforcement of rules or sanctions for failure to follow them, but, in ICT professional societies, these rules seem like toothless tigers. There are no real sanctions for failing to follow the rules. Professional societies can have a significant roar without any associated bite.

We see the attitudes of many computer practitioners toward attempts to promulgate rules of behaviour clearly in Lucas (2008). A member of the Australian Computer Society (ACS) complained about the rules because they wanted them to apply to everyone practicing ICT. 'The ACS can discipline its members for breaches of its ethical code but that is no barrier to employment and it has no effect on the vast majority of workers in the industry who are not ACS members' (Lucas, 2008).

This failure of universal application of principles and regulations is a common complaint of honest, hard-working computer practitioners. It should be noted that these are not complaints about the importance of following such rules. Rather, they are complaints about the fairness of their being held accountable, as members of the ACS: 'Why should I follow the rules if those outside the ACS are not bound by them?'. Another motivation for the complaint is that the practitioners realise that following these standards will reduce some of the harms caused by software and improve the lot of humanity and, as such, everyone ought to follow the rules. The same need to 'enforce' compliance is perceived in ICT governance. The primary motivator for following the rules is the addition of teeth — sanctions — for not following the rules. As we have seen this is done with some compliance standards like Sabanes-Oxley and ISO Standards.

The limitation of the ICT governance approach

ICT governance is repeating some of the same mistakes made by software developers in their attempts to address the software crisis and I believe ICT governance is heading for a similar set of problems as those faced by software developers. The development of systems software has to develop an interface between the technology of computing and the nature of the enterprise. The nature of this interface must be guided by a consideration of the impact of the system on a broad range of stakeholders. Software developers focus on the nature of the software and how to reduce errors in the programs. They focus on a limited set of stakeholders in the system: developer, customer, sponsor, and vendor, and pay limited attention to those who will be impacted by the deployment of the software. ICT governance, likewise, has an internal focus on the business side of the software system and a narrow a view of the stakeholders as those with a financial interest in the system (Agle et al, 1999; Weill & Ross, 2005). We can see some of the consequences of this narrow approach by looking at the ICT treatment of 'software risk'.

The narrow stakeholder focus in software risk

Although the need for high quality software is obvious to all, despite efforts to achieve such quality, information systems are frequently plagued by problems (Ravichandran, 2000). A narrow approach to risk analysis and understanding the scope of a software project and information systems has contributed to significant software failures.

Informaticians have been evolving and refining techniques to moderate the risk of developing software products that do not meet the needs of clients. The risks include: missed schedule, going over budget, and failing to meet the system's specified requirements. In spite of this attention to risks, a high percentage of software systems are delivered late, over budget, and do not meet requirements, leading to software development still being characterized as a "software crisis" and leading to a general mistrust of software systems.

Risk management generally consists of an iterative series of steps, similar to the ones shown in Figure 1.

The context referred to in the top box—the context in which the project is being developed—includes the organisational structure, and its competitive and political position, as well as its risk management structure.

The risk identification process identifies potential negative impact on the project and its stakeholders. AS/NZS 4360-1999 lists potential negative areas of impact such as

> Asset and resource base of the organisation, Revenue and entitlements, Costs, Performance, Timing and schedule of activities, and Organisational behaviour.' (AS/NZS, 1999: 39)

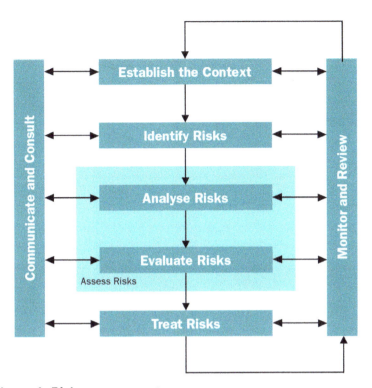

Figure 1: Risk management

Source: AS/NZS, 1999, 16

The risk analysis process divides the identified risks by their severity and the likelihood that they will occur, producing a given level of risk. This level of risk is generally determined using statistical analysis or calculations with fault trees and event trees. A typical calculation is 'Risk exposure', a metric derived by multiplying the anticipated costs by the probability of the event occurring.

Two forms of exposure are commonly calculated. The first method, using quantitative risk analysis, provides quantitatively expressed assessment of the negative consequences of an event as the outcome of an event; for example, 'A delay of one day will cost $3000 in sales'. The second method, qualitative risk analysis, is often used to address risks which are not readily quantifiable, other than by describing the broad degree of risk; for example, 'The delay will upset our distributors causing significant loss of goodwill'. Generally, 'qualitative analysis is often used first to obtain a general indication of the level of risk ... or where the level of risk does not justify the time and effort for a quantitative analysis ...' (AS/NZS, 1999: 14). Its primary role is to characterise and identify the impact of a risk generally asserted in terms of dollars.

Surprisingly, in standard risk methodologies, the qualitative risk approach typically looks at quantifiable data, which can be easily prioritised and facilitates analysis. These standard methods of risk identification and risk characterisation have been used extensively in software projects.

The Association for Information Systems (AIS) defines 'system quality' in terms of currency, response time, turnaround time, data accuracy, reliability, completeness, system flexibility and ease of use (AIS, 2005). Even after using these generic models of risk analysis, information systems have been produced which have significant and negative social and ethical impacts. The risks of these impacts are not traditionally included in the tripartite concept of software failure — over budget, late, or not meeting stated functions. The extended stakeholders in developed software are all those who are affected by it, even though they are not directly related to the use or financing of a system. The political candidate who is not elected because of a difficult voting machine interface is a stakeholder in the development of that voting machine. The person who suffers identity theft because of a flaw in the security for an information system is a stakeholder in that system. The developer's obligations to these stakeholders are not included in the generic concept of software failure.

These systems may have been a success in terms of being developed on schedule, within budget, and delivered on schedule, but were a failure because they failed to take into account the conditions in which they were used. The user interface, which met specifications, had a significant impact on the lives of others. For example, the system that was used to record dosages of paediatric medicine correctly handles negative interactions of dosages, but was awkward to use in emergency situations, resulting in three medication errors out of every 100. (Walsh, 2006)

Contributing factors

Two interrelated factors related to system stakeholders contribute to these professional and ethical failures being overlooked. First, limiting the consideration of system stakeholders to just the customer/client, software developer and those who have a financial stake in the system ignores the needs of other relevant stakeholders.

Some have realised that the focus on technical risks is too narrow, but, unfortunately, the risk focus only expands to other internal issues that are related to the development of the system. For example, Thiagarajan Ravichandran writes 'Research in software quality has focused largely on the technical aspects of quality improvement, while limited attention has been paid to the organisational and socio-behavioural aspects of quality management' (p. 119, 2000).

A second factor arises from limiting the scope of software risk analysis just to technical and cost issues. A complete software development process requires 1) the identification of all relevant stakeholders and 2) enlarging risk analysis to include social, political, and ethical issues. A complete risk analysis requires a process to help identify the relevant stakeholders and broaden the scope of risks anticipated.

To meet the goal of quality software, developers focus on the particular risks including that they perceive as a threat to a project, such as budgets, timelines and suitability of the product. This focus may mean that other critical aspects of the product, such as the use of easy-to-read fonts or back-up systems, are not given adequate consideration.. Nevertheless developers use Risk Exposure to help them focus on the most critical risks. The use of easy to read fonts or an easy to use back up system may be ignored in an effort to get a product out in time or produced at a lower cost.

The risks that are addressed are those with the highest Risk Exposure. All consequences are given dollar values. Even qualitative risks are turned into a numerical hierarchy. The resulting risk of the September 11th disaster was calculated in terms of the number of deaths that occurred on that day or lifetime dollar earnings potential of those who died.

The negative effects that need to be addressed in risk analysis include both overt harm and the denial or reduction of goods. An automated surgical system which randomly moved inches instead of centimetres, thus hurting patients, would have a negative effect; just as a pay-phone system which disabled all usage, including emergency numbers, without an approved credit card, would also have a negative effect. These stakeholders, patient and someone hurt in a fire, are not normally considered.

The scope of a project must identify all stakeholders to eliminate the possibility of negative effects.

This extended domain of stakeholders includes: users of the system, families of the users, social institutions which may be radically altered by the introduction of the software, the natural environment, social communities, informatics professionals, employees of the development organisation and the development organisation itself.The design of many of the USA's voting machines correctly counted votes but made it difficult for people to enter their votes.

Modifying the approach to stakeholders

The response of software developers to the ICT crises was internal and focused on the technology and how to be a good technician. Analogously, ICT governance is going through a similar, though much shorter, life cycle. As initially developed, ICT governance focuses on the governance within an organisation; 'evaluating and directing the plans for the use of ICT to support the organization and monitoring this use to achieve plans. It includes the strategy and policies for using ICT within an organization' (AS8015, 2005). The focus is clearly internal with statements like "to better understand their obligations and work more effectively to maximise the return and minimise the risks to the organisation from ICT" The internal focus suffers from the same problems as 'over the wall' software development.

The role of professional organisations

Professional organisations have been involved in the development of these ICT governance standards. For example the ACS was involved in the development of AS8015 and its narrow approach. Both ICT and the business sector have contributed what they view as solutions from their own sectors — object-oriented software design and financial flexibility of corporate systems — which model single-mode solutions. They have difficulty enforcing these solutions and worry about system ownership. They are each concerned that their side of the wall has control of the system and, of course, they also differ as to which stakeholders primarily need to be considered when developing these systems.

The professional organisations have the ability to resolve this. One of the roles a professional organisation needs to play in resolving these problems is to broaden the range of stakeholders considered in the current concept of ICT governance. ICT professional organisations have codes of ethics and codes of practice that address each of the ICT governance problems discussed above. Their codes of ethics require the ICT professional to consider a broad range of stakeholders, including all those whose lives are affected by the ICT project and the way in which it was implemented. The code of ethics is very useful as a model of governance but is ignored because these codes are viewed as less enforceable than other types of regulations. They are considered worthless because they like toothless tigers.

Professional computing codes of ethics can address the ICT governance problem

To see how codes can be useful we need to look at some of the general functions of codes and how they are managed in professional organisations.

Functions of codes

Codes of ethics are developed by professional organisations for a variety of reasons. They serve a variety of functions and are frequently directed at several audiences. At the simplest level, a code of ethics is a statement of the obligations of individual computing professionals in the conduct of their profession. The code will generally embody a moral commitment of service to the public. Sometimes they are used to clarify expectations and appropriate behaviour of professionals. More positive functions of codes of ethics include: making a statement to members or aspiring members of the profession about shared commitments and agreed upon rules, sensitising members to new issues, and providing guidance to individual members when they are confronted with ethical situations.

Codes are also used to win public confidence and stave off external regulation. Some codes are disciplinary in order to convey a sense of self-regulation by the profession. They also help set expectations for employers and clients about dealing with members of the profession and socialise novices in field. Codes can express and strengthen the community orientation of the group. The general nature of some codes makes it difficult for the general practitioner to apply their directives in concrete cases.

Some confusion about codes of ethics arises from a failure to distinguish between closely related concepts about codes which direct the behaviour of practicing professions. The less restrictive codes can be primarily aspirational, in that they provide a mission statement for the profession. There are also codes of conduct which describe professional attitudes and some professional behaviour. Codes of practice are specific and closely tied to the practice of the profession. They are the easiest to use as a basis for legal action. Because practicing professionals deal with human affairs, the underlying ethical principles are the same across professions. Studies have shown that most codes are a hybrid of these three types of code (Berleur, 1994).

The goals of a code of ethics could be ranked from the most benefit to society to most benefit to the individual member of the profession. These goals include:

1. Inspiration and guidance for ethical conduct.

2. Support for those seeking to act ethically by appealing to the public system of ethics established in a code.

3. Education and shared understanding (by the professional and the public) of standards of practice.

4. Deterrence and discipline for specific actions by sanctioning code violations.

5. Protection of the status quo by stifling dissent and state only minimal levels of ethical conduct.

6. Promotion of business interests by forbidding competitive bidding.

One way to evaluate codes of ethics is to examine which of these functions assumes prominence in the code.

Codes also indirectly educate the public at large about what professionals consider to be a minimally acceptable ethical practice in that field, even as practiced by nonprofessionals. The Swiss Information Society Code of EthicsComputer Code (SIS Code 2005) suggests that the responsibility of a national and/or professional society is to be in charge of making the public aware of the society's guidelines It also advocates regularly publishing information about code violations as a means of informing the public about what is to be expected of a computing professional. Some codes also include the responsibilities of the ICT profession itself.

Two common problems for codes in computing are that they need to be able to address a rapidly changing environment and there are difficulties in enforcing them. Turning a code into law makes it static and eliminates some of the other important functions of codes of ethics.

Recent codes have become more specific about ICT workers' responsibility to society and a broad range of stakeholders (Gotterbarn, 1996). The Canadian Information Processing Society (CIPS) code of ethics and standards of conduct addresses a diverse audience. The CIPS code has imperatives for six audiences: colleagues, clients, students, the public, myself, and the employer and management. By separating the client and the employer it avoids the possibility that the interests of the client and the employer may not be identical. This code starts from the belief that a set of ethical obligations — professional ethics — is in part based on the high social impact of the profession; because of the broad and significant impacts of computing the computing professional owes a higher order of care to their clients. Because of the nature and impact of computing, a higher level of care is required. Consistent with this, many codes advocate the avoidance of negative consequences of professional activities. In the Association of Computing Machinery (ACM) code there are general statements dealing with responsibilities in the event of negative consequences. For example, section 1.2, which deals with the responsibility for negative consequences, states that a person is obligated to undo or mitigate negative consequences as much as possible. This is clearly a shift from earlier codes, which were designed to protect the computing professional. These, and other sections of the ACM code, are designed to protect society. Some codes limit corrective responsibility to merely fixing one's own mistakes. In the ACM code, however, even if the negative consequence were the fault of the customer's incorrect use of a product, the member is still responsible. The code first protects society and then the professional on the basis that the development of a computer system requires a consideration of all stakeholder's rights. For example, section 3.5 states

> *Articulate and support policies that protect the dignity of users and others affected by a computing system.* Designing or implementing systems that deliberately or inadvertently demean individuals or groups is ethically unacceptable. Computer professionals who are in decision making positions should verify that systems are designed and implemented to protect personal privacy and enhance human dignity.

The ACM and the IEEE have adopted a 'Software engineering code of ethics and professional practice' (SECEPP). This code has been adopted by numerous ICT professional organisations and it clearly points to the developer's responsibility to a range of stakeholders. The Preamble states

> These obligations are founded in the software engineer's humanity, in special care owed to people affected by the work of software engineers, and in the unique elements of the practice of software engineering. …
> In all these judgments concern for the health, safety and welfare of the public is primary; that is, the 'Public Interest' is central to this Code.

But, as we have seen, simply mentioning a broad range of ethical obligations does not satisfy those who want to see codes enforced by severe sanctions. A Code such a SECEPP can still be enforced even if the Code does not include mention of specifice sanctions. A large aerospace firm enforced its ethics regulations and fired 28 software engineers for violating the ethics policy. The enforcement of the policy struck many as toothless since the fired software engineers will easily get other jobs and the firm and their staff suffered because of a staff shortage. It sometime looks as if professional organisations actually do not want to enforce the code. There is a problem that some codes seem vague and are neither easily enforceable nor useful in making decisions. There is the already mentioned problem of jurisdiction only over an organisations membership. How can you enforce a code on people who are not members of your organisation and have not agreed to be bound by the code? Professional organisations can leave their codes as hollow statements or do something to help them have an impact.

Two major difficulties with codes for professional ethics committees

It is useful to trace the development of the ACM's code of ethics because it is similar to the pattern occurring in the development of most codes and it maps a difficulty with ICT governance. The two major themes running through the code's development are the questions concerning enforcement — by whom and how — and currency — how to address changing technology (Berleur, 2008).

The early years: Ethical standards in search of identity

On 11 November 1966, the ACM adopted an ethics standard. They adopted a set of guidelines called 'Professional conduct in information processing' (Parker, 1968). The pattern of concerns and development of this ethics standard is similar to the patterns of many such developments. The 1966 discussion revolved around questions of: whether information processing was really a discipline, whether it was a single discipline that could have a single standard, what types of effective enforcement it could have, was it meaningful to merely expel miscreants from membership in the ACM, who would determine when to enforce the standard and, a narrowly US-centric concern, that the enforcement of a professional standard by the ACM might alter the ACM's tax status as a scientific society. At least two trends from this early approach continue through the ACM's development of ethical standards: there is recognition that the rapid and unanticipated changes in the profession will require modification of the ethical standards at some level and that agreement on enforcement is difficult to manage.

The 1966 standard handled both the enforcement and the change issue in the same way. The recognition of change was addressed by calling the first ethics document a 'guideline'. The label was also intended to address the enforcement issue — 'the ACM Council has wisely adopted ethical rules as a guide to members rather than a code to be enforced'. As a result, there is no enforcement function directly related to the code/guidelines. Approaches to the issue of change have been constant while there has been a significant change in the approaches to enforcement. The society's means of addressing these issues of enforcement and technological change dictate the role of the ACM's committees related to ethics.

1970–92: From guidelines to standards and the need for enforcement

Just four years later, a change was made to article 3, section 4 of the ACM constitution that stated 'demonstrating a lack of integrity' was a reason for being 'admonished, suspended, or expelled' and gave authority to the ACM council to impose these sanctions. The amendment also mandated the development of a code of professional ethics. This led to the development of a code with detailed ethical statements, the violation of which is easier to determine and which are easier to enforce.

There was an obvious flaw in the implementation of this approach: in order to obey the rules, you need to know what they are. This education problem is not unique to the ACM. According to Lucas (2008), more than 30 per cent of those questioned are unaware of the ACS code. Perhaps it is education that needs to be emphasised, more than enforcement.

There is also a conceptual problem with the approach. A basic problem with a precise checklist-approach to ethics arises particularly when it is applied to a technical field. The computing field changes and advances very rapidly. What was considered best practice 20 years ago may be dangerous now. Early medical practices are now considered very dangerous. A precise ethics checklist, which is easy to enforce, is out of date almost as soon as it is off the press and what it advocates may be inconsistent with current best practice.

The ACM code was adopted in 1974, but an enforcement procedure was not approved until 1978 (Smoot, 1981), reflecting the continuing uneasiness about enforcement (Perspective, 1981). During this time, the ACM's committee on 'Professional standards and practices' was responsible for services to individual ACM members who faced ethical problems such as whistleblowing, product reliability and safety issues, and employment problems. The adoption of the 1974 code, and later adoption of a policy designating the enforcement method as the sole responsibility of ACM council, was a significant change from the ACM's original ethics guideline.

Addressing and enforcing ethical issues related to ACM members was now the sole responsibility of the governing organisation of the ACM — the ACM Council. According to the ACM constitution article 6 section 8 'a member may be admonished, suspended or expelled for demonstrating lack of integrity' by a three fourths vote of Council after a hearing'. This still left open the nature of the hearing.

1990: The emphasis on guidelines returns

In 1990, adopting the insights from 1966 that ethical guidelines for computing need to change to address unanticipated changes in the profession, Ron Anderson proposed that a code like the current ACM code, which had a structure that listed possible ethical violations, needs to be revised. He further argued that the 'ACM needs a revised organisational structure for an ongoing review, reformulation, interpretation, and application of its Code of Ethics and Professional Conduct'(1990). The ACM council supported this request and, two years later, on 16 October 1992, a code of ethics and a suggestion for a revised review standard was presented.

This 1992 code was developed over a two-year period, during which there were multiple drafts and reviews by ACM members. The ACM approved a new code of ethics which de-emphasised enforcement and emphasised education of members, of prospective members and of the public. The code's use as an aid to decision-making was also emphasised. The code, which is still in use today, has a two-level structure. It consists of 24 ethical imperatives each of which has an associated guideline illustrating the application of the imperative in computing. The imperatives are divided into four sections. The first section gives a set of general moral considerations, the second identifies additional ethical principles which apply to computing professionals, the third section pertains to organisational leaders, and the final section deals with issues of general compliance with the code.

It was envisioned that the high level imperatives would be constant and that the lower level clauses would require updating when technology and practices changed. In addition to the approval of the code, a committee on professional ethics (COPE) was also established to meet the need for revision and nurturing of the code. The charge for the COPE committee was to 1) promote ethical conduct among computing professionals by publicising the code of ethics and by offering suggested interpretations of the code; 2) plan and review activities to educate the membership in ethical decision making on issues of professional conduct; and, 3) review and recommend updates, as necessary, to the code of ethics and professional conduct and its guidelines.

These changes should have addressed the 1966 concerns about enforcement and code revision. The nature of the code, emphasising voluntary compliance and consisting of aspirational and normative imperatives rather than disciplinary imperatives helped reduce the concern about sanctions and enforcement. The structure of the code, with fairly constant imperatives and flexible guidelines, helped address the state of flux within the computing profession.

COPE was a committee with a charge but without a structure. Many of the items described above were not addressed in the establishment of COPE. The president of the ACM appointed the chair of COPE, who determines its structure and guides its activities. The outreach functions of COPE are straightforward and modelled to some extent on the work outlined when the code was first passed. COPE members present papers and participate in computer ethics workshops. They write articles that offer interpretations of the code (Miller, 2003). Other professional societies, the ACS and German Gesellschaft für Informatik, for example, have used the original case studies developed when the 1992 ACM code was passed (Anderson, 1993). COPE is currently working on a specific set of examples related to internet issues because the World Wide Web only achieved prominence after the general code was approved in 1992.

The role of COPE extended in 1999 when the ACM and the IEEE–Computer Society jointly developed and adopted the SECEPP as a standard for a sub-specialisation of computing. COPE's domain now includes both of these codes.

COPE has helped with the translations of the ACM's codes by professional organisations that want to adopt them. The SECEPP has so far has been translated into nine languages.

Other computing organisations have adopted SECEPP. For example, in September 2006, the Association of Software Testing resolved to adopt the ACM code of ethics as a series of principles to guide and govern practice among its membership. The ACS has also adopted the SECEPP (ACS, 2004).

In meeting its charge, COPE is also involved in the design of posters of the code of ethics, which are distributed to member organisations, design of web pages, and includes a commitment to the ACM as a separate item on membership renewal forms. COPE has primarily limited its education function to the membership of the ACM and has only reached out in terms of getting the code included in appropriate textbooks and conducing workshops a computing conferences.

The original charge to COPE is merely a starting point. The absence of a fixed structure, including a regular schedule of meetings, has led COPE to function in response to external requests. The committee is one of the ethical focal points within the ACM. COPE members are asked to review many of the ethics articles submitted to the *Communications of the ACM*. It also responds to ethics complaints that are forwarded to it by ACM headquarters. These complaints vary from the trivial to very significant, such as the development of a plagiarism policy that is consistent with the ACM's codes of ethics. In many cases, committee members are not knowledgeable in the domain of the ethics problem and need to bring in other committees who have a better understanding of the situation.

The absence of a charge which involves COPE in all such issues means that on occasion some very significant complaints and ethical issues do not have their ethical component addressed adequately. For example, there was a significant issue raised by one ACM member on his website — regarding religious discrimination — that was never brought to the attention of COPE (Complaint, 2003). One of the reasons for the omission is that COPE only serves in an advisory role.

Without a clearly defined structure, it is sometimes difficult for COPE to achieve its goals. Another problem is the separation of the primarily proactive and advisory functions given to COPE, and the enforcement functions that are given to the ACM council. Often, this structure contributes to a perception that the role of COPE is less important.

On one hand, the lack of a clearly defined structure makes it difficult at times to achieve its goals. On the other hand, the absence of a defined structure has the virtue that, when an unanticipated issue arises, such as the relation of the code to a plagiarism policy, COPE can be involved in those situations without having to wait for a formal meeting.

Every professional organisation ought to have an ethics committee for the promotion of the code of ethics. There are several things required for such a committee to function effectively:

1. When an ethics committee is established both its charges and structure should be specified.

2. All ethics issues should be passed through the ethics committee.

3. The method of updating a code of ethics needs to be clearly defined by either the national/professional society or by its ethics committee. This method should be as rigorous and cautiously entered into as the original creation of the code.

4. The structure of the ethics committee should not limit the issues it can address.

5. Ethics committees should have a regular venue in the society's publications to help promote a proactive approach to ethics. This should be accompanied by an annual ethics award that is included in a national/professional society's repertoire of awards.

The ethics committee and ICT governance and the software crisis

An ethics committee promoting the code of ethics amongst the ICT community is helpful in addressing the three common difficulties identified in ICT governance, and in addressing the software crisis. A major role of the ethics committee is public education. The codes of ethics focus on the professional's responsibility to a broad range of stakeholders. This broad focus lessens the significance of the controlling influence of the discussion of 'who owns the system' and emphasises the question of the system's consequences, for all concerned. The breadth of concern makes clear that single-mode solutions are inadequate for systems with broad stakeholder responsibility. A code advocates a quality of action and not a particular technology. A constant reminder of the social and ethical impacts of ICT systems makes clear that simply focusing on inline documentation in a software program is an irresponsible approach to building something with the impact of an electronic voting system.

But codes have no teeth

Even though the concerns about the constant flux of the computing profession and enforcement of a code's imperatives have been addressed by modifications to the structures of the codes of ethics, there remains the underlying concern of how to get everyone to follow the code. Codes are not the teeth of an organisation. They do not contain the due process and sanctions within an organisation, but they do describe the conscience of a profession.

Within organisations generally there is limited enforcement of a code of ethics. Codes get some teeth when they are used by a professional organisation to make decisions. The ACM code of ethics, for example, was used in developing the ACM policy and procedures on plagiarism based on the imperative that 'Respecting intellectual property rights is a foundational principle of the ACM's Codes of Ethics'. The ACM publications board defined the procedure for reporting alleged plagiarism, for investigating the allegation, and managing confidentiality during the investigation. If the offending paper has already been published, the 'ACM will post a Notice of Plagiarism based on the investigation on the ACM Digital Library's citation page of the plagiarising paper and will remove access to the full text.' This response gives teeth to the charge of plagiarism.

Codes of ethics do not have teeth and they do not define the disciplinary action for a code violation. Codes are not self-referential: organisations have bylaws and the code is a bylaw. The due process and sanctions for violating the code is defined outside the code. Codes are the mind and conscience of a profession. The profession is what nutures the code and gives the code teeth.

The original concerns of the ACS member about those outside the ACS not being bound by the standards remain. In a recent case, major sections of a student's masters thesis were copied and submitted to a conference where the original author's supervisor was present. The conference was not an ACM-sponsored event and the plagiariser was not an ACM member, so ACM processes for dealing with plagiarism did not apply. But, that does not prevent the ethics committee from informing the reviewer and the conference about how the ACM understands professional standards and the action they would take with members.

Codes of ethics impose accountability on the professional organisation

A code of ethics is like a Swiss Army knife, serving many important and useful functions. It is a statement to members about the ethical stand of an organisation and profession, a conscience of the profession, an announcement to non-members of what the profession standards for (although most often stated in terms of the actions of individuals), it imposes functions on an ethics committee to educate the membership, and it imposes responsibilities on the professional organisation itself.

One of the functions of an ethics committee is to help its own professional organisation understand the importance of the role played by ethical standards. Sometimes, professional organisations lose sight of these responsibilities or get distracted. In 1972 the IEEE set up an ethics task force in response to the fireing of engineers reporting ethical problems in the development of the Bay Area Rapid Tansit system. Because the existence of the committee was not publicised, no case was referred to it until 1978 when, following advice of its existence being sent to the membership, it received notice of 11 cases to address. In 1990 the IEEE set up hotlines and sent copies of the code out with membership renewals. With each renewal, ACM the members agree to adhere to the ACM code. The IEEE hotline provided a direct channel for IEEE members to get help when they were faced with painful professional dilemmas. Unfortunately, and for a variety of reasons, the hotline was discontinued.

In 2003 the ACS established the committee on computer ethics (CCE) to promote the development of computer ethics policies (in Australia). The committee is charged with working with ACS Special Interest Groups, to help develop policies for government, and to promote the importance of computer ethics in the ICT community.

Micro-macro-ethics confusion

The distinction between macro and micro ethics is important for these committees and for the use of codes of ethics in addressing the ICT governance/ software crisis problem. The lack of attention to this distinction is a problem that pervades many codes of ethics, ethics committees and professional organisations. Generally, they all focus on the responsibility of the individual computing professional.

The general view is that codes and regulations are about the behaviour of the individual member and not about the organisation, or the profession as a whole. Micro-focused codes of ethics talk about 'You' and not specifically about the professional organisation or about the profession.

Complainants about an absence of sanctions tend to view codes as being primarily about the ICT person's behaviour. Of public sector workers in Australia, 23 per cent did not see their work as related to a larger whole (Lucas, 2008). It is the ICT individual who is asked to be ethical.

This narrow understanding of the scope of a code of ethics and professional standards is a problem and affects their work. Individuals who 'view themselves as NOT part of other systems but as separate ICT folk' (Lucas, 2008: 23) are taking a an exteme form of micro-ethics approach where enforcement of a narrow set of rules is primary and they lose sight of the positive contributions of ethical development, lose sight of the contribution 'doing it right' makes to the quality of life.

Unfortunately, few codes have sections dealing with macro ethics. Sometimes the need for a macro consideration is not clear in an ethics committee's terms of reference. The terms of reference for the ACS Committee on Compuiter Ethics is correct in asserting that it should promote 'the value and importance of Computer Ethics within the wider Australian, as well as regional and international, ICT community'. The standards are not just for ACS members, but for the ICT community at large. They also recognise the responsibility to society as a whole in the charge 'to advise the Society and the ICT community on 'best practice' in relation to Computer Ethics;'

This awareness and promotion of the responsibility to the whole profession, not just to ACS members, addresses the software development standard. It shows a major function of professional ethical standards being beyond the punishment of miscreants.

This macro understanding justifies an approach to encourage adherence to the code. If the standards become common knowledge, the public knows what ought to be done and those who do not follow the standards will receive less business, provided the failure to follow the standards is made known, as in the Swiss model. For example, there was some concern about the lead content in the paint of some children's toys sold in the USA. Citizens were made aware a) of the danger of the levels of lead and b) that toys exceed these levels of lead. In spite of a lack of government testing on all toys, when the knowledge spread that X toy was below this safety standard the toys are no longer purchased.

The ethics committees need to clarify and formally state those principles that are important to the profession as a whole. This need was supported by the comments in Lucas (P.57) which indicate the importance of training, but the commentators see ethics solely as an individual responsibility

> Punitive measures [when asked about them] are also useful but you can't take action unless you communicate your expectations. The message is:

> communicate your expectations, police them, and then maintain them. If there are no consequences people will not be motivated to behave ethically. So you need to put the sign post up (i.e. 80 km/h), communicate it and police it.

> Needs to be incentives, and the value of ethics for the business to be recognised.

> Organisations need to communicate their ethical expectations from staff and then ensure that they are aware of them.

> Management should encourage staff to take their ethics training so that they do not have an excuse when it comes to expectations.

> Needs to be more promotion of ethics. Frequently overlooked. Essentially it is up to individual, but training helps.

Conclusion

Codes of ethics and regulations can have teeth. If professional organisations are committed to elevating professional practice, and to the standards in their codes, then they should publish a list of expelled miscreants who violate those standards. Maintaining silence in the light of such violations is inconsistent with the content of the codes of ethics. The ACS takes a different view. ACS reprobates are not named but 'describe[ed] in general terms (to protect the privacy of those involved) the breaches that have occurred and the actions that have been taken by the ACS in respect of those breaches.' The important stakeholder is not the reprobate. The important stakeholders are everyone affected by that person's actions, and the profession which maintains silence about unprofessional actions.

Committees need to be aware of:

1. A focus on micro ethics contributes to single-mode solutions and mistakenly ignoring obligations to a broad range of stakeholders. It also is used to

justify ignoring a sense of responsibility for the whole project, for the whole profession.

2. The worst model of leadership is the whip. The committees must educate about the 'what' and the 'why' of regulations.

3. The rules are guidelines with a purpose. Codes and ICT governance are not checklists, which will be out of date shortly after they are written.

4. There needs to be as much if not more emphasis on and financial support for education thenis given to methods of enforcement. Enforcement/encouragement can take a variety of forms.

5. Take the codes seriously — publish the list of those ejected for violation of codes of ethics. Be the 'better business bureau' for software quality and list if there are open complaints

Codes of ethics and regulations may be toothless tigers, but they can still be heard. Their message is important to address the problems at the interface of ICT governance.

References

ACS , 2004 *Image* See <http://www.acs.org.au/news/060404.htm> accessed 2008.

Agle, BR, Mitchell, RK & Sonnenfeld, JA, 1999,', Academy of *Management Journal*, vol 42, no 5, October, pp 507–25.

AIS , 2005, 'Definition of system quality', <http://business.clemson.edu/ISE/html/system_quality.html> 2005.

Anderson, R, 1990, 'A rationale for the proposed revision of the association for computing machinery's code of professional conduct', <http://www.southernct.edu/organizations/rccs/oldsite/resources/research/comp_and_priv/anderson/references.html>

——, 1995, 'The ACM code of ethics: history, process, and implications,' *Social issues in computing*, McGraw-Hill, New York, pp 48–72.

——, Johnson, D, Gotterbarn, D, & Perrolle, J, 1993, 'Using the new ACM code of ethics in decision making', *Communications of the ACM* vol 3, 2 February.

Association for Computing Machinery (ACM), 2006, 'Policy and procedures on plagiarism', revised 2010, <http://www.acm.org/publications/policies/plagiarism_policy?searchterm=Policy+and+Procedures+on+Plagiarism>

Berleur, J, 2008, *Ethics of computing committees: suggestions for functions, form and structure: To Promote Discussion Inside the IFIP National Societies*, Berleur, J., Burmeister, O., Duquenoy, P; Gotterbarn, D. ; Goujon, P, Kaipainen, K; Kimppa, K, Six, B.; Weber-Wulff, D.; Whitehouse, D. (Eds)., IFIP Press, Laxenburg-Austria, 2008, ISBN 978 3 901882 24 3. The account of the ACM Code of Ethics development is based on my report to this group.

—— & d'Udekem-Gevers, M 1994, 'Codes of ethics, or of conduct, within IFIP and in other computer societies', *13th World Computer Congress*, pp 340–48.

Bowern, M, 2006, 'Ethics: part of being a professional', *Information Age*, June/July p 51-52.

Cockburn, A, 2004, *Crystal clear*, Addison-Wesley.

Dijkstra, EW, 1976, *A discipline of programming*, Prentice-Hall, Englewood Cliffs, NJ.

Gotterbarn, D, 1996, 'Software engineering: the new professionalism', in C Myer (ed), *The professional software engineer*, Springer-Verlag, New York.

——, Miller, K & Rogerson, S, 1998, 'Software engineering code of ethics', viewed 27 January 2007, <http://www.acm.org/serving/se/code.htm>

Halstead, MH, 1977, *Elements of software science, operating, and programming systems series*, vol 7, Elsevier, New York, NY.

Highsmith, J, 2002, *Agile software development ecosystems*, Addison-Wesley, Boston.

Lucas, R 2008, 'ETGovICT briefing paper for presenters', Centre for Applied Philosophy and Public Ethics, Australian National University, Canberra.

Martin, MW & schinzinger, Roland 1989, *Ethics in engineering*, 2nd ed, McGraw-Hill, New York.

Miller, K & Gotterbarn, D, 2003, 'Computer ethics in the undergraduate curriculum: case studies and the joint software engineer's code', *Small College Computing Conference Journal*.

Parker, D, 1968, 'Professional conduct in information processing', *Communications of the ACM*, vol 11, 3 March.

Paulk, MC, 1995, *The capability maturity model: guidelines for improving the software process*, Addison-Wesley Publishing Company, Reading, MA.

Perspectives on the Professions, Centre for the Study of Ethics in the professions at Illinois Institute of Technology, <http://ethics.iit.edu/perspective/v1n1%20 perspective.pdf>

Ravichandran, T, 2000, 'Total quality management in information systems development: key constructs and relationships', *Journal of Management Information Systems*, vol 16, no 3, pp 119–37.

Ross, DE, 2003, 'My complaint against the ACM — a leading technological society condones employment discrimination against some of its own members', <http://www.rossde.com/acm.html>

SIS Code, 2005 Swiss Information Society Code of Ethics. <http://www.s-i.ch/ fileadmin/daten/si/SI_Code_of_Ethik_V1.pdf>

Software Engineering Ethics Research Institute, 1999, 'Software engineering code of ethics and professional practice (5.2)', <http://seeri.etsu.edu/Codes/ TheSECode.htm>

Smoot, 1981, http://ethics.iit.edu/perspective/v1n1%20perspective.pdf

Software Testing, <http://www.associationforsoftwaretesting.org/about/ governance/>, Resolution on adoption of the ACM code of Ethics.

Standards Australia, 2005, *Corporate governance of information and communication technology*, Australian standard AS8015.

Standards Australia & Standards New Zealand, 1999, *Risk management*, Australian and New Zealand standard 4360.

Walsh, K,, Adms, W., Bauchner,H., Vinci,R. 2006, 'Meidcation Errors Related to Computerized order Entry for Children', *Pediatrics,* vol 118, n 5 pp 1872-1879.

Weill, P & Ross, JW, 2005, 'IT governance on One Page', MIT Sloan Working paper nimber 4517-04 <http://papers.ssrn.com/sol3/papers.cfm?abstract_ id=664612>

14. Business benefits from keeping codes of ethics up to date

Michael Bowern and Oliver K Burmeister
Australian Computer Society, Charles Sturt University

Introduction

The greater the extent to which a code of ethics is kept up to date, the better is the case that it benefits business to abide by that code. In 2010 the Australian Computer Society (ACS) updated its code of ethics after an extensive review, including national focus groups with members, and international input. Its last update, in 1985, predates most of the recent industry advancements. It was undertaken prior to advances and innovations, such as object-oriented techniques, the Internet and nanotechnology, of the last two decades. Industry currency of the ACS code is important if it is to have endorsement and adherence by Information and Communications Technology (ICT) professionals.

Almost every professional society has a code of ethics. Sometimes it is called a code of professional conduct, or behaviour, or something similar. In each situation it is a standard for the personal conduct of members of that professional society. In the ICT industry there are some professional societies that do not have such a code and, for the majority that do have one, questions have been raised about their efficacy.

Are such codes enforceable? Does the public hold ICT professionals accountable, based on such codes? Are codes simply weighty statements that serve no practical purpose? Why should the business community pay attention to such codes? This chapter attempts to answer questions such as these in two ways.

One way is by making a case for having a professional code of ethics. Many books have been written on ethics and, within many of these, the purpose of such a code is defined. That is not what this chapter means, however, by making a case for a code of ethics. Instead, a business case is proposed. A few years ago, environmental issues were given lip-service, but essentially ignored, whereas, today, companies see a business case for taking environmental issues seriously. Similarly, a business case can be made for creating and enforcing adherence to an up–to-date code of ethics for the ICT industry.

The second way such questions are answered is through making the case for renewing codes of ethics on a regular basis. Just as ICT changes over time, so a code of ethics in ICT should be kept current. Codes should stay current and relevant to professionals working in the industry. Some aspects of a code may not change frequently, others should. For instance, since its inception, the ACS code of ethics has had 'Honesty' as a guiding principle. The way in which this concept has been interpreted, however, in professional ICT behaviour over the past 40 years has changed. It is not envisaged that a future ACS code will call on ICT professionals to be dishonest, but the way in which honesty is practiced in the workplace can subtly change over time.

This chapter begins by making a business case for the ACS code of ethics, and then goes on to reflect on the current process of renewing this code. For the latter, the case is made that such codes ought to undergo a major review at least every decade, with minor reviews possible in the intervening years.

A business perspective

In Australia the ACS, in collaboration with the Australian Research Council, funded a major industry review of ICT ethics (Lucas & Weckert, 2008). That report has many findings, one of which, the most pertinent to this chapter, revealed that ACS members have a poor understanding of issues relating to professionalism and ICT ethics. The contention presented here is that concerns such as these are best addressed by making a business case for professional/ ethical behaviour.

Richard Lucas and John Weckert used evidence from surveys and interviews to show that many people in the ICT industry do not see themselves as ICT 'professionals'. The term 'professional' is poorly understood in the ICT industry, even amongst members of the ACS.

This is an important finding, because it shows that there is a need for the ACS to raise the profile of professionalism. Arguably, the ACS has already taken significant steps in this regard. The Lucas and Weckert report showed, however, that these efforts have not, to date, yielded the desired results. New ways of marketing professionalism need to be created. Professionals have special responsibilities and public accountabilities, be they in ICT or another industry such as medicine or law.

Inherent in the promotion of professionalism is the promotion of a code of professional behaviour and, in this regard, the ACS code of ethics is the focus of this chapter.

Professionalism defined

The Australian Council of Professions (ACP), of which the ACS is a member, is a peak body for professional societies in Australia. Trading as Professions Australia, it states:

> A profession is a disciplined group of individuals who adhere to ethical standards and hold themselves out as, and are accepted by the public as possessing special knowledge and skills in a widely recognised body of learning derived from research, education and training at a high level, and who are prepared to apply this knowledge and exercise these skills in the interest of others.

> It is inherent in the definition of a profession that a Code of Ethics governs the activities of each profession. Such codes require behaviour and practice beyond the personal moral obligations of an individual. They define and demand high standards of behaviour in respect to the services provided to the public and in dealing with professional colleagues. Further, these codes are enforced by the profession and are acknowledged and accepted by the community (ACP, 2009).

Using a detailed definition of 'profession' avoids the confusion of 'professional' being used as an adjective as well as a noun and, more importantly, identifies a number of special conditions which help to distinguish a 'profession' from an 'occupation'. From the definition above, a professional is a member of a disciplined group of individuals. The ACS is one such group, which represents the people working in the ICT industry, but, unlike doctors and lawyers, Australia does not require ACS members to be registered or licensed, and membership of the ACS is voluntary.

In 2000 the ACP recognised members of the ACS as professionals. It was the first time this had happened outside software engineers being able to claim to be professionals through membership of the Institute of Electrical and Electronics Engineers (IEEE). At the time the ACS promoted this development widely and significantly raised its membership numbers. However, the Lucas and Weckert (2008) report shows that the momentum was not sustained and that, even today, some ACS members do not understand what professionalism means.

The business case

The ACS believes that the society's code of ethics has two main roles. First, to promote professional behaviour by members of the society. Second, to promote to the government and business sectors, the value of employing people who practise the values of the ACS code. This section presents a business case for the second role and, in so doing, reflects the community expectations identified in the previous definition of a profession.

A decade ago, very few businesses realised that being eco-friendly could deliver a commercial competitive advantage; nowadays, that proposition is uncontentious. Promoting professionalism should be positioned in a similar way, so that it quickly becomes commercially desirable, and uncontentious, for professional people in the ICT industry to behave in a manner consistent with a code of ethics.

If a business makes clear that being eco-friendly meets its corporate objectives, it follows that it will recruit and value staff who help it to conduct its activities in that way. Similarly, if a business makes clear that being professional has marketable commercial advantages, it will give priority in its recruiting to people who manifest this value, and will send a signal to the community that employing professional people has value.

The other side of the coin is for the community to demonstrate that professional conduct is valued to the point where preference will be given to those businesses and individuals who behave in that way.

To support the previous points, the ACS needs to promote the value of professional conduct to both consumers and suppliers. That requires the society to devise examples and scenarios to illustrate how the code of ethics works in particular situations, and highlights how resolving an issue in an ethical way confers tangible benefit on the consumer. The outcome ought to be that consumers of ICT demand that suppliers practise what the ACS code of ethics requires.

Professional principles would apply to the teams and organisations developing systems that are critical, in terms of safety, the environment, and other social aspects. At one level, this approach would ensure a reduction in the public's regular infuriation with technology, by supplying products and services that meet consumers' reasonable expectations without faults or failures, or undisclosed constraints, expenses or risks.

Professional principles would be promoted as the reference point for safe and professional practices, and would be publicised on the ACS website, and other sites dedicated to this issue.

From an ACS internal perspective, it should develop a marketing strategy that involves corporate professional partnerships in which each corporate partner undertakes to encourage its relevantly qualified staff to practise its code of ethics for the benefit of its customers, and to join the ACS.

The aim is for ACS membership to become a proxy for certification to which an individual has committed, and who practises the type of conduct that is espoused by their professional society. By creating a demand for people who conduct themselves in that way, the value proposition of ACS membership commensurately increases.

The ACS should be working with national, state and local governments, and other areas of the public sector, to help them lead by example and make compliance with the ACS code of ethics mandatory in the ICT areas of government agencies, whether or not their technical staff are ACS members.

The approach in the Australian pharmaceutical industry

Some ideas and direction for the ACS could be gained by considering aspects of the Australian pharmaceutical industry, in which there is regulation of some of the products and activities of that industry.

In Australia there is a federal government requirement that the manufacturing processes of pharmaceutical products comply with standards addressing good manufacturing practice (GMP), or similar quality requirements. Regulation is based on risk mitigation, so complementary medicines (such as vitamins) are not regulated to the same extent as prescription medicines, or devices that are invasive to humans and/or include software, for example automatic drug-delivery systems. These may also be regulated against other product-specific standards.

The benefits to pharmaceutical businesses of compliance with their professional code are well recognised. These benefits include risk management and mitigation, and consumer confidence in products through compliance with professional and product standards.

Pharmaceutical businesses enforce their standards within and beyond their industry. A condition of registration of every new medicine is that its marketing and promotional literature complies with a specific code of conduct. This code of conduct is maintained by an industry organisation, the members of which include the manufacturers of many Australian pharmaceutical products. The code covers all aspects of marketing, including the relationships with health professionals such as doctors, dentists, and pharmacists. There is an industry committee to investigate complaints, identify breaches of the code, and impose fines or other disciplinary actions. The government condition attached to product registration makes some requirements of the code of conduct applicable to companies that are not members of the industry organisation.

Obviously, there are differences between the ICT industry and the pharmaceutical industry, and the purpose of this section is not to argue for government regulation of all ICT products. As a starting point, however, it is advocated that those people and organisations engaged in the development and maintenance of public-sector ICT projects should be required to observe the ACS code of ethics. Also, the ACS disciplinary committee should have jurisdiction to hear and determine complaints against both ACS members and non-members in those circumstances.

The benefit to business

As described below, a code of ethics typically consists of a) a code of conduct — how a person conducts him or herself in an ethical manner, and b) a code of practice — how a person carries out his or her specific tasks and responsibilities. In a business, such a code of ethics should not exist in isolation, it should be part of a business integrity system, in which the code is supported by policies, staff training, and disciplinary processes. In turn, the integrity system should relate to the systems for quality and other management functions used by the business. These other systems could describe how the business addresses the management of finance, risk, occupational health and safety, and the environmental aspects of its operation. It is preferable that these are not individual systems, but are integrated into one business management system. In effect, the code of practice is complementary to this business management system.

In the 1980s many businesses recognised that by adopting the techniques of quality management and process improvement they would improve their profitability, and gain a competitive advantage through increased customer satisfaction and improved products and services. It has been argued that quality-management activities will maximise the long-term value of a business (Foley et al, 1997). Value means more than just financial value. Other values could be a safe and healthy work environment, the provision of adequate training and tools to employees, an eco-friendly workplace, and a work culture that recognises a job well done. Likewise, a focus on the quality of products would be of value to suppliers, who might then focus on the quality of their own products, and to customers.

A business which instils these types of value by focusing on its quality processes is behaving ethically, and reflects the existence of a code of practice for quality management. If a business gains benefits by following good quality-management practices, it would also see that it could benefit by following codes of practice for occupational health and safety, and the environmental aspects of its business, for example. By establishing codes of practice for work activities, a business has already adopted one half of its code of ethics. The next step is to adopt a code of conduct, which is available from the ACS for ICT businesses.

Renewal of the ACS Code of ethics

A code of ethics has several major functions. These include the requirement to articulate ethical standards of the profession; to educate practitioners and the public about ethical obligations; and to provide guidance to resolve ethical quandaries (Burmeister, 2000; Burmeister & Weckert, 2003; Coldwell, 2008). In order to fulfil these functions satisfactorily it is important that the code is reviewed on a regular basis to keep it relevant, and up to date.

What is included in a code of ethics?

There are numerous terms in the literature for professional codes of 'ethics'. This chapter follows the guidance of the International Federation for Information Processing (IFIP) (Berleur et al, 2004), in discussing two types of codes.

The first type, is a code of 'ethics/conduct', which has a set of high-level statements, concerning such issues as honesty and integrity. This code governs 'how the person to whom it applies conducts him or herself in an ethical manner' (Berleur et al, 2004: 11). To avoid confusion, this type will be referred to as the code of conduct in the rest of this chapter.

The second type of code refers to a 'code of practice' for professionals, which 'governs how the person to whom it applies carries out his or her work technically' (Berleur et al, 2004: 11). This code includes a set of detailed statements related to the professional's particular occupational environment. These statements of practice are more specific and more likely to change over time, than are the conduct statements.

Typically, a professional society would define both types of code for its members in a single document. One example for ICT is the standard for teaching and practicing software engineering, published by the ACM and IEEE-Computer Society (CS) (SECEPP, 1999). In this chapter, reference to a 'code' or a 'code of ethics' means a document that includes both types of code, covering conduct and practice. Other references will be made to the specific type of code under consideration, for example a 'code of practice'.

To benefit business, professional codes should address current business challenges

The development of any technology can provide a range of new business opportunities, and this is particularly so with respect to ICT. Just as businesses periodically re-evaluate their strategic direction, so professional societies need to ensure the currency of their codes to the business communities they serve.

A simple example of this could be if a number of companies developed technology and software to use the wide range of personal information on social networking sites in ways that were not originally intended. The ACS code would apply to the professionals in these organisations who were members of the society. Some of these members might feel that their code did not sufficiently cover the ethical issues of information security and privacy arising from this new business approach, and so they make the ACS aware of their dilemma. The ACS should then review the relevant areas of the code in the light of this new use of technology. They might decide that the information security and privacy provisions in the code were sufficient, but that it needed additional material on whistleblower action and protection.

In a domain that changes as frequently as that of ICT, a change in work practice, or the advent of some new technology, should not of itself constitute grounds for changing a code of ethics. Examples of the types of change that could warrant reviewing a code include the growth of the Internet, the increased use of outsourcing and off-shoring, and the arrival and impact of micro/nano computing.

Codes, however, cannot be reviewed with every change in technology that occurs.

To avoid ineffective continuous code reviews, procedural guidelines are needed within a professional society that ensure regular review of its code or codes. How frequently should such a review be undertaken? Given the logistical difficulties involved, and based on observations of previous changes to codes in Australia and the United States, Bowern et al (2006) proposed that codes ought to be reviewed at least every 10 years. The ACM code underwent significant reviews in 1992 and again in 1998, but not since. The ACS code was reviewed in 1975, and again in 1985, but not since then. The review procedure also ought to allow for reviews that are determined by major technology changes. The advent of a biological computer implanted in the brain might be the sort of thing that justifies immediate code review rather than waiting a prescribed period of time.

Changes to codes of ethics have been motivated by a desire to include their expanded sense of professional responsibility and behaviour occasioned by some technological advances (Gotterbarn, 1996). Other changes have been made to include an aspirational component in the code (Burmeister, 2008).

Why aspirational? To appeal for ethical behaviour when enforcement is not possible. In medicine and law, a breach of the code of ethics can mean loss of ability to practice one's profession. Not so in the ICT industry. An ACS member can be held accountable through the ACS disciplinary procedures. But 80 per cent of ICT practitioners in Australia are not ACS members. Also, a member facing ACS disciplinary procedures could simply resign. Then s/he can continue to behave unethically; they simply can no longer claim to be an ICT 'professional', which is a right that the ACP (2009) has restricted to use only by ACS members. This ACP restriction, however, is not legally binding. There is nothing directly that the ACS can do to prevent non-ACS members calling themselves ICT professionals.

Instead, the ACS should raise the standard of what it means to be an ICT professional in Australia. The ACP recognition provides the ACS with an ideal foundation from which to redefine ICT professionalism. Given that the ACP requires professionals to adhere to a code of ethics, this further provides the ACS with the opportunity to define ICT professionalism in Australia in terms that closely align with its own code of ethics, further strengthening its position in Australia as the arbiter of professional behaviour. The Lucas and Weckert (2008) report clearly showed that the efforts of the ACS to do this have not been successful to date. The ACS has to devise new strategies to market the benefits of its professional code to the ICT business community.

Gotterbarn (2000), writing about experiences in the development of two codes of ethics in the United States, found that a critical issue is that of specificity; that is, how prescriptive and detailed the code should be. One approach to specificity is in the clauses of the code itself. Generally, the code of conduct does not change over time, but changes in the activities and products of the ICT industry practice mean that the application of the code, as seen in the code of practice, is different. The task of maintaining the code is simpler where the code of conduct is separate from the code of practice, and it is much harder where there is no clear distinction in the document between conduct clauses and practice clauses.

The relevance of the ACS code of ethics to business

The joint ACM/IEEE software engineering code of ethics has been endorsed and adopted by over 50 companies in the United States. One reason for its acceptance may be that its most recent review included wide involvement of the ICT business community. In contrast the current review of the ACS code was only carried out among ACS members. About six per cent of ACS members are from academia, a tiny percentage are student members or retirees, and the rest are all members working in the ICT industry. Although the majority of participants in the review were from the business community, they participated as individuals, not directly as representatives of their respective businesses. In some of the focus groups, however, the comments made were specifically relevant to businesses.

If an ICT company decides to adopt the business case for an ethical approach described above, it will be necessary to define the company's policy on ethics, adopt an appropriate code of ethics, provide training in this code, and start to develop an ethical culture in the company. The ACS code of ethics would be a good starting point for these activities. For additional support in these activities, the company could become a professional partner with the ACS. This initiative by the ACS provides companies that become partners with a number of benefits, including professional development support, recruitment of qualified professional people, recognition in the ICT industry, and attractive professional indemnity insurance rates. Currently, the ACS has over 170 professional partners.

Structure of the ACS code

An earlier review of the renewal process for the ACS code of ethics (Bowern et al, 2006) noted that it comprises two sections. The first section is a simple policy statement plus a declaration of six values and ideals, which are then expanded into the second section, a detailed set of 37 statements or 'Standards of conduct'. This code of ethics is defined as part of the ACS national regulations, to emphasise the importance of the code, and to ensure its prominence in the ACS body of documentation.

Supplementary to the code of ethics is the code of professional conduct and professional practice. This code was developed to provide more practical guidance in the day-to-day activities of ICT professionals. It is not part of the national regulations, which means that it is easier to amend and update. Changes to the national regulations require a vote by the national congress, followed by a vote by all members of the ACS. This has implications for future amendments to the codes. Figure 1 illustrates these various codes.

National regulations			Code of professional conduct and professional practice	
Clause				
4.	ACS code of ethics		Introduction	
4.1 and 4.2	Policy statement			
4.3	Values and ideals			
4.4 to 4.10	Standards of conduct	There is overlap between the standards of conduct and the code of professional conduct	Section A	Code of professional conduct
There is no identified code of professional practice in the national regulations			Sections B to L	Code of professional practice

Figure 1: The 'conduct' overlap

Source: Bowern et al, 2006

The code of professional conduct is intended as a guideline for acceptable personal conduct for each IT professional practicing in the industry and, as such, it is complementary to the values and ideals and the standards of conduct. There is some overlap between the standards of conduct and the codes of conduct.

All reviews of the ACS code of ethics, including the current one, have been limited to the left side of Figure 1. That is, they are limited to the values and ideals and their associated standards of conduct.

The code of professional practice is intended as a guideline for acceptable methods of practice within the ICT industry. The guideline is generic and addresses a range of aspects of product life cycle, and acquisition, development, implementation and support processes. The code of professional conduct and professional practice has not been updated since its adoption by the society.

Engineering organisations in Australia tend to have their own values and obligations, although some of these companies may also practise the professional code of the Institution of Engineers Australia. Similarly, the six values and ideals of the ACS code of ethics could form the basis of the values and obligations of ICT businesses in Australia. That would, in effect, create closer ties between business values and those of the ACS code, and possibly bring some businesses in closer contact with the ACS, as mentioned previously in relation to the professional partners program.

Guidance and education

Further work is needed to identify how the ACS code can best provide guidance and education to its members. Current attempts by the ACS Committee on Computer Ethics CCE to accomplish this are mainly through better communication and publicity of the code to ACS members.

Some members of the CCE have produced a set of case studies related to each of the clauses in the ACS code, which have been publicised to members via the ACS website, and used for teaching in some tertiary institutions. Also, since late 2004, the CCE has arranged for a regular column in *Information Age*, the ACS bimonthly magazine for members and other professionals in the industry. The column seeks to promote the professional code, and discuss the ethical aspects of current ICT news items. The case studies are one source of material for these articles.

Advice should be provided on how the code would apply to the wide range of ACS members, some of whom are not directly involved in systems development; for example, ICT professionals dealing directly with customers, such as some empirical software engineers. If the ACS code is to cater to the widest possible interpretation of 'ICT professional', it must be examined for its applicability to all aspects of the profession. An excellent example has been set through the ACM/IEEE software engineering code, that has been adopted by the ACS for its members who are software engineers (Davidson, 2004). The adoption of the software engineering code is another contribution to the specificity of supplementary ACS codes.

Updating the ACS code of ethics

The review process for the ACS Code of Ethics, which was initially described in the Bowern et al 2006 report, has now been completed. Following international advice, as detailed in the earlier report, an extensive consultative process was followed, particularly with ACS membership. As with the 1985 review, this involved focus groups at each of the ACS branches across the country. The Western Australian branch was the exception at this stage, as this branch was reserved to provide a final review of the completed, new code. A preliminary report was submitted to the ACS management committee, which foreshadowed the recommendations of the review.

The final step in the renewal process was the drafting of the new document, which meant carefully capturing members' views about individual clauses in the code of ethics. Contradictory suggestions were resolved, and some clauses were reworded to support the updated code. This was completed in late 2009, after which the Western Australian branch focus group provided a further opportunity for ACS members to have input into the review process, before the new code was formally submitted to ACS management.

The new code of ethics was first presented to all ACS members in the April/May 2010 edition of *Information Age*.

Structure of the new ACS code of ethics

The various ACS codes described previously, namely the code of ethics, which comprised the values and ideals and the standards of conduct, and the separate code of professional conduct and professional practice have now been combined and rationalised. The new code of ethics is accompanied by two complementary codes: the code of conduct and the code of professional practice.

The code of ethics

This is part of the ACS constitution, and describes six values to which professionals in the ICT industry are expected to adhere. When the final approval is given to the new code, the formal procedures for amending the constitution will be implemented and the existing code of ethics will be replaced with the new text.

The code of conduct

This is an expanded guide to how the code of ethics should manifest itself in the way members conduct themselves, and replaces the earlier material on professional conduct. This code is not part of the constitution and is, therefore, able to be amended by the ACS management committee on a majority vote. This is a public, and more explicit statement of how the ACS expects its members to conduct themselves.

The code of professional practice

This is a separate document that is yet to be reviewed and amended, but, for the time being, the existing document of that name remains in use, pending its replacement. The committee on computer ethics will take on the task of this review. The code of professional practice will have more practical examples of how the code of ethics is intended to operate in specific workplace situations.

Summary of changes made

All of the responses from the focus groups were considered and applied, although the specific recommended wording was changed in some cases, for consistency with other text, and to avoid duplication.

Environmental issues did not figure at all, let alone prominently, in the 1985 review. Now, environmental aspects have emerged as an issue that ACS members are concerned about and wish to have addressed in their code of ethics. Clauses on the environment were suggested in every focus group considering the current six ideals. This led the review team to consider the possibility of adding a new 'ideal' to do with environmental issues. No new clauses have been included, but the environmental consequences of using ICT have been mentioned in the preamble to the code, and in the value addressing public interest.

Similarly, in the 1985 review, issues related to women in the industry and as members of the ACS were not prominent, whereas, in the current review, one focus group session was entirely devoted to issues raised by women in the ACS, through the Director ACS-Women, concerning aspects of professional practice, particularly in relation to discrimination in the workplace. This focus group proposed five new clauses for the value relating to professionalism. Three of these have been combined in the new code, to maintain the spirit and intent of the original wording. The other two clauses were already covered by an existing, less specific clause.

Every clause was reviewed, and revised or deleted, as required. Other clauses were added to reflect the changed values. Other changes include:

- A preamble has been added to the code, and each value has an introductory paragraph.
- For consistency, only 'ICT' or 'information and communications technology' has been used.
- 'I must' has been changed to 'you will'. This makes the wording consistent with a regulation, it stresses an imperative, and is consistent with other international codes.
- 'Public' and 'public interest' have been used instead of 'community' as these words are used in Professions Australia's definition of a profession, and in other international codes. Also, some focus groups had concerns about the use and meaning of 'community'.
- 'Values and ideals' are now just 'values'. Some values have been changed to be actual values, and the new values are now:
  - the primacy of the public interest
  - the enhancement of quality of life
  - honesty
  - competence
  - professional development
  - professionalism

After each of the last two major revisions of the code of ethics, the ACS has published a one page summary of the values and ideals, known as the 'one-pager'. An outcome of this renewal process will be a new version of the ACS values for distribution to all members. Given the above arguments for a business case, producing a new one-pager serves a dual purpose. Not only does it easily convey the main ideals upheld by the code of ethics, but it can also serve businesses. That is, it can be prominently displayed by businesses as a way of publicly aligning themselves with the professional values of the ACS, the peak professional ICT body in Australia. This link would be even stronger if, as suggested above, ICT businesses develop their organisational values and obligations to professionalism in close alignment with those of the ACS, and considered the benefits of the professional partner scheme.

This review process is significant to the wider ICT community as well. Just as the review process drew on similar work from the United States and the

United Kingdom, so it is expected that the process will help inform other societies working through similar issues. The latter will be achieved through disseminating information about the process of the review through working groups of IFIP, a process that has already started, given four members of the CCE are IFIP working group members. More particularly, three are also members of the special interest group within one IFIP working group that is charged with the responsibility for developing frameworks on ethics, which has as one of its responsibilities the directive to help member societies to formulate their codes of ethics (IFIP SIG 9.2.2).

Conclusion

The greater the extent to which professional codes are kept current and relevant, the greater the business case for endorsing and promoting those codes.

In order to raise the profile of a code of ethics, one approach is to see the benefits to the organisation from a business perspective. As seen above in the pharmaceutical industry, such benefits include risk mitigation and the enforcement of professional and product standards. This argument can apply to many businesses, especially those that already have documented processes for running their operation. It has been argued above that an ethical and quality approach is consistent and complementary; they are two sides of the same coin.

Each professional society needs to ensure the currency of its code of ethics, through regular reviews, including major reviews every 10 years, and minor reviews taking place as required by significant technological advancements. A major review should involve an extensive consultation process, as advocated by IFIP, which claims that the 'process used to develop a code is as important as the code itself' (Berleur et al, 2004: 13). Revisions of the ACS code of ethics should ensure consistency between the various supplementary ACS codes; and rationalise the way that part of the code is incorporated into the national regulations to mandate its use, and the way that other parts of the code can be more easily updated.

The ACS is the guardian of professional ethics and standards in the ICT sector, and is committed to ensuring the beneficial use of ICT for all Australians. The code of ethics is the linchpin that enables the ACS to promote the value of professional conduct of its members to both consumers and suppliers. The ACS must take a leadership role in the marketing of the benefits of its professional Code to the ICT business community.

Acknowledgement

The authors acknowledge the work of Philip Argy, past president of the ACS, who developed the original business case for the ACS code of ethics.

References

Association for Computing Machinery (ACM), 1999, *Software engineering code of ethics and professional practice, version 5.2*, viewed 5 January 2009, <http://www.acm.org/about/se-code?searchterm=software+engineering+code+of+ethics+and+professional+practice>

Australian Council of Professions, 2009, Australian Council of Professions, viewed 5 January 2009, <http://www.professions.com.au/>

Berleur, J, Duquenoy, P, Holvast, J, Jones, M, Kimppa, K, Sizer, R & Whitehouse, D, 2004, 'Criteria and procedures for developing codes of ethics or of conduct', IFIP Press, Laxenburg, Austria.

Bowern, ME, Burmeister, OK, Gotterbarn, D & Weckert J, 2006, 'ICT integrity: bringing the ACS code of ethics up to date', *Australasian Journal of Information Systems*, vol 13, no 2, pp 168–81.

Burmeister, OK, 2000, 'Applying the ACS code of ethics', *Journal of Research and Practice in Information Technology*, vol 32, no 2, pp 107–20.

——, 2008, 'Introduction', in D McDermid (ed), *Ethics in ICT: an Australian perspective*, Pearson Education Australia, Frenchs Forest, NSW, pp 1–25.

—— & Weckert, J, 2003, 'Applying the new software engineering code of ethics to usability engineering: a study of 4 cases', *Journal of Information, Communication & Ethics in Society*, vol 3, no 3, pp 119–32.

Coldwell, J, 2008, 'Professional ethics and responsibilities', in D McDermid (ed), *Ethics in ICT: an Australian perspective*, Pearson Education Australia, Frenchs Forest, NSW, pp 277–302.

Davidson, P, 2004, 'ACS and IEAust jointly adopt software ethics, practice code', *Information Age*, April/May, pp 47–50.

Foley, K, Barton, R, Busteed, K, Hulbert, J, & Sprouster, J, 1997, 'Quality, productivity and competitiveness: the role of quality in Australia's social and economic development', Wider Quality Movement.

Gotterbarn, D, 1996, 'Establishing standards of professional practice', in C Meyer, ed, *The responsible software engineer*, Springer Verlag.

——, 2000, 'Two computer-related codes', *Perspectives on the Professions*, vol 19, no 1.

Lucas, R & Weckert, J, 2008, *Ethics and regulation in the ICT industry*, report for the Australian Computer Society, Centre for Applied Philosophy and Public Ethics, Charles Sturt University, Canberra.

Section VII

ICT and society

Ethics first or ethics last?

In the first chapter of this book the Hon Michael Kirby showed the difficulties of regulating new technologies, information and communications technology (ICT) included. These difficulties, we argued, highlight the need for a high level of professionalism in the ICT industry and sections two to six all focused on aspects of ICT professionalism. This section brings to the fore an important role for the ICT professional that has been in the background in at least some of the earlier papers but is discussed explicitly here; the ICT professional's role in decisions regarding the development and use of technology in society. Technical expertise is essential in this decision-making, and those in the industry are the ones with a responsibility to provide it. A topical example in Australia is internet-content control. ICT practitioners are not the only ones with expertise regarding what content should or should not be allowed, but they have the knowledge about what is technically possible in managing content, and what the likely technical consequences will be for particular actions.

Ethical considerations of technology in society raise methodological questions of how these considerations should proceed in this context, and who should be involved. Our view is that it is essentially an interdisciplinary activity. ICT practitioners, ethicists, social scientists, lawyers and policy makers, are all necessary if real progress is to be made. Another important point relates to the proper subject matter of ICT ethics, something that is raised in both papers in this section. According to Bernd Stahl and Simon Rogerson, 'A central problem of the ethics of technology is that it [the ethics] tends to arrive too late' and, in a similar vein, in the following paper, Philippe Goujon and Catherine Flick state that 'the strong push for technology development too often obscures the need for any deep ethical consideration before a technical project is funded, developed and deployed'. The question is, should ICT ethics be reactive, proactive or both? That is, should ICT ethics just respond to existing problems, try to anticipate problems or both respond and anticipate? A distinction needs to be drawn here between software engineering and the more general development of ICT. In software engineering, it is not new to hear calls for ethics to be considered early in the development process (see Gotterbarn's discussion of SODIS for more on this (Gotterbarn, 2008)), but, in ICT development in general, as in most technical development, ethical questions have generally been left until problems appear. This is our focus in this section.

Ethics, as we have just seen, can either react to the technology; that is, the ethics is done last, after the technology has been developed, or it can be proactive; that is, done first, before the technology is developed. This ethics first, ethics last approach raises, or seems to raise, what is commonly known as the Collingridge dilemma (Collingridge, 1980). If the ethics is done before the development, it is difficult to predict what the impacts of it will be and, if done after development, it is difficult to control the impacts. Marvin Croy formulates the dilemma this way:

> Either a technology is in a relatively early stage of development when it is unknown what changes should be made, or a technology is in a relatively late stage of development when change is expensive, difficult and time-consuming.
>
> If the former, then control is not possible.
>
> If the latter, then control is not feasible.
>
> Therefore, either controlling technology is not possible, or controlling technology is not feasible. (Croy, 1996)

It is not possible because prediction is so unreliable and is not feasible because once the technology is developed change is difficult.

While this poses real difficulties, it is not the dilemma that it initially appears to be. Some prediction can be informative and some control after the development is possible. Furthermore, it is possible, to some extent, to slip between the horns.

The difference between reactive and proactive active ethics can be illustrated through a brief look at the history of the ethics of information technology (IT). Problems in IT ethics are not different or new, in the sense that they are different in kind from other ethical problems. Moral philosophy has been studied systematically at least from the time of the ancient Greeks, and the ethical issues in computing are part of this tradition. What is new and different is that the development and use of computers has raised old questions in new and different ways (Johnson, 2001), often creating what James Moor calls *policy vacuums* (1985). The work of the computer ethicists, then, is to develop policies to fill those vacuums. One example in computing is hacking. Breaking into someone's computer account is in some ways like breaking into someone's house, but there are interesting differences. It is a logical rather than a physical entering. Another is unauthorised copying of software, which is a bit like unauthorised copying of a book, and a bit like taking a television set, but there are significant differences. There are also questions relating to work and the loss or creation of skills, which arise in a unique way. This way of doing computer ethics is essentially reactive (Johnson, 2001: vii–viii), and this is an essential role of applied ethics in general.

When a proactive, or ethics-first, approach is taken, the emphasis is different. One is much more likely, and in fact it is necessary, to think carefully about what is wanted from the technology, and that involves thinking about what sort of life one thinks is a good one. This approach means taking action that will guide the development of the technology in a particular manner. This proactive stance also highlights a more positive view of applied ethics. Ethicists are frequently seen as playing only a negative role, always criticising and attempting to hinder development. While this view is to some extent correct, it is not the only thing that ethicists should be doing. Technology clearly has a positive role. In many ways our lives are much better because of various technologies.

The argument here is that the ethics first model and the ethics last model are popular but poor solutions to a false dichotomy (see Weckert, 2007). ICT ethics is not something one can complete satisfactorily, either first or last, but something that needs be done continually as the technology develops and as its potential consequences become better understood. Ethics is dynamic in that the factual component on which it relies has to be continually updated. Norbert Wiener outlined this approach in 1960. Talking about automated machines, he writes:

> To be effective in warding off disastrous consequences, our understanding of our man-made machines should in general develop *pari passu* [in step with] with the performance of the machine (1960: 1355)

What he is suggesting is that if we wait until the technology has been developed, it may be too late to avoid 'disastrous consequences'. Predicting is hazardous and he does not suggest that these machines should not be developed on the grounds that they may produce undesirable consequences. Rather, the understanding must develop in step with the development of the machines. Similarly, the contention here is that understanding of the ethical questions must develop as the technology develops. This will be dynamic; partly reactive and partly proactive, continually returning to the technology to understanding how it is developing and what its actual or likely consequences are.

Ethicists must work with scientists and technologists to predict future problem areas. This is of course fraught with danger given the uncertain nature of prediction, but if care is taken, it is a useful and important activity. In many cases it is at least plausible that ethical problems will arise and that we can know something about what they will be like. Take the case of radiofrequency identifier device (RFID) chips. It is known that they have been developed much further than the passive chips currently used, for example, in libraries, and research is continuing. A new passive RFID chip is under development that will be able to be printed onto clothing, and paper, with a range eventually of close to 300 metres (Williams, 2010).

It is clear that developments in nanotechnology, particularly nanoelectronics, will lead to further miniaturisation of chips and that readers and other sensing devices will become more powerful and sensitive. Given these developments and uses of the chips in other contexts, for example, warehouses, it is certainly plausible that libraries and other industries will extend their uses in ways that are increasingly threatening to privacy, and make them much more than fancy barcodes (Thornley, et al, 2011). This is not inevitable of course. We are not proponents of technological determinism, the view that technology will move on regardless of what we do. Given the extent to which the values of efficiency and productivity seem to override other values in the assessment of technologies and their uses, however, the pressures to extend their uses will be intense if it is seen to aid efficiency. Resistance might also be weak. In other areas privacy is eroded, a little at a time, in a way that is almost unnoticeable. Imagine if, about 20 years ago, before paying for groceries with credit cards was popular, the supermarkets had employed people to sit and watch all customers as they paid, and made a record of their purchases and their names and addresses. Most of us probably would have been a little concerned about this. Now it all happens automatically for those who pay by credit card and most people do not give it a second thought. This suggests that what the eye doesn't see doesn't really bother most people and what happens gradually also goes largely unnoticed.

It is here that proactive ethics comes in to play. The argument is not that RFID chips should not be used because of future dangers. The question to be asked is, is it morally responsible to use the chips in certain extended ways? And how can the technology be designed to mitigate the dangers? What kind of regulations need to be put in place to control the use of the chips? These are all legitimate questions for the proactive ethicist and their answers should feed into policy decisions about the future use of RFID chips.

In both papers of the section a central concern is with the relationship between ethics and technology and both argue that ethics has a role to play early in the process of technological development. Bernd Stahl and Simon Rogerson, the current and immediate past directors of the Centre for Computing and Social Responsibility at De Montfort University, acknowledge the difficulty of forecasting future developments, but, through an examination of European Union projects, isolate a number of ICT trends. These are: increasing computational power and decreasing size, new types of interfaces, network connection anywhere and anytime, and increased development of virtual places. These trends suggest ethical issues that require addressing or at least, serious consideration. Philippe Goujon, a continental philosopher and researcher, and Catherine Flick are critical of the sharp divide that exists between ethics and technology, and argue for ethical reflexivity; that is, the constant reassessment of ethical norms in response to the technology, to economics, to politics and to gaps in regulation. The context of development is important and the non-neutrality of ICT requires emphasis. Inadequacies in both current risk assessment and technology assessment are highlighted. Both papers relate their conclusions to ICT governance.

References

Collingridge, D, 1980, *The social control of technology*, St Martin's Press, New York.

Croy, M, 1996, 'Collingridge and the control of educational computer technology', *Society forPhilosophy and Technology*, vol 1, no 3–4.

Gotterbarn, D, 2008, 'A practical mechanism for ethical risk assessment — a SoDIS inspection', in HT Tavani & KE Himma (eds), *The handbook of information and computer ethics*, Wiley, Hoboken, NJ, pp 429–71.

Johnson, D, 2001, *Computer ethics*, 3rd edn, Prentice-Hall, Upper Saddle River, NJ.

Moor, J, 1985, 'What is computer ethics?', *Metaphilosophy 16*, pp 266–75.

Thornley, C, Ferguson, S, Weckert, J & Gibb, F, 2011, 'Do RFIDs (radio frequency identifier devices) provide new ethical dilemmas for librarians and information professionals?', *International Journal of Information Management International Journal of Information Management*, vol 31, pp 546–55.

Weckert, J, 2007, 'An approach to nanoethics', in G Hodge, D Bowman & K Ludlow (eds), *New global regulatory frontiers: the age of nanotechnology*, Edward Elgar.

Wiener, N, 1960, 'Some moral and technical consequences of automation', *Science*, vol 131, pp 1355–58.

Williams, M, 2010, 'Nano-based RFID tag, you're it', viewed 13/9/2010, <http://news.rice.edu/2010/03/18/nano-based-rfid-tag-youre-it/>

Biographies

Dr Catherine Flick is a lecturer in computing and social responsibility at the Centre for Computing and Social Responsibility, De Montfort University. Her research interests lie in the area of ethics in ICTs, particularly in social media, user experiences, and emerging technologies. Apart from her research into informed consent, some of her previous work has been in the areas of online child protection, ethical governance of emerging technologies, anonymous computing, and trusted computing. She has a computer science and computer ethics background, and gets excited about developing ethical value centred technologies from the ground up. She is an active member of the British Computing Society and the International Federation of Information Processing Working Group 9.2 and SIG 9.2.2.

Prof Dr Philippe Goujon Dir. de recherches: Habilité à diriger des recherches (HDR) — director of the Legit (Laboratory for ethical governance of information technology). He has authored articles and books on artificial life, self-organisation, thermodynamics, the complexity, biotechnology, genomics and also on the connection between science, techniques, and society. He has directed international conferences concerning governance, ethics and information technologies and edited books concerning the relation between governance ethics rationality and communication technology. He is a partner in four European projects the Etica project ('Ethical issues of emerging ICT applications), EGAIS project ('The ethical governance of emerging technologies — new governance perspectives for integrating ethics into technical development projects and applications'), and Consider project (Civil society organisations in designing research governance), and the project goldenworker (scientific expert). He is in charge of the ethical training for the European Commission. He is an expert for the European Commission and the French government.

Prof Simon Rogerson is professor emeritus at the Centre for Computing and Social Responsibility and member of the Department of Information Systems at De Montfort University. He joined the university in 1983, following a successful industrial career which culminated in being the computer services manager for Thorn EMI. At DeMontfort he has lectured, researched and consulted in the management and ethical aspects of computing. He has published and presented papers internationally about these issues. In 1995 he became the director of the newly formed Centre for Computing and Social Responsibility and conceived and co-directed the ETHICOMP conference series on the ethical impacts of IT. He is a member of the Parliamentary IT committee in the United Kingdom, a fellow of the Institute for the Management of Information Systems and a fellow of the Royal Society for the encouragement of Arts, Manufactures and Commerce.

Prof Bernd Carsten Stahl is professor of critical research in technology and director of the Centre for Computing and Social Responsibility at De Montfort University, Leicester, United Kingdom. His interests cover philosophical issues arising from the intersections of business, technology, and information. This includes the ethics of ICT and critical approaches to information systems. From 2009 to 2011 he served as coordinator of the European Union (EU) FP7 research project on 'Ethical issues of emerging ICT applications', ETICA (<http://www.etica-project.eu>) and from 2012 to 2015 he served as coordinator of the EU FP7 research project 'Civil society organisations in designing research governance' (CONSIDER, <http://www.consider-project.eu>).

15. Ethical issues of emerging ICT applications — a Euro-landscape[1]

Bernd Carsten Stahl and Simon Rogerson
De Montfort University

Introduction

A central problem of the ethics of technology is that it tends to arrive too late. In many cases ethical issues are only recognised when the technology is already on the market and problems arise during its widespread use. Ethics can then become a tool to clean up a mess that might have been avoidable. It is probably not contentious to say it would be desirable to have ethical input at the earlier stages of technology design and development. Indeed, there are ethical theories and approaches that explicitly aim at an early integration of ethics into the technology life cycle (van den Hoven, 2008). One central problem of this type of approach is that the future is unknown. By definition we do not know with certainty what will happen in the future and an ethics that relies on future development needs to be able to answer the question of how it decides which technological developments to pursue. Ethics has traditionally not been well equipped to deal with issues of uncertainty (Sollie, 2007) and, in particular, future uncertainty.

The present chapter aims to contribute to this discussion. Its approach is to identify likely scenarios of future information and communications technology (ICT) developments that are grounded in empirical facts. The idea is thus to strike a balance between unavoidable speculation when talking about the future and factual grounding necessary for academic research. This chapter should be understood as a first step in identifying future developments in ICT. The chosen approach is to concentrate on an identifiable and relevant regional and policy area, namely the European Union (EU). It aims to give a high-level overview of the European landscape of emerging information and communication technologies. Its purpose is to come to an understanding of the ICTs that are likely to develop in the next 10 to 15 years, with a view to understanding which

---

1   This chapter is part of the research project, 'Ethical issues of emerging ICT applications' (ETICA) (GA no 230318) funded by the European FP7 Science in Society strand. An earlier version of this chapter was presented as BC Stahl & S Rogerson, 2009, 'Landscapes of ethical issues of emerging ICT applications in europe', in, *Proceedings of the eighth international conference of computer ethics: philosophical enquiry*, 26–28 June 2009, Corfu, Greece.

ethical issues we can expect and how we may best prepare to meet them. This will lead to policy recommendations for the EU as well as advice for individuals and organisations involved in technology development. The chapter is intended to provide the grounding necessary to develop empirical work. It will develop categories of ICTs and ethical issues that will be used to investigate specific ICT research projects in order to assess whether and how ethical issues are currently taken into consideration and how policies need to be developed.

The ETICA project, outlined in this chapter, can be described as replacing or complementing the existing feedback ethical analyses which review implemented ICT and consider the reactions of those who are impacted by such ICT. ETICA is a feed-forward approach by which we are trying to provide an ethical direction for future landscapes. In doing so, we hope to reduce the likelihood of ICT which damages individuals, society and the environment and increase the likelihood of real beneficiaries. This is done through establishing policy scaffolding in advance of ICT development and subsequent implementation.

Conceptual basis

Before we move to a detailed discussion of the European landscapes of technology, it is important to outline briefly the context of this chapter. This will start with a description of the seventh framework program. We then outline our concept of ICT ethics. Finally, we discuss some of the policy background that informs the EU's view of technology, and gives reasons for the plans and resource allocations that are meant to shape the development and use of technology. The section finishes with some considerations concerning the methodology of this chapter.

The seventh framework program for research and technological development (FP7)

The EU has a long tradition of fostering research and development through so-called 'framework programs'. The current seventh framework program (FP7), which runs from 2007 to 2013, has a total budget of over €50 billion. The majority of this money is and will be spent on research grants, predominantly in Europe. Research to be co-sponsored by such grants is chosen on the grounds of calls for proposals and following a competitive peer-review process. Given that there are national research-funding mechanisms in many European countries, the European framework funding has the additional characteristic of being centred on international collaboration. The European Commission (EC) names two main aims of the framework program (European Commission, 2007: 7): 'to strengthen the scientific and technological base of European industry [and] to encourage its international competitiveness, while promoting research that supports EU policies.' One aspect of FP7 is that it is meant to contribute to the European Research Area (ERA) (<http://cordis.europa.eu/era/concept_en.html>), which aims to overcome the weaknesses of European research caused by its fragmented and dispersed nature.

In order to meet the broad objectives of FP7, the program has been divided into four categories: co-operation, ideas, people and capacities. Each of these is then subdivided into further categories and sub-programs. The core of FP7 is the co-operation program, which is used to fund collaborative projects involving partners from at least three European member states. This program has been further divided into 10 key thematic areas:

- health
- food, agriculture and fisheries, and biotechnology
- information and communication technologies
- nanosciences, nanotechnologies, materials and new production technologies
- energy
- environment (including climate change)
- transport (including aeronautics)
- socio-economic sciences and the humanities
- space
- security.

The ideas program aims to support 'frontier research', and funding is based on scientific excellence, without the need for cross-border collaboration. The people program supports researcher mobility across Europe and the capacities program aims to strengthen the research capacities of Europe. This chapter will concentrate on the co-operation program and, more specifically, on its ICT sub-program. This is justified by the particular emphasis on ICTs of the project. While it stands to reason that ICTs will be developed in other areas of FP7, the ICT work program is specifically focused on them. In addition, the ICT program is the largest of all sub-programs with a budget of over €9 billion over the lifetime of FP7 (<http://cordis.europa.eu/fp7/budget_en.html>).

A final word of justification of the choice of concentrating on European ICT research program is necessary, given that this chapter aims to investigate the global phenomenon of ethics in ICTs. In addition to the practicalities of this chapter being a result of a European research project, one can also easily argue that the EU is one of the most important economic and political entities internationally, and that its research policy has the potential of shaping future technical and economic standards. With a population of around 500 million and a gross domestic product (GDP) that represents about one third of the world's GDP, it has significant international power. The European view of ICT is important because it is developed in intercultural discourses with scientists and researchers worldwide. It shows the ways that policy makers perceive the role of ICT. At the same time, it has the potential to shape future developments. This refers to the funding available via FP7 but, maybe more importantly, to the many ways in which it is necessary for the EU to set policy that can shape the way technology is designed or used. While this chapter and the underlying project are thus concentrating on a particular region, we believe that our findings should be of interest more generally and are likely to be transferable at least to a considerable degree.

ICT ethics

Ethics can be defined as the philosophical study or reflection of morality (Adam, 2005; Weil, 1969). In everyday language, and even in much academic writing, this distinction is not always observed (Forester, 1994; Weckert, 1997). The distinction between social norms and their reflection is important to observe, however, if one wants to come to a measured understanding of normative issues and their ethical evaluation. Ethics as the reflection of morality can have different tasks. There is a distinction between descriptive ethics, normative ethics and metaethics (Marturano, 2002).

Here, the term 'ICT ethics' is used to denote ethical issues that arise from or in conjunction with ICT. Work in ICT ethics can be distinguished along the lines of the earlier distinction of ethics in general, namely in descriptive, normative, and metaethical. Scholars from different disciplines undertake the different types of investigation. Descriptive ICT ethics work is typically done by researchers with a technical, social science or information systems leaning (Moores and Chang, 2006). Normative and, in particular, metaethical work is frequently undertaken by scholars with a background in philosophy (Bynum, 2006; Floridi, 2006; Introna, 2002; van den Hoven, 1997).

Research in ICT ethics is often multidisciplinary and attempts to come to a broad understanding of the subject at hand. Much research is focused on specific issues and problems. Among the most prominent, one can find issues such as privacy (Brown, 2000; Introna, 2003), intellectual property (Burk, 2001; Syme and Camp, 2002), access and digital divides (Rooksby and Weckert, 2006), data quality (George, 2002), but there are many others. It often overlaps with related discourses in neighbouring disciplines; for example, computer law (Poullet, 2004).

Much work in ICT ethics engages with the normative question of how normative problems can be addressed in an ethically sound way. A typical approach that tends to be taken is the adoption of a behavioural guideline, policy or code (Siau, Nah & Teng, 2002). Some of the most important professional bodies have taken this approach, for example the British Computer Society (<http://www.bcs.org/server.php?show=conWebDoc.1587>), or the Association for Computer Machinery (<http://www.acm.org/about/code-of-ethics>). Codes of ethics can raise as many problems as they solve (Fairweather, 2000; Ladd, 1985). Alternative forms of governance are, therefore, discussed in this chapter.

The aim of this chapter is not to champion any of the applications or approaches but to develop a framework that captures the work currently going on with a view to providing a more holistic understanding of research questions and expected future developments.

Policy aims

Current public policies, in particular EU regulations, are pertinent to issues of ICT ethics and influence the outcomes of our chapter. Normative perceptions and their ethical evaluation strongly influence what democratic governments perceive as issues to regulate. In current EU policy there are several areas where normative and ethical issues of ICT are addressed. ICT research has been identified as one of the three pillars of the 'i2010 — a european information society for growth and employment' initiative of the EC. i2010 is renewing the Lisbon agenda and relies heavily on ICT to realise efficiency and economic gains (<http://ec.europa.eu/information_society/eeurope/i2010/introduction/index_en.htm>).

The EU, furthermore, views ICT as an essential tool in addressing its demographic challenges. In its green paper 'Confronting demographic change: a new solidarity between the generations' (European Commission, 2005), the EC has outlined the challenges the EU is facing. The demographic development continues to be a main area of concern for the EU (European Commission, 2006). Three general trends combine to create the problem of decreasing population: continuing increases in longevity, continuing growth of the number of workers over 60, and continuing lower birth rates. The EU intends to address the resulting problems with a variety of strategies. Among them there is the aim to use ICT to allow older people to remain an active part of society, but also to allow them to remain independent in their homes. This has economic implications for health and social care, but, more importantly, it is a matter of the quality of life for EU citizens.

The aims of the European ICT policy are broad and arguably contradictory. The aim of increasing competitive advantage, for example, can lead to the use of ICT to replace traditional workplaces. Wiring companies and creating digital infrastructure can have the unintended result of facilitating outsourcing, thus further limiting the stated European aim of creating employment. To some degree, the question of the net effect of technical development on the labour market is an empirical question. The ICT policies, however, can also be contradictory in other aspects. The inclusion of disadvantaged groups in social processes is an ethically relevant aspiration. At the same time, evidence from the literature on digital divides suggests that the provision of technology can exacerbate existing barriers to social participation. The EU is aware of this and digital inclusion, with all its implications, is high on its list of priorities. An interesting question remains, however, whether general policy aims and the ICT research agenda that is investigated in the present chapter are consistent.

The first section of the ICT work program 2009 summarises the policy aims behind the EU ICT research initiatives (European Commission, 2008a: 4) as:

> Improving the competitiveness of European industry and enabling Europe to master and shape future developments in ICT so that the demands of its society and economy are met. ICT is at the very core of the knowledge-based society. Activities will continue to strengthen Europe's scientific and technology base and ensure its global leadership in ICT, help drive and stimulate product, service and process innovation and creativity through ICT use and value creation in Europe, and ensure that ICT progress is rapidly transformed into benefits for Europe's citizens, businesses, industry and governments. These activities will also help reduce the digital divide and social exclusion.

Methodological considerations

While this chapter is fundamentally of a conceptual nature and explores possible and likely futures to allow the development of more detailed research agendas, it nevertheless needs to be grounded in a shared social reality to gain acceptance of the variety of audiences who have an interest in ethical issues of emerging ICTs. In order to provide a transparent and shared account of likely developments, the empirical basis of the chapter was based on a content analysis of a range of sources. Primary among these were documents created by the EU with regard to policy planning, in particular of the seventh framework program. In order to supplement and contextualise these, other sources on ICT, its future developments and ethical issues were considered. The content analysis was conducted by reading the documents with a view to the following items: applications of future technology, artefacts, ethical issues, governance structures, and others. The findings of the analysis were stored in a mindmap for easier reproducibility, and then used for summarising the findings below.

European landscapes

This section shows the major areas of technological development in ICT as well as ethical and governance aspects related to it. It is broken down according to the main items used for the data analysis: trends, applications, artefacts, ethical issues, and governance structures. The first attempts to provide an overall view of where ICT is going. The two sections on applications and artefacts relies heavily on the most recent call for ICT projects at the time of writing this document, which is the 'FP7 ICT call 4', 19 November 2008, with a submission deadline of 1 April 2009 (European Commission, 2008a). This is the document that explains in most detail the aims and objectives of the ICT work program and, thereby, gives an exact view of what European policy makers believe to be desirable and realistic. Further documents are drawn upon where necessary.

Figure 1 represents a higher level view of the relationship among the main concepts that constitute the landscape as derived from the ICT call 4 of FP7. The relationship diagram shows there are at least three views of the derived Euro-landscape: the technology worldview; the technology supplier view and the political view. At least two relationship types exist; drivers (shown as solid arrows) and feedback (shown as chequered arrows). There might be a fourth view, related to user/victim/beneficiary. This fourth view is not easily located in the diagram. It may be better conceptualised to be on a contextual level, which is invisible in this abstract diagram.

In the following subsections we describe the individual categories depicted in the relationship diagram (Figure 1) in more detail.

ICT trends

Those who have tried to forecast the next technological advances are usually incorrect. ICT has a track record of unpredictability in the specific nature and consequent impact of these future advances. The only certain thing is that there will be always be significant advances and these will always impact upon society and its people. Several general ICT trends can be seen, however, even though the specifics are unpredictable. These trends influence the overall strategic approach, for example, to national and European research funding and to societal acceptance or rejection of technology. John Vaughn (2006: 8–14) suggests that there are four key ICT trends.

- **ICT trend 1: Ever-increasing computational power plus decreasing size and cost**

  The move towards more computational power, with decreased size and cost, can make possible improved and entirely new types of technology and new application opportunities.

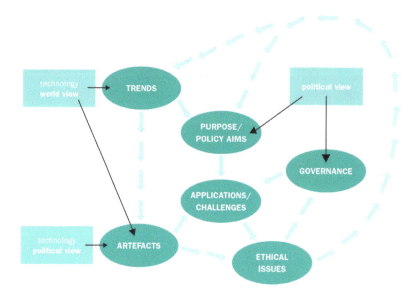

Figure 1: Euro-landscape relationship diagram

Source: Authors' research

- **ICT trend 2: Technology advances enabling new types of interfaces**

  The human interface is one of the most important determinants of whether a technology product can be used by people regardless of their skill, experience, affliction or disability. For example, advances in interface technology are creating new opportunities for better assistive technologies, more accessible mainstream technologies, and entirely new concepts for controlling both. Some of the more innovative interfaces include augmented reality, hands-free operation, voice control and direct control from the brain.

- **ICT trend 3: Ability to be connected anywhere, anytime with services on demand**

  The latest innovations such as wireless electronics, location awareness, wearable technology and implantable technology point towards a society with widespread connectivity. This allows people to think about communication, control and presence in entirely new ways.

- **ICT trend 4: Creation of virtual places, service providers and products**

  Web technologies have provided people with new ways of doing things hitherto not thought of or not possible. Such technologies have fostered the development of entirely new social, commercial, and educational concepts.

The evolution of ICT through such trends could impact upon everyone both positively and negatively. This is explored in detail by both Roe (2007) who uses a SWOT (Strengths, Weaknesses, Opportunities, Threats) analysis and Vaughn (2006) who considers opportunities and barriers. Luciano Floridi considers such trends at a meta level and argues that 'in information societies, the threshold between online and offline will soon disappear, and that once there won't be any difference, we shall become not cyborgs but rather *inforgs*, i.e. connected informational organisms' (2007). If this is the case, then the ethical dimension of ICT becomes the ethical dimension of society per se.

Applications and challenges

In this chapter we distinguish between applications and artefacts that may give rise to ethical issues. This distinction is not reflected in the ICT call 4 (European Commission, 2008a), but it can be plausibly made. By applications we mean areas where ICTs can lead to solutions or applications. Artefacts, on the other hand, are software, hardware, or related items that can be used for particular purposes. There is often a relationship between applications and artefacts and, in many cases, artefacts are built specifically for particular applications. Artefacts can usually, however, be used in different ways and different artefacts can be used for the same applications. Since ethical issues can arise in a number of ways, including the non-intended use of artefacts, we believe that the analytical distinction between applications and artefacts is helpful to our overall aim of identifying ethical issues of emerging ICT applications.

The applications that the EU views as relevant for the next 10 to 15 years are reflected in the main challenges. These challenges are divided into two groups: 'overcoming technology roadblocks and reinforcing Europe's industrial strengths', and 'seizing new opportunities and applying ICT to address Europe's socio-economic challenges'. The first group contains those challenges that can be seen as technological in nature, which seems to imply that their social and economic context is less important or maybe unproblematic. The first one of the three technical challenges has to do with 'pervasive and trustworthy network and services infrastructure.' Its content is based on the perception that current network infrastructures, in particular the Internet, is problematic and needs to be replaced soon. The second technical challenge aims at context-aware and easy to use technologies. These are perceived to be a key technology that can further policy objectives in a number of ways. The work program, therefore, calls its second main challenge that of 'cognitive systems, robotics and interaction'. The third and final technical challenge has to do with 'electronic components and systems'. These are seen to be crucial for the development of the next generation of technologies and, therefore, as a central basis for further innovation in products and services. It is interesting to note that for all of the three technical challenges the work programme is silent on their expected consequences and link to policy goals. This implies a pervasive belief that technological progress is desirable because of its knock-on effects, the hope that it will lead to successful products, higher competitiveness and thereby to wellbeing and employment.

The second group of challenges, the socio-economic ones, are more immediately and more visibly linked to the European policy goals. The four challenges meant to address 'Europe's socio-economic challenges' are aimed at specific areas where technology is perceived to have a crucial role. The first one is the area of 'libraries and digital content'. Under this heading, one can find research aimed at digitising libraries and cultural heritage. It also includes a section on technology-enhanced learning and one on intelligent information management. The second challenge addresses issues in relation to sustainable and personalised healthcare. This one is linked to the increasing costs of sophisticated healthcare that are set to further spiral because of the changing European demographics. The challenge is split in three main groups, one on personal health systems, one on patient safety and one on virtual physiological humans, which covers simulations of humans for training and research purposes. The third challenge centres on ICT for mobility, environmental sustainability and energy efficiency. Among the aims here, one can find a range of aims related to efficiency, mobility, environmental protection and distribution of energy. The fourth and final challenge on 'ICT for independent living, inclusion and governance specifically' aims at developing applications for ICT related to ageing, accessible and assistive ICT, as well as ICT for governance and policy modelling.

Together these seven challenges represent the applications that the EC sees as central to advance its policy agenda. They set the boundaries for the type of research that will be funded under the seventh framework program. They are therefore likely to have an influence on the technologies that will become viable and widespread in the next decade. It is clear that this is not an exclusive list and that there are other development agendas from private organisations, such as nation states or NGOs, that are similarly worth exploring. For our purposes, however, the EU policy is of central interest and we will therefore concentrate on these applications.

Artefacts

While the applications and challenges are relatively easy to identify and list, the same cannot be said for the artefacts that are envisaged to attain the policy goals. A detailed reading of the call 4 document displays a range of artefacts that are considered possible solutions to a variety of problems. In addition to physical artefacts, there is a strong emphasis on processes and procedures that may lead to products or services. Rather than try to identify all of the artefacts, this chapter will briefly discuss some of the more speculative ones or ones that recur as specific artefacts to be emphasised.

The most notable of such artefact is related to the future of networks and, in particular, the Internet. This is the next generation of Internet protocols, 'Internet protocol version 6 (IPv6)'. Mentioning of IPv6 recurs throughout the document. More importantly, the promotion of IPv6 is named as one of the strategic priorities of European ICT research policy.

In addition to IPv6, the call document goes on to enumerate a number of ICT artefacts that are currently of speculative, but that are seen as bearers of great potential that deserve to be developed. Interestingly, these artefacts are not linked to the challenges discussed in the preceding section and are, therefore, not clearly identifiable as contributors to the policy aims. Instead, they form a separate part of the call document, which is listed under the heading of 'future and emerging technologies'.

Given that the aim of the present chapter is to provide a framework for the investigation of ethical issues of emerging technologies, these emerging technologies are of particular interest. As they are at a more exploratory stage, their conceptual and physical form are currently still uncertain, but the technologies suggested render it clear which way the development is expected to take. The first set of such emerging technologies has to do with high-speed data processing and it is listed under 'Concurrent terra-device computing'. The next set of technologies is based on 'quantum information foundations and technologies'. 'Molecular-scale devices and systems' are suggested as a further important research area. Another predominantly technical area is that of 'bio-chemistry-based information technology'. The attempt to use cross-disciplinary research in order to improve ICTs is furthermore developed in the 'brain-inspired ICT'.

In addition to these technical challenges, there are also application-driven emerging technologies. The first, 'human-computer confluence', explores new modalities for individual and group perception, actions and experience in augmented, virtual spaces. There is also an area of self-awareness in autonomic systems, which aims at an improvement of the interaction between computing artefacts and their environment. Environmental concerns are reflected in the research towards zero-power ICT.

To some degree one can see reflected the distinction between purely technical considerations, which at this stage are not yet application-oriented, and those that are specific to particular issues. An interesting question that will guide our further research is whether this more or less specific-outcome focus of the artefacts raises particular ethical issues.

Ethical issues

The seventh framework program (Decision N°1982/2006/EC), Article 6 (1§) states that 'All the research activities carried out under the Seventh Framework Programme shall be carried out in compliance with fundamental ethical principles.' The same decision also states that 'the opinions of the European Group on Ethics in Science and New Technologies [EGE] are and will be taken into account' in research under the eventh Framework Programme. The emphasis on ethics is based on the recognition of the potential impact of ICT on human rights as established by the European convention on human rights (<http://conventions.coe.int/Treaty/en/Treaties/Html/005.htm>) and developed by the 'Charter of fundamental rights of the European Union' (<http://www.europarl.europa.eu/charter/pdf/text_en.pdf>). Such general considerations are complemented by other more specific statements, notably the extensive guidelines on addressing ethics, which are included in the guides for applicants for FP7.

Despite this high-level recognition of the relevance of ethics to ICT, it is worth exploring in more depth what is meant by ethics in the context of the EU ICT research program and how it is to be addressed. It is easy to follow the EU policy assumptions that ICT has important ethical aspects and promises solutions to pressing social and ethical issues. At the same time, ICT can raise a host of new ethical questions.

The interesting question for the present chapter is how these general ethical concerns are operationalised, and whether there is any guidance on the type of ethical problems that should be considered. There are several documents that offer guidance on how to recognise and address ethical issues. A helpful distinction to categorise different ethical issues is the distinction between ethical issues as arising out of the research process and ethical questions arising from research content. In its 'ethical guidelines for undertaking ICT research in FP7' (European Commission, 2008b) the EC lists a number of substantive issues that may result from emerging ICT. The first problem concerns the autonomy and privacy of potential users. Researchers are reminded that a responsible approach is required and that compliance with European and national legislation is required. Further substantive issues are those connected to specific technologies, such as implants and wearable computing, which have been elaborated by the

European Group on Ethics. E-health is seen as a further area worthy of specific warnings as it poses particular problems to privacy and security. The same is true for nano and bio-electronics.

The same concerns that are included in the ethical guidelines are reflected in annex 4 of the guidance for applicants, which also forms a part of the proposal form. This annex is a checklist that covers informed consent, privacy, and ICT implants. Additional issues, which are not further explained, are research on human embryos / foetuses, research on animals, research involving developing countries, and dual use of ICT for military or terrorist purposes. The points on this list are elaborated in the 'Ethics for researchers' document (<ftp://ftp. cordis.europa.eu/pub/fp7/docs/ethics-for-researchers.pdf>).

Questions of the social consequences of widespread use of particular technologies, which, in areas such as e-health, could have foreseeable consequences, are not elaborated. The documentation does not address, for example, how one can recognise terrorist applications and how to address such issues. One could argue that such substantive ethical issues of emerging technologies are beyond the scope of concrete current guidance and that this type of issues should therefore be covered by procedures that allow researchers to be alerted to ethical questions. A look at the procedural guidance shows, however, that it is not geared to capturing such issues either. The two main aspects of procedural guidelines are to ensure informed consent and to comply with legislation. Both are well-established ways of dealing with issues arising from the process of doing research. Informed consent, in particular, is the cornerstone of ethical conduct of medical research. It is open to question, however, whether it is sufficient to deal with ethical issues arising from emerging technologies. It is interesting to note that the guidelines do acknowledge that there are likely to be hitherto unrecognised and emergent ethical issues resulting from advances in ICT research. Due to the apparent reliance on procedural ethics, it is important to ask which procedures are envisaged in the governance structures of projects.

Governance structures

The most immediately visible aspect of governance has to do with ethics review of projects. Ethical review is described as one aspect undertaken by the panel of experts that undertake the scientific evaluation of a project. The panel of experts will identify a project as requiring special attention if 'projects raise sensitive ethical issues or when applicants failed to address ethical issues in an appropriate way.' ('Ethical review procedure, <http://ec.europa.eu/research/ science-society/index.cfm?fuseaction=public.topic&id=130>). All projects thus identified, as well as all projects dealing with human intervention or human embryonic stem cells, will undergo ethical review. This ethical review will be conducted by a panel of experts and it aims to determine whether the project follows the standards of ethics of FP7. Projects that are found to be in violation of such fundamental ethical issues are then excluded from funding.

Such external governance of projects is described in some depth, but there is little guidance on internal governance of research projects. There must be some explicit ethical governance, for example, in order to ensure that the procedural human research issues, in particular informed consent, are addressed according to standards. This will presumably require some sort of ethics committee but the exact form of such a committee is not clear. Specifically with regards to dual use, the 'Ethics for researchers' document recommends the recruitment of an advisory board, which can advise the project consortium on societal, political, and legal aspects of potential applications, on exploitation and dissemination strategies. In addition to such external guidance, the 'Ethical guidelines for undertaking ICT research in FP7' state that 'activities may, if appropriate, include specific tasks or a specific work package that explicitly addresses ethical concerns (in terms of the research, its conduct and outcomes) and outlines how ethical issues raised by the proposed research will be handled'. Further guidance on how such work packages are to be defined, which membership is desirable or how they are to be integrated in the project is not given.

Summary

It is easy to imagine that there are further general categories of relevance to be explored for a better understanding of the ethical issues related to emerging ICTs. For our current purposes of charting a landscape of such ethical issues, the outlined categories offer enough of a differentiation to allow for a detailed picture of issues that can reasonably be expected to develop in the medium term future of 10 to 15 years. Table 1 below summarises the issues enumerated in this section. It is easy to see that there are numerous possible combinations of trends, applications, artefacts, and ethical issues, which allow questions of how they relate to policy aims or which type of governance structure would be likely to be able to address them. This table provides a high-level summary of the landscape of emerging ICT ethics and can be used as a basis for further research.

Table 1: Summary of emerging EU ICT research landscapes

Trends	Purpose/policy aim	Applications/ challenges	Artefacts	Ethical issues	Governance structure
Ever-increasing computational power plus decreasing size and cost	Economic growth	Network infrastructure	Physical artefacts	Research process	External governance
Advances enabling new types of interfaces	Employment	Cognitive systems, robotics	Processes and procedures	Informed consent	Ethical review as part of scientific review
Ability to be connected anywhere, anytime with services on demand	Demographic challenges solutions	Components, systems, engineering	IPv6	legal compliance	External advisory board
Creation of virtual places, service providers and products	Social/political inclusion	Digital libraries and content	Concurrent terra-device computing	Research content	Internal governance
		Healthcare	Quantum information foundations and technologies	Privacy/data protection	Work package on ethics
		Sustainability	Bio-chemistry-based information technology	ICT implants/wearable computing	Informed consent procedures
		Inclusion	Human–computer confluence	E-health related issues	
			Self-awareness in autonomous systems	Nano- and bio-electronics	
			molecular-scale devices and systems	Research on animals	
			Brain-inspired ICT	Research involving developing countries	
				Research involving human embryos or foetuses	
				Dual use (military or terrorist applications)	

Source: Authors' research

Conclusion

This chapter aims to provide an overview of current social, political, and technical developments with a view to provide a framework for further research. It has identified current EU policy with regards to ICT research, which will have manifest effects when current research and development projects come to the stage of market entrance, 10 to 15 years from now. To be useful, the framework itself needs to be expanded and applied as the basis of further research.

This chapter draws on a particular document from a particular source to identify relevant concepts that allow us to understand which emerging ICTs we can expect and which ethical issues these may raise.

It is, therefore, beyond question that, in order to come to a better understanding of the subject area, a richer understanding of the field is required. This is true with regards to the identification of emerging ICTs, and probably even more so with regards to ethical issues. Ethics in real life is always a question of context and situation, which general descriptions cannot capture. This leads to the question of how such contexts can be incorporated. The answer decided upon by the ETICA consortium is that it can only be done as a second step, once a general understanding of the landscape of emerging ICT has been gained. The reason for this decision to concentrate on the abstract first is that a consideration of substantive ethical issues at an early stage is epistemologically impossible and practically infeasible. The epistemological issue arises from the fact that any technology can be applied in an infinity of context, some foreseen, most not. In each of these contexts the technology may raise a number of different ethical issues. It is thus not possible to give a representative overview of these issues. A more abstract level of analysis is therefore required before more detailed cases can be subjected to analysis.

This approach does, however, raise new issues. Primary among them is the question of how a general understanding of emerging ICTs and their ethical issues can be developed. The approach chosen by the ETICA consortium in order to address this question is to do an analysis of a number of sources that focus on high-level visions of emerging technologies. The EU FP7 ICT call analysed here is one example of such visionary documents but, in order to have a more representative understanding, further documents will need to be included. These will include other European documents but also policy documents from other areas of the world; for example, the United States, Australia and Japan. The justification of the choice of such high-level views is that, collectively, they are likely to give a good understanding of what the political leaders of this world envisage as future technologies. Such a vision is likely to influence factual developments through the means of research funding and legislative developments.

At the same time, an understanding of what is expected to happen in actual research should be sought via an analysis of the vision documents produced by leading research institutions around the globe. Such institutions, in many cases, have mission statements or comparable documents that give an insight into the type of technologies on which they are working, which gives a view of what is likely to become reality in the medium-term future.

The sources of the ETICA data analysis can thus be represented as follows:

Figure 2: Delimitation of data sources

Source: Authors' research

The delimitation of data sources, as suggested in Figure 2 is justified by the fact it will give a good view of what is intended and envisaged by organisations that are in a position to enforce their view of the world. It should not be misunderstood as aiming at a true or verifiable prediction of the future. As with any future-oriented research, this project is not able to do this under any circumstance. What it can do is aim to provide plausible possible futures to enrich discourses concerning desirable futures and possible ways to get there. In order to spawn such discourses, one needs to give thought to what aspects of the sources to analyse and in what way, so that the outcome of analysis is of relevance and produces novel insights.

The decision to concentrate on documents to investigate which technologies are emerging and which ethical issues can be expected betrays a constructivist assumption of the project. Technology is not a neutral and independently existing entity but it gains meaning through the medium of discourses. Analysing such discourses can, therefore, help to understand the formation of the meaning of technologies. At the same time, there is a potential infinity of emerging technologies and their applications.

It was, therefore, decided to follow the phenomenological idea of identifying the essences of phenomena. This essence is then going to be subjected to an ethical analysis. The immediate question following this statement is going to be: how do we identify the essence of an emerging technology? The answer, from the point of view of a researcher interested in emerging ICTs and ethics, is that the essence of the technology is the way in which it changes the world and the relationship of humans to the world. Central questions to be asked of texts, therefore, include the way the technology is represented in the text, which includes sub-questions of the expected social impact, the representation of society, social values and anthropological views elaborated in the text. Of similar interest are critical issues; that is, social, legal or ethical issues outlined in the text. To avoid falling into the trap of either social or technological determinism, not only the social representation of the technology will be covered but also the artefacts that it produces; that is, its technical constraints or capabilities. These aspects will be collected together with higher level categories, such as fields of technology, target audiences, fields of application and, wherever possible, application examples. In addition, metadata on the sources will be collected.

As a result of this, a grid of analysis will be developed that will allow an analysis of a rich field of emerging technologies, which can be analysed from a variety of perspectives. Once the analysis is finished, it will, for example, be possible to compare predicted technologies on the basis of the type of source they are derived from, or investigate whether particular types of application lead to particular types of predicted critical issues. The analytical grid will lend itself to a relatively simple input of further data; for example, from surveys from project coordinators. The analytical grid will contain sufficient information to allow a normative ethical analysis of particular technologies. And, finally, it can be used to determine whether interesting clusters of technologies, issues, or descriptions emerge that warrant further attention. In this way the analysis will allow a targeted description of emerging ICTs, a sound linkage of these technologies with ethical issues and a justified basis for policy advice.

The earlier analysis of a particular policy document has thus shown how relevant categories of analysis can be distilled from an existing source. The categories enumerated here (see Table 1) are not identical to the categories of analysis enumerated in this section, but they are encouragingly similar. As a result, the exercise shows that it is reasonable to expect to find information on the relevant categories in policy-related documents.

The current chapter therefore provides a good starting point of further data collection and analysis, provides a sample of such analysis and shows how further research can proceed. As a result, the ETICA project will be able to produce findings that go beyond restating the obvious without falling into the trap of unjustified concreteness in the light of an unknowable future. It will give an indication of possible futures and thereby contribute to the aim of allowing a more proactive feed-forward approach to ICT policy.

References

Adam, A, 2005, *Gender, ethics and information technology*, Palgrave Macmillan, Basingstoke.

Brown, WS, 2000, 'Ontological security, existential anxiety and workplace privacy', *Journal of Business Ethics*, vol 23, no 1, pp 61–5.

Burk, DL, 2001, 'Copyrightable functions and patentable speech', *Communications of the ACM*, vol 44, no 2, pp 69–75.

Bynum, TW, 2006, 'Flourishing ethics', *Ethics and Information Technology*, vol 8, no 4, pp 157–73.

European Commission, 2005, 'Confronting demographic change: a new solidarity between the generations', COM(2005) 94 final: Green Paper, viewed 10 December 2008, <http://ec.europa.eu/employment_social/news/2005/mar/comm2005-94_en.pdf>

——, 2006, 'The demographic future of Europe — from challenge to opportunity', COM(2006) 571 final, viewed 10 December 2008, <http://ec.europa.eu/employment_social/news/2006/oct/demography_en.pdf>

——, 2007, 'FP7 in brief: how to get involved in the EU 7th framework programme for research', viewed 26 November 2008, <http://ec.europa.eu/research/fp7/pdf/fp7-inbrief_en.pdf>

——, 2008a, Work programme 2009; Cooperation; Theme 3, Information and communication technologies (European Commission C(2008)6827 of 17 November 2008), viewed 26 November 2008, <http://cordis.europa.eu/fp7/dc/index.cfm?fuseaction=UserSite.FP7DetailsCallPage&call_id=185>

———, 2008b, *Ethical guidelines for undertaking ICT research in FP7 (Annex 5) to ICT*.

Fairweather, NB, 2000, 'No, PAPA: why incomplete codes of ethics are worse than none at all', in G Collste (ed), *Ethics in the age of information technology*, Centre for Applied Ethics, Linköpings Universitet, Linköping, Sweden.

Floridi, L, 2006, 'Information ethics, its nature and scope', *ACM SIGCAS Computers and Society*, vol 36, no 3, pp 21–36.

———, 2007, 'A look into the future impact of ICT on our lives', *The Information Society*, vol 23, no 1, pp 59–64.

Forester, T, 1994, *Computer ethics: cautionary tales and ethical dilemmas in computing*, 2nd ed, MIT Press.

George, RTD, 2002, *The ethics of information technology and business*, WileyBlackwell.

Introna, LD, 2002, 'The (im)possibility of ethics in the information age', *Information and Organization*, vol 12, no 2, pp 71–84.

———, 2003, 'Workplace surveillance "is" unethical and unfair [opinion]', *Surveillance & Society*, vol 1, no 2, pp 210–16.

Ladd, J, 1985, 'The quest for a code of professional ethics: an intellectual and moral confusion', in D Johnson & J Snapper (eds), *Ethical issues in the use of computers*, Wadsworth Publ. Co, pp 8–13, retrieved 14 December 14 2008, <http://portal.acm.org/citation.cfm?id=2569.2570>

Marturano, A, 2002, 'The role of metaethics and the future of computer ethics', *Ethics and Information Technology*, vol 4, no 1, pp 71–8.

Moores, TT & Chang, JC, 2006, 'ethical decision making in software piracy: initial development and a test of a four-component model', *MIS Quarterly*, vol 30, no 1, pp 167–80.

Poullet, Y, 2004, 'The fight against crime and/or the protection of privacy: a thorny debate!', *International Review of Law, Computers & Technology*, vol 18, no 2, pp 251–73.

Roe, PR (ed), 2007, *Towards an inclusive future: impact and wider potential of information and communication technologies*, a COST 219ter report. COST, Brussels.

Rooksby, E & Weckert, J, 2006, *Information Technology and Social Justice*, Information Science Publishing.

Siau, K, Nah, FFH & Teng, L, 2002, 'Acceptable internet use policy', *Communications of the ACM*, vol 45, no 1, pp 75–9.

Sollie, P, 2007, 'Ethics, technology development and uncertainty: an outline for any future ethics of technology', *Journal of Information, Communication and Ethics in Society*, vol 5, no 4, pp 293–306.

Syme, S & Camp, LJ, 2002, 'The governance of dode: open land vs. UCITA land', *ACM SIGCAS Computers and Society*, vol 32, no 3.

van den Hoven, J, 1997, 'Computer ethics and moral methodology', *Metaphilosophy*, vol 28, no 3, pp 234–48.

——, J, 2008, 'Moral methodology and information technology', in K Himma & H Tavani (eds), *The handbook of information and computer ethics*, WileyBlackwell, pp 49–68.

Vaughn, JR, 2006, *Over the horizon: potential impact of emerging trends in information and communication technology on disability policy and practice*, National Council on Disability, Information and Communication Technologies, a COST 219ter report, COST, Brussels.

Weckert, J, 1997, *Computer and information ethics*, Contributions to the study of computer science (175), Greenwood Press, Westport, Conn.

Weil, E, 1969, *Philosophie morale*, Librairie Philosophique J Vrin.

16. Ethical issue determination, normativity and contextual blindness: Epistemological studies of the limits of formalism in ethics and their consequences for the theory of governance

Philippe Goujon and Catherine Flick
Université de Namur, De Montfort University

Information society, technical development and ethics

The impact of techno-scientific developments on societal evolution and lifestyles no longer needs to be demonstrated. In particular, the last half of the twentieth century has witnessed a considerable acceleration of the integration of technological elements into the means of economic production and social life in general. The profound transformations that have taken place in the last few decades equally involve energy, transportation, construction, telecommunications, administration, medicine, pharmacy and agricultural sectors. These transformations are closely linked to techno-scientific developments and particularly to stunning developments in information and communications technologies (ICTs). The information society emerging in the contemporary period, however, can no longer simply be considered as a result of technical mutations. Up to now, this ongoing global phenomenon that is technological, economic, political and cultural, is in search of social and political projects, references and reaffirmed values. We are faced with the task of building networks that are based on a cultural model incorporating clear collective choices, so that the principles of democracy are transferred on line — hopefully without loss — in the future. The knowledge society is the embodiment of a change in civilisation whereby science and technology have become omnipresent and are developing at a disconcerting rate.

More and more people, in particular through their work and their responsibilities, are questioning the relationships between knowledge and power, between science, technology and society, and the notion of governance whereby decision-makers impact on the public. In this fast-changing context, the relationships between government and citizens, between industry and consumers, and between designers and end-users is no longer the same. New governance models must be established and socio-economic progress must be rethought, providing a stronger consideration of the sustainability issue.[1] Consequently, different relationships and a positive dialogue between the scientific community, society, society's decision-makers and end-users are required, as underlined in the Council of the European Union Decision of 30 September 2002 (2002/835/EC).[2] This document emphasises the need for 'structuring the European Research Area' (2002–06), adopting a specific program for research, technological development and demonstration: 'Today, and even more in the knowledge-based society of tomorrow, science and technology have a ubiquitous presence throughout the economy and in everyday life. If they are to realise their full potential in securing a continually increasing quality of life — in the broadest sense — for Europe's citizens, new relations and a more productive dialogue between the scientific community, industry, policy-makers and society at large, as well as scientists' critical thinking and responsiveness to societal concerns, will be needed.'(12)

The evolution of ICT is driving our society towards situations and applications where humans interact so deeply with 'intelligence', pervasively distributed among them, that, at some point, we will reach a divide where a fundamental choice will be presented to us: to develop a 'utopian' environment where all humans will have access to an empowering and accessible techno-environment ('ambient intelligence (AmI) environment') or head towards a 'dystopian' environment where Bentham's panopticon[3] will become an increasing reality due to thousands of sensors, computers and networks that will track every human movement. David Lyon called this new form of 'cooperative surveillance', 'synopticon' and 'scopophilia' (1993; 2006). The notion goes beyond Bentham's panopticon and was interpreted by Zygmunt Bauman (1995) as a significant trend of the globalisation process.

In a world that is becoming increasingly homogenised, any capacity for questioning is likely to be stifled by the rational constraints extended to all

---

1   See, for example, European Commission (EC) (2002), *Science and society: action plan*, European Commission, Brussels.

2   Council of the European Union, Decision of 30 September 2002 adopting a specific program for research, technological development and demonstration: 'structuring the European Research Area' (2002–2006): (2002/835/EC). English version: <http://eur-lex.europa.eu/LexUriServ/LexUriServ.do?uri=OJ:L:2002:294:00 44:0059:en:pdf>

3   The famous Panopticon was designed by Jeremy Bentham, British jurist and philosopher, towards the end of the eighteenth century. It is a type of prison, also known as the 'Inspection House', which enables an observer to watch all the prisoners without their knowledge. This essential principle of construction is reflected in the Greek neologism, pan — meaning 'everything', and opticon — concerning 'vision' and, as such, the word is meant to express 'the all-seeing place'.

fields. This brings us back to the question of the possibilities and the place of ethics within this framework of corrective regulation. The problem of the real action of norms in the context of technical development is crucial and curiously largely ignored, and needs to be scientifically taken into account if ethics is to be effectively integrated in technical development and more generally in the relationship between science and society. The need for serious attention to the problem of allowing the conditions for the development of a responsible information technology is strong. It is based on the conditions for developing what is sometimes referred to as value-sensitive design, which recognises that any technology and/or artefact (ie ICT) embeds values into the technological design, research and development.[4] It assumes that human values, norms and moral considerations are imparted to the things we make and use and it construes information technology (and other technologies for that matter) as a formidable force which can be used to make the world a better place, especially when we take the trouble to reflect on the ethical aspects in advance. A wider approach in evaluating emerging technologies should include not just the legal dimension (often referred to in the ICT field as 'compliance') or the social acceptance, but also the ethical dimensions which imply a normative horizon that can't reduce itself to the mere social acceptation.[5]

Conditions for a transformative critical room[6]

What we propose is to review the state-of-the-art in respect of the ethical analysis of ICT developments. Practically, the main problem consists of the absence of background: the strong push for technology development too often obscures the need for ethical consideration before a technical project is funded, developed and deployed. Some efforts have begun to consider ethics and ICT in the AmI domain that adopt different approaches such as analysis from scenarios or 'ethical review' panels (set up after the project has started; eg MINAmi).[7] These consist of 'ethical experts', who may come from a separate community:

---

4 It is highly problematic, however, in that there is no mechanism developed for the choosing of the relevant values: simply relying on the emergence of values once the technology or context is inspected (Friedman, Batya, Kahn, Jr, Peter H, & Borning, Alan, 2006, *Human–computer interaction and management information systems: Foundations*, ME Sharpe). Value-sensitive design, then, relies on the identification of direct and indirect stakeholders, the benefits and harms for each (particularly looking at the different levels of competence for each group, for example, the elderly may not be so proficient with the use of technology). For that reason one shouldn't confuse value-sensitive design and ethics. A value-sensitive design approach does not by itself ensure that the values that will be embedded in the technical artefact are ethical values.
5 A society can, in fact, reach a consensus on something that is ethically unacceptable.
6 The concept of transformative critical room is from Crutzen (2003). These are characterised as those interaction worlds where actions of questioning and doubt are present, which have the potential to change habits and routines, where the 'change of change' has a differentiated potential.
7 MINami (Micro-Nano integrated platform for transverse ambient intelligence applications, an FP6 project).

the 'technical' community is typically separated from the 'ethical' community.[8] One of the reasons for this separation is the specialisation of high-level studying (eg, within universities) where the 'technical' study plan of engineers is different from that applied in 'humanities' subjects of other faculties.

An interdisciplinary approach is strongly required.[9] Indeed, some technical universities in Europe are introducing areas of study such as 'Science, society and technology' for engineers, or 'Computer ethics' for computer scientists, but these courses are not always included as a mandatory component. Scientific projects can raise ethical questions, evidenced by the rules and procedures of the European Commission for the systematic ethical evaluation of projects submitted for funding in FP7 (Seventh Framework Programme). The elaboration of ethical standards, however, is made uneasy by the common dividing line which separates the justification of norms and their application. These two levels ought, however, to be integrated if one considers the issue of ethical universality, which has to be newly analysed within the context of a multicultural Europe (Drenth et al, 2006). The reflexive articulation of ethical norms and cultural contexts raises many problems, the first of which is the problem of the conditions of an ethical reflexivity.[10]

This is natural because the researchers and technical developers of ICT systems focus mostly on the technical and economic challenges before them, and are not usually aware of potential ethical issues because they see ethical considerations and analysis as an obstacle to the technical and economic development. In short, the problem we must first analyse is not so much the problem of determining solutions to ethical issues as to settle the conditions for raising ethical questions, and for a new approach that authorises a real reflexivity which allows for a questioning of the integration of ethics in complex technical systems. The obligations set out by economic constraints, interests concerned with the influence of experts, the general impression of the ineluctability of the technical projections, social requests, and the needs for the consumers make it increasingly difficult to define the conditions of a critical perspective respecting the moral autonomy requirements for thought.

---

8   For example, noted in the 'Human report ethical audit' produced in December 2004 (edited by P Goldie, SA Döring and WP10 members, <http://emotion-research.net/deliverables>. Humaine (Human-Machine Interaction Network on Emotions) was an IST FP6 project.

9   There is some recognition of the interdisciplinary problem, which has resulted in the PEACH project ('FP6 coordination action on presence') that includes a working group on social impact, legal issues and ethics. The approach taken is to analyse social impact scenarios and raise and address potential ethical issues <http://www.peachbit.org>

10   Reflexivity may be defined as the capacity of actors and institutions to revise basic normative orientations in response to the evolution of economic, techno-scientific or political systems and to shortcomings in current modes of regulation. This reflexivity is not given, however, and needs to be specifically integrated into governance approaches.

We do not mean that existing criticisms don't allow for a certain reflexivity. We support, nevertheless, that these criticisms, generally, show a tendency to restrict ethics to a *categorical field* by sacrificing the existence and tension in the name of flattering pragmatism, being satisfied with ad hoc answers to artificially isolated specific contexts and being conditioned by the reigning instrumental rationality. An example of this would be the use of legal regulation or best practices documents that are decontextualised and not reliant on ethical principles, but which instead aim to avoid litigation or censure.

The danger and problem is to limit the debate to the scientific perspective alone (hence the importance of expertise, and the tendency among politicians to favour traditional, 'top-down' governance of activities in which risks are involved) and, shunning an approach based on technology assessment (TA), debate the meaning and the ethical, cultural and social stakes. Instead of initiating an inclusive debate on the nature of the different forms of knowledge and visions of the world, discussions limit the debate by adopting a positivist and, more often than not, reductionist approach that leads to cognitive closure, where those involved are unable to consider other potential issues, or to 'step outside the box'. Hence, we ask the question: how can we elicit the cognitive opening-up required for a genuine reflexivity that would allow us, as Jean Ladrière (1984) puts it, to extract the existential and the political meaning from the objective meaning? How can we decompartmentalise ethical discourse and make it play a more important role in the joint construction of technologies? How can we transcend the neo-classical, technological approach (the cost-benefit approach for Slovic)?

The economists' answer appeals to the industrialists, for it confirms their practices and habits. Positivism has become so potent these days that the project of emancipation through reason is, for the most part, either rejected or ignored. In its place appear instructions on how to increase one's power over social processes that have been reduced to the status of objects. Hence again — even if, as Ulrich Beck has stressed, this is a perspective that needs qualifying — there is significant difficulty in controlling the rampant growth of technological innovations politically. More often than not, political institutions make do with a regulatory and financial framework within a dynamic system accompanied by positive feedback that leads to a self-fulfilling positive feedback loop.

In this context, expertise, be it philosophical or scientific, becomes the indisputable new source of normativity,[11] and the problems revealed are confined to a scientific perspective alone — which means that the problems taken into

---

11 Normative is contrasted with its antonym, positive, when describing types of theories, beliefs, or statements. A positive statement is a falsifiable statement that attempts to describe ontology. A normative statement, on the other hand, is a statement regarding how things should or ought to be. Such statements are impossible to prove or disprove, thus forever banishing them from the world of the scientific.

account are confined to the realm of strict scientific rationality, and democracy is confiscated. One must be wary of any theory which tries to objectify the world we experience (emotion or attitude, as in the MIAUCE[12] project — Multi modal Interaction Analysis and exploration of Users within a Controlled Environment) in order to predetermine the form of the world we share. There's a risk that the possibility of genuine reflexivity will be stifled by a technological and scientific rationality imposing its value system with, as a result, a dismissal of the prestige of moral reason.

This dismissal of the prestige of moral reason found credit in 'progress' of all kinds that were acquired from scientific work and technological discoveries. This progress, tested only slightly in many fields, maintained the idea (or the myth) of a possible emancipation with regard to morals. Modern thought, thus, by no means needs morals, since it is, in itself, an action, basing humans in knowledge and technical rationality.

What is at stake is of importance, indeed various sectoral ethics tend increasingly to reinforce social differentiation characteristics of modernity by proposing an internal, and specific, framing of moral problems, with the risk of exclusion of other external and alternative framings. As a consequence, ethics is disconnected from the design of technological devices and the lack of a concrete grid of assessment concerning the embedding of ethics in technological development makes this issue important. Briefly, this is the European situation: some ICT projects investigate ethical aspects, but ethical considerations are not a 'matter of course' in the development cycle. In some cases, ethical experts are brought in at some stage in the project to assess the ethical implications. As a result ethics is often an 'add-on', a sort of accessory and instrumentalised guarantee, and not properly integrated nor understood in its methods and objectives, which are clearly very different from the method and objectives of sciences and techniques.

Consequently, there is a need for the inclusion of ethical consideration before, during and at the end of technical and scientific projects, so that the technology 'incorporates' and tackles the ethical side (within its whole concept and implementation). The risk is that by not analysing the conditions (institutional, rules, cognitive) for the effective integration of those considerations in the context of a technical project, the ethical considerations will be excluded from the technical rationale and treated as a totally separate domain. The consequence of this separation is a loss of impact, and an undermining of the integral role of ethics in the application of technology. This is understandable since the technology can only be limited to the set of its rules (that is, objectivity, and the technical rationality which frames its vision and conception).

---

12   MIAUCE is a European Commission FP6 project.

Ethics is never in the answer (always conditioned) but well into this dynamic movement of questioning, before the action and on a border, which separates our subjective existence (with its presuppositions, preferences, convictions and hidden motivations) from the constraining externality (economic constraint: political, hierarchical, technical, and ideological). The ability to position ourselves with a moral freedom is fundamental, as we are then able to question ethical feasibility and conditions.

These issues are at the basis of this problem if we desire to take into account the fundamental changes that affect our world.

A world in the midst of change

What is at stake at a time of globalisation and the expansion and domination of technology, market forces and calculating rationality, is the capacity of societies to regulate themselves. It is also the possibility of taking a reflexive relationship into account in the emergence of collective action based on individual behaviour in institutions of self-regulation.

The feeling that there is something automatic about the growth of technology and economic and administrative rationalities is all the more widespread since their political origins are masked, reinforcing the impression that the prospect of democratic supervision of social sub-systems — the possibility, that is, of collective autonomy in the conditions of a radicalised modernity aimed at dominating nature and mankind — is barely credible, if not to say irrational.

Yet, the growing desire shown by citizens today to be involved in a certain number of decision-making mechanisms is tangible. This is notably the case with regard to situations perceived as presenting risks: people are loath to run risks that they have not been warned about beforehand, that they have not accepted or that have not been debated in a democratic manner. The questions this raises extend well beyond the field of science and technology and are essentially bound up with political and ethical issues, as well as with technological, economic and financial stakes. It has to be accepted that, in order to face these challenges and the risks and dangers attendant upon the spectacular growth of the techno-sciences, we need to rethink our modes of governance in science and technology.

Arguments involving a semantics of risk or danger are commonplace in modern societies and now dominate discussions of governance. Though the risks that might be mentioned are heterogeneous in nature (bio-technological risks, epidemiological risks, or challenges relating to geopolitical security), a number of common features can be identified. The scale of each risk is virtually global, both in terms of the threat itself and of the solutions which might be envisaged.

Each involves complex phenomena, funds of knowledge that are in conflict with one another, developments in science and technology and substantial public funding if the challenge is to be met; each presents potentially gigantic mortal dangers and, above all, involves much public debate about current assessments of the risks involved. These risks pose challenges in the realms of politics and governance. To understand them is essential, all the more so since supervision of the norm is no longer limited to vertical, state-type supervision alone but today passes through far more diversified and complex channels. These questions relating to the role of governance in a context that is uncertain and involves risks cannot seriously be addressed, therefore, without posing the question of normativity and of the links between society, norm and law.

Social complexity and judicial norm

Today, the State has *de facto* forfeited its monopoly on social regulation and the production of judicial norms. Faced with the problem of regulation and social order in a society characterised by the multiplication of sources of normativity, the failure of policies designed to reinforce the means available to the welfare state and the failure of deregulation, Jurgen Habermas (1992; 1997) and Günther Teubner (1995) have each put forward a solution: self-regulation for the former, proceduralisation for the latter.

Both take as their starting point the difficulty of achieving collective autonomy in a modernity radicalised in its project of dominating nature and mankind. Social subsystems must be integrated by 'proceduralisation'. This means that the law should identify itself with what Teubner calls the 'structural judicial preconditions for a self-regulating society' (1995: 89). Similarly, Habermas favours a process-based model of law and, in so doing, means to give a new legitimacy to the legal system.

If one takes as given the functional differentiation of society and the multiplication of subsystems, each employing its own rationality and each running the risk of closing in on itself, the question arises not only of communication between these subsystems, but of where the legal system stands in relation to them. As Teubner sees it, 'the law cannot monopolise the epistemological authority pertaining to other forms of knowledge and alone assume full responsibility for all the constructions of reality. ... It cannot, however, delegate all epistemic authority to the other forms of social discourse,' (1995: 202) with the attendant risk of conflicts emerging between these different systems.

Faced with conflicts of this kind, there is a very real danger that, in many instances, positivism will end up presenting itself, in cultural terms, as the only genuine form of knowledge, the only activity endowed with meaning, and will forgo, as Habermas has remarked, 'any reflection on the role it plays in the social context, that of a legitimation of technocratic domination or of vested interests'[13] (Gentil, 1997). The activity of the state is no longer geared to achieving practical ends, but to finding technical solutions to crises affecting the system. Hence the growing importance of experts and the depoliticisation of civil society. Hence, also, the force of technocratic ideology, an unacknowledged domination that arises from the fact that it hides behind technological rationality. The risk approach can, in many cases, be a strategy to divine more profound issues.

The limits of the risk assessment approaches

The perception of risk cannot be disassociated from a type of informal assessment of technology which combines socio-economic and ethical stakes and the question of the relationship (whose relationship?) to institutions. In the absence of certainty, precaution means coming down on the side of procedural strictness. When the 'truth' of a situation and the 'reality' of a risk cannot be established, it is the strictness of procedures, and of the actors involved in drawing up, implementing and overseeing those procedures, which becomes the dominant value. *The absence of certainty does not prevent us from analysing and managing a situation strictly.*

To limit the debate to the question of risk alone, regardless of the problems relating to risk assessment, has an important consequence: when all is said and done, it limits the debate to the scientific perspective alone (hence the importance of expertise, and the tendency among politicians to favour traditional, 'top-down' governance of activities in which risks are involved) and, in shunning an approach based on TA, shuns a debate on the meaning and the ethical, cultural and social stakes of biotechnology and genetic engineering in the case of genetically modified organism (GMOs) or on the representation of ICT. Instead of initiating an inclusive debate on the nature of the different forms of knowledge and vision of world, discussions limit the debate by adopting a positivist and, more often than not, reductionist approach that leads to cognitive closure, even if the framework is masked by an appearance of deliberative democracy the implementations of which (particularly citizens' conferences), disconnected from decision-making bodies, serve only to mask traditional forms of governance 'from the top down'.

---

13   J Rivelaygue, *Leçons de métaphysique allemande*, Vol. II / Kant, Heidegger, Habermas, p 459, quoted in Gentil (1997).

The autonomy of the technique in question and the legitimacy of ethics

It is undeniable that 'the scientific method', as a considered and self-checking specification of the rational method, is at the base of the common dynamics which crosses the properly scientific field as well as the technological field. What this dynamics tends to generate, is an autonomous reality, an intermediary between nature and properly human reality, a kind of 'third world', of which the formal structure is given in conceptual architectures of science and the concrete figure in the equipment of all kinds which constitute around us, the extremely dense network of what one calls the technical world. Nevertheless, if there is autonomy of the technical sphere, this autonomy is, however, only an expression of an ideology, that of the engineer, or a society which justifies, by subjecting its various components to scientific and technical rationalities, its incapacities to find political and social remedies for its problems. In a world full of doubt, characterised by its complexity, the technique becomes, or tends to becoming, in spite of the undeniable suspicion which weighs on it, the supreme reference of the truth and the pragmatico-economic effectiveness. The myth of progress, after having been identified with science, coincides with the technique, accompanied with a decline of the question of the truth in the name of a pragmatism which is more than ready to respond to all the economic and industrial requirements of the context.

In this arena, ethics is reduced to playing the role of mere justification and legitimisation of what was *a priori* decided on according to economic, technical and industrial justifications, without being able to exercise its normative specificities and characteristics. This is a consequence of the fact that, conditioned by the technical framing, we have forgotten that technology is not neutral. It is a manner of thinking, of making, and of transforming the world which cannot be disassociated from policy or subjacent choices of ethics. From this point of view, data processing and ICTs are not neutral and *reflect*, in their constitution and their use, expectations of the society and are influenced by their socioeconomic context. For example, technological innovations that focus on satisfying needs of individuals also come to fulfil a function dependent on the cultural features of the society in which they fit. As in any society, these needs can be seen as negative or positive, for instance, perhaps negatively to control, or more positively to align the political, social and institutional goals with the individuals' desires (Lyon, 1993). On this view, the organisational life has to become increasingly rationalised and controlled and organisational control will be 'less and less apparent and increasingly powerful' (H Isaac and Mr Kalika, 2001; WJ Orlikowski, 1991; JR Barker, 1993).

These characteristics also apply in a broader social sense, with the result that the introduction of ICT also shapes social practice: 'information technology has become a constitutive technology and partly constitutes the things to which it is applied. It shapes our discourses, practices, institutions and experiences in important ways' (van den Hoven, 2007: 68).

Non-neutrality of ICT and AmI

This recognition of the non-neutrality of ICT should not be surprising, after all, technology is designed with a purpose guiding the technical direction. The implications of the technology, however, in its capacity to affect and change social practices are not so easily seen. The context in which the development takes place — the culture, language, discourse — already provides the framework for the resulting artefact. Consequently, although an expert in another field (eg ethics) can advise on possible impact, the advice is reinterpreted within the context and cultural knowledge of the development team, and thus does not extend the understanding to a wider view. That is, 'learning' has not taken place. Without this deeper understanding brought about through learning, we are all bound in some way to narrow horizons that need an external mechanism to be opened to wider horizons. In the case of ICT and AmI, more generally, the construction of the social legitimacy (and not just acceptability) of science and technology requires that the subjacent metaprinciples of normative nature be considered along with techno-scientific actions. The technological artefact forms the system and is also a form of organisation and perpetuation of social reports, which are a manifestation of the dominant ways of thinking and behaviours. The technological artefact is always an historical and social project, and this project reflects the intentions which nourish a society, the interests which dominate it and the values which guide it.

> Technological progress in all these fields has contributed to the shaping of the AmI vision, but at the same time, progress in the diffusion and acceptance of past and current ICTs was mutually important. In 10 years, GSM has grown in Europe to more than 300 million subscribers. In many European countries, penetration rates of mobile phones are above 70% of households. And SMS has given a considerably boost to the mobile services market during the last years. Internet access from home has increased in the EU15 to 40% in June 2002 (against 28% in October 2000) (Eurobarometer 2002). If the increased diffusion and usage of computers, the Internet, mobile phones, etc. had not happened, technological progress in these fields would have slowed down. And acceptance of these technologies is enabled by demographic and social

trends such as the emergence of individualism, diversity, mobility, and choice of personal life styles, affecting the structure of groups and community and the ways we live and work. (Punie 2003, 17)

In the case of the vision of AmI, at first glance it seems that it is based on technological progress in the fields of microelectronics, communication networks and interfaces. Nevertheless, it is also driven by socio-economic factors that go beyond the technologies alone, including but not limited to economic neoliberal rationality, rationalisation of the production, demographic conditions, impact of terrorism, and consumerisation of communication.

For a critical perspective

As we have underlined, every technological artefact is a construction which rests on some *a priori* foundation (social, political or economic) and, even if it is partially suggested by preliminary information on the behaviour of the objects, it reinterprets the latter starting from its own categories. The political impact of a technological artefact cannot thus be assigned to this artefact alone, but must be allotted to the techno-speeches which diffuse it, give it a specific meaning and envisage for it specific usages. Data processing and innovations related to ICT, if they seem to be binding to individuals, come, actually, to satisfy a need and to fulfil a function largely dependant on the cultural features of the society in which they fit.

It is only to the condition of recognising the non-neutrality of ICT that one can start to change its cognitive framing and consider ethical and societal issues. Without this propaedeutic step, one can just interpret the world and technology within the restricted cognitive fields allowed by its framing (in our case the technological framing). The result of this is to either negate any justification of ethical and societal considerations or to instrumentalise them and consider them as a means to obtain a sort of ethical guarantee and label. This latter issue is a problem with all ethical guidelines: they don't take into account the issue of their application, and so, most of the time they have no effect at all. The mechanism which consists of providing the answer expected by a given context (such as economic or industrial contexts) poses ethical questions since the justified context itself becomes the justification of the social function of ethics.

This recognition of the non-neutrality of ICT nevertheless returns a realistic ambition of relativising instrumental rationality and aiming at political and societal control, which means also its rehabilitation into the world of social and cultural life. The immediately urgent issue is to correct the manner of approaching ICTs aiming at applying approaches which dissociate the social approaches from the technological ones, and political approaches from economic and ethical. Too often the accepted responses are only the economic, political and institutional constraints. It is undeniable, for example, that policy-makers are fascinated by technology. Positivism continues to influence our political leaders, who are in a state of utter disarray in relation to the complexity of our world. A symptom of this tendency is the call to use technology in all areas of public policy — education, health, environment, administration — to solve the problems which affect our society, and afterwards use an appeal to ethics to justify the decisions that were already taken (for example, the European ethical assessment of technical projects).

Thus we need to recognise that the possibility of a critical perspective doesn't mean we can ensure its reality, and indeed many factors can prevent the effectiveness of its realisation.

Ethics in trouble

We must acknowledge here that the background we use, and build on, refers to the contextual pragmatics and the theory of governance developed by the Louvain school,[14] as it offers a solid, theoretically founded framework which ensures the above requirements. This theory relies particularly on opening up the process to many stakeholders, who reflect and assess the ongoing feasibility for the governance tools used throughout the full lifetime of the project in order to ensure efficiency. The justification for relying on this theory is that it directly addresses the problem of the conditions for the effectiveness of norm expression. It does this from a theoretical perspective and using an applied set of studies.

As it is currently, ethics is restricted to a categorical field by a pragmatic use of ad hoc answers to artificially isolated contexts which are conditioned by the reigning instrumental rationality. Deontological codes and ethical guidelines are good examples of this because they relate to issues that are, most of the time, sectorially defined with a theoretical approach that is chosen by 'experts' employing a decisionist process (which refer to an a priori chosen principle, with their own specific framing, determined by their field of expertise. This sectorial ethics reinforces the social differentiation characteristics of modernity by proposing both internal and specific framings for moral problems. The risk, here, is that the possibility of genuine reflexivity will be stifled by a technological and scientific rationality imposing its value system. Sectorial ethics tends to also propose an internal and specific framing for moral problems, risking the exclusion of other alternative framings. For example, in ICT particularly, developers will solve problems using technological means (such as websites

---

14   For more details, refer to Lenoble & Maesschalck (2003; 2006); Coppens & Lenoble (2000); Maesschalck (2001).

offering privacy controls) rather than questioning whether the technology should raise such problems in the first place (exclusion of social or cultural framings).

In this context, expertise, be it philosophical, scientific or economic, becomes the indisputable source of normativity, and the problems revealed are confined to these expert perspectives alone. As a result, ethics is often an accessory without proper integration nor understanding of and respect for its methods and objectives, which are very different from the methods and objectives of the science or technological development.

Consequently, there is a strong need to investigate and reflect on the governance conditions that allow for an ethical reflexivity inside the project and for the effectiveness of that reflexivity, so that ethics does not reduce itself to a means to satisfy the conditioning constraints of the context, such as economic, political, or scientific constraints. If reduced as such, it transforms the question of the ethical acceptability into the question of the social acceptance of the technology, which is easily addressed by sociological enquiry. Such governance conditions also need to allow for the inclusion of ethical consideration before, during, and at the end of technological and scientific projects. This is so that the technology incorporates and is confronted with the ethical side throughout its conception, development, and implementation stages.

Within a regulation framework, for ethical integration to be successful, it firstly needs to be considered for inclusion in that framework, and to be accorded a certain status which acknowledges and respects its fundamental epistemological and methodological characteristics and requirements.[15] For this we need to consider the cognitive structures, the stock of knowledge, the habitus, based on

---

15 Agreeing with Ladrière, 'The ethical dimension [...] manifests itself when it exercises and founds an authority under an unconditional and, in a certain way, intemporal form. Speaking of a historicity of ethics introduces different variations and relativity' (translation by the authors). Nevertheless, as Ladrière recognises, there exists a historicity of ethics and that, for a very simple reason, is related to the fact that historicity is a fundamental constitutive dimension of the existence in the meaning where it can't be reduced to a mere contingent proprietary but, on the contrary, must be understood as a structural requirement. Ladrière, in this, drew from the lessons of contemporary phenomenology, according to which, the historicity is a dimension of the existence of the dimension underlying the historical consciousness by which it is left as a historical form. Ethics, itself a constitutive dimension of the existence, is necessarily affected by this other constitutive dimension which is the historicity specific to the human condition. As a result, the ethical dimension must necessarily inscribe itself in the historic effectiveness; it cannot deploy itself concretely in the 'historic flesh'. There is no ethical effort that does not model itself on the historical effort. But to this reverberation of the historicity on ethics, and to this historicity of ethics, should be added, according to Ladrière, an 'ethicity' of the historical, that is to say, an ethical determination of the historicity (understanding essentially as the projection into the effective real history of the ethical dimension). If there is a historical determination of ethics (historicity of ethics: the first level of this comes from the fact that ethics evolved, and the second that it acquires its effectiveness only through the historical forms that it takes; this explains why the ethical field is an open, not closed, field), there is also, reciprocally, an ethical determination of historicity (ethicity of the historic). Such a crossing constitutes what Ladrière called 'the dual relation'. J Ladrière, 'Philosophie politique et philosophie analytique', in Ladrière & Van Parijs (1984: 222).

our background, which founds our relation to the world. The background, or the 'stock of knowledge' (from the Husserlian analysis), gained from the analysis of the experience, is a corpus of preliminary knowledge and is necessary for the act of interpreting the world.

In this context, the 'stock of knowledge' takes the form of an individual's interpretation schemata (after Kant (1964)) or the cultural inheritance (according to Alfred Schütz (1943, 1951, 1954, 1959)). The concept of 'stock of knowledge' can be defined as a concept whose meaning is broader than the concept of the 'reserve of experience'. It gives an account of the preliminary knowledge that is also from social origins. The stock of knowledge of the socialised individual has a basis of knowledge that is composed of three parts: lived experiences, common knowledge (traditional or scientific knowledge), and knowledge related to corporality, skills, and other practical things. The processing of the stock of knowledge as preliminary knowledge (foreknowledge) is used in theories of interpretation. It is not difficult to see a parallel between the cultural knowledge in Schütz's (1951; 1954; 1959) conception and the Gadamerian (1976) conception of prejudices. After the rehabilitation of the prejudice, the former receives an important function in understanding practice, and becomes one of the fundamental elements of the hermeneutic circle.[16] The interpretation practice always begins (according to Gadamer) from a preliminary knowledge. Concerning the action, to quote Ladrière (1993), 'the action mobilises itself, always under the inspiration, be it tacit or not, of an informing interpretation'. The action inscribes itself concretely in a class of events which anticipates and which is conditioned by its comprehension of the insertion process within its context. This understanding rests on a background that refers to the principles and presuppositions, related either to a community, or, for example, more globally to the dominating value of the society. The prejudice or the preliminary knowledge allows us to go within the circle of interpretation. From our perspective, it's precisely the a priori frame of contextualisation of practical meaning in reference to the actual possibility that needs to be destabilised, so as to allow the opening of the cognitive closure resulting from the framing effect.

To talk about background here is not an unnecessary detour. Through exploring these concepts, the normative discourse can be confronted by practical logic that may neutralise, by practical anticipation, the normative production's sought consequences. We can also conceive of the possibility of an internal distancing of the value systems that block all possibility of alternative understandings and all possibility of discussion of common norms (again, the framing effect). It

---

16 According to Spinoza, in order to understand extremely dense written material, one needs to keep in mind the historical background as well as the person who wrote it (that is, their history, attitudes, culture, outlooks, and bias). In order to understand the parts of the text, one must understand the whole, but the whole can only be fully understood based on the parts. The hermeneutic circle is the moving back and forth between these contexts, between the whole of the text and its parts (Ramberg & Gjesdal, 2005).

allows us ultimately to think about opening the closed cognitive doors, and to think about the access to the contextuality of the normative production. Marc Maesschalck (1998) writes that

> the contextual efficiency has a meaning only within the strategic plan as information on the appropriateness of the mean or movement. Concerning the justification of the action, it has an effective relation only to itself on the model of a hermeneutics of the strategic judgement of application. We lose from that fact all possibilities of an immediate reference to the specific coherence of the context as an environment of the practical application of normativity if not as an environment that configures the relation to the norm in general. (1998)

The urgency is also not to limit ethics to questions of the institution and the decontextualised (or a priori contextualisation referring to a predetermined context) application of the standards and principles. Indeed, such an ethics is purely a decisionist one. Its reference immediately limits the problem of the adequacy of the ethical representation to the context of action and also ignores the origin and legitimacy of this representation — a practical way to automatically justify it and instrumentalise all ethical considerations and, also, to reduce the complexity of the context. This problem is all the more important from an ethical point of view, because the theoretical image of humans conveyed by a sphere of activity deserves, as underlined by Maesschalck, 'to be questioned according to its effective impact on the culture and the traditions. The extrapolation of a model of activity can become totalitarian when it claims to redefine the relation of a human with its horizons'. The function of knowledge in modern sciences is, in an increasing way, conceived in relation to the system of work. These sciences generalise and rationalise the technical capacity of humans on the *objectified* processes of nature and society, with the use of techniques of administration increasingly more effective and generalised.

Once the knowledge is reduced to the applied science, this type of science and rationality monopolises the rational behaviour. The positivist conscience, consequently, dogmatically combats any theory referring to the practice differently from those improving the possibilities of the practical applications. That means that the practical (moral) questions become the object of a decision and are confined to the irrational field, unless they are not, in their turn, subjected to the only technical criteria of instrumental rationality. This results in 'blocking any possibility of discussion on common ethical standards while imposing, preliminary a manner of understanding the world in general, a meaning of life determined usually inextricably by traditions, solidarity and personal identifications'. The explanation is that the directives required by the action are divided into a rational determination of techniques and strategies and into an irrational choice of systems of values.

The consequence, for the economy governing the choice of the means, is a total freedom characterised by a 'decisionnism' in the choice of the highest goals (values associated with the action). However, as Ladrière underlines it, recalled by Maesschalck,

> the decisionism is radically insufficient, because one cannot be satisfied to seek and pose principles of action and from that justify the action. It is also a question of permanently assessing the 'ethicity' of the lived situations, i.e. of their particular relationship with the ethical requirements, themselves included/understood ideally like realisation of the human one.

It is the fundamental reason why what is ethically at stake is not so much to find an answer but to make room for ethical thinking, that is to say an attitude that can localise the issues raised by technology and have a theoretical tool with which to face them. Without that ethical standing relying on ethical references, any answers are useless since the cognitive tools to understand and apply them are not present (the cognitive framing being in that situation still the technical framing). The ethical step cannot be limited to provide the elements to establish the justification of a decision. The ethical interrogation refers to the construction of a human order and questions the way humans are perceived and treated,[17] a construction that is most of the time absent from the TA process.

---

17   Ethics concerns convictions that are not explained in a logic of compromise. A consensus can respect the interests of all the parts in question without being, by its object or its purposes (industrial, economic, scientific), in conformity with the ethical requirements. The ethical provision precedes the consensual effort and is its only guarantee. But this provision is not obtained by negotiation, it refers to the intentionality of the actors, their relation to the real world, their conceptions of existence and on which values those conceptions are founded. Without such an image of human responsibility in relation to a real world, the stake of the ethical reflexion disappears. In the field of subjectivity, the ethical interrogation is not satisfied by transformation of the individual forces into social autonomy. It seeks to exceed the responsibility concerning the measurable to open a reflexion concerning the responsibility with regard to the non-measurable; that is, with regard to the destiny of humanity, life, and our world. If we think that it is impossible for us to measure what we are and to measure what is life, a particular dimension of our responsibility must also concern the irreducible dimensions of the subjects, of life, dimensions which are not calculable. Compared to the life given, ethics is a concrete answer where a figure of humanity with its specific features is at stake. It thus answers partially in the concrete world to the challenges of a future for the human world. It cannot elude the specific fields of activities. Ethics answer the injunction of reality when it questions the rational choices for the management of the limits of our capacity to answer. It refers the human action to a destiny which exceeds it, that of the life which is 'carried out' in it. Ethics must assumes a normative authority which wants to be free of any contextual constraints, otherwise ethics would be subject to the reign of instrumental rationality, and would transforms itself in its justification for objectives that are not ethically founded but may be, for example, economically driven. Refusing to be subject to such constraints we assume, in this chapter, deliberatively the normative authority of ethics.

TA expertise and ethics

All forms of TA processes involve an expert committee designed to give input on the potential impacts of the technology. Many of the more modern forms of TA have stakeholder input, using one or more of many tools available to gauge the concern of users, shareholders and interest groups. The inclusion of participants from outside the expert community and from the groups that are likely to be affected by the technology is important for not only the identification of norms, but the construction of the contexts within which the technology will function. The norms, however, constructed by both the experts and the stakeholder participants are in no way required to be *ethical* norms and, in fact, they are most likely to be societal norms and expectations of the target groups. Of course some of these may correspond to ethical norms, but there is no definite requirement within TA for the explicit establishment of normative ethical horizons.

The ethical approaches that are used in TA depend greatly on the context, though it is safe to say that ethical approaches are not usually appealed to as such, yet some are underlying the motives for carrying out the TA. Consequentialism features highly in traditional TA, such as those identified in 1980 by Joseph F Coates:[18] '[Technology assessment] emphasises those consequences that are unintended, indirect, or delayed', or by Marvin Cetron and Lawrence Connor in 1972:[19] 'Technology assessment is an attempt to establish an early warning system to detect, control, and direct technological changes and developments so as to maximise the public good while minimising the public risks'. Later on, however, more deontological approaches were underlying the ideas for incorporation of the public and other interest groups in the discussions regarding technology and the potential impacts of it on society. Normative TA processes involve a particular focus on assessing technology against moral principles such as beneficence, respect for autonomy, justice, and harm prevention (Verbeek, 2006). A virtue ethics-based approach is rarely found in TA (however implicit), because it relies on assessment of the processes and practices involved in directing technology rather than the technology itself.

In terms of reflexivity, TA processes rarely involve any such reflexivity within their own procedures. With each type of TA comes a structured approach that is followed and includes the choosing of stakeholders and experts and methods of conducting focus groups. TA could, however, be used as part of a reflexion process within a project, allowing for a learning operation to take place and

---

18   Joseph F Coates, 'Technology assessment: here today, gone tomorrow', in *Technological forecasting & social change* 49, 1995, pp 321–23.
19   Marvin J Cetron & Lawrence W Connor, 1972, 'A method for planning and assessing technology against relevant national goals in developing countries', in Marvin J Cetron, Bodo Bartocha, *The methodology of technology assessment*, Gordon and Breach Science Publishers, New York, 1972.

then the project to adapt to the findings of the reflexion, and to determine the conditions for effective integration of norms into the technology being developed. Real-time TA does attempt to integrate some sort of reflexivity into its approach, by assessing the technology throughout its lifespan of design and implementation, but this is limited by the primitive nature of early technology and the problem of choice of stakeholders. These limitations restrict TA to being a tool to incorporate into an overall reflexive approach on a larger view of the technology and its ethical impact on society, rather than for it to be used as the point of reflexivity.

From contextual blindness to ethical governance

Ethics is in the position of questioning before an effective action, and is on the border that separates our subjective existence from constraining externality. Paul Ricoeur defines ethics as 'the aim of a good life with and for the other, in just institutions' (1990: 202). It is a teleological conceptual framework, therefore, with the aim for a social, collective, just way of living together in a 'good life'. In the cultural realm, ethics can be perceived as a legitimisation process of this pluralism of conceptions of 'good life'. There is a diversity of ethical demands that reflects an autonomisation of the action spheres in the social life and individuality sphere. These ethical responses to these demands align themselves according to the professional, social, and cultural contexts in which they are formulated and used.

There are a number of philosophical responses to the conception of ethics:[20]

1. Analytical: ethics is conceived as a theory of principles from which a preferable interaction can be formed (Rawls). It responds to deontological preoccupations.

2. Pragmatic: ethics is related to consensual procedures that allow the institution of common norms and their collective control (Habermas), related to the choices of society and of ethical decisions in conflicting contexts.

3. Contextualist: ethics is a way to rationalise the values of a culture (Charles Taylor), and allows formalisation of the normative pretention contained in cultural opinions.

---

20   As quoted by Maesschalck, 'The last book of Enrique Dussel proposes an analytical landscape of the main tendencies of contemporary ethics and questions their negative relation to the material dimension of values.' (E Dussel, 1998, *Ética de la liberación en la edad de la globalización y de la exclusión*, (Maesschalck, 2001).

4. Semantico-symbolic: ethics is a disposition to respond to the absolution of freedom in relative acts (Ricoeur, 1990), and is directly related to the subjective aspiration for the respect for human life.

So, contemporary ethics keeps its diversity, but the diversity of tendencies hides some of the specificity and the consequences of such specificity. For example, analytical ethics (Rawls) and pragmatics ethics (Habermas) postulates a semantic indetermination. Semantic ethics (Ricoeur) finds that the legitimisation process of the norm doesn't belong to ethics, but that the real ethical question is the response to the injunction of the suffering of others. Contextualist ethics (Taylor) proposes a realistic ethics of the way that the subject assumes its cultural space.

The semantico-symbolic approach appears to be a connection between ethical efforts and life in itself, with the openness of the structure constitutive of the subjection to the injunction of historical reality. The problem is, however, that the actor needs to voluntarily submit itself to ethics: ethics' main challenge is this confrontation of irrational human behaviour with rational choices for the management of this behaviour.

Thus, if ethics is to assert a normative authority, which is a fundamental characteristic of ethics, since ethics concerns the relationship between what is and what should be, it is necessary to question the condition for its effective integration in the context of technical projects. The conditions for effective integration relate directly to the normative conditions, that is, the relationship between the norm and the context. Once knowledge is reduced to the rational field, limited to sciences and applied sciences, the practical questions are perceived as belonging to a different realm, which is seen as not having the same properties as the scientific and technical realm, and become the object of a decision. They are then confined to the irrational field, unless they are subjected to the only technical criteria of instrumental rationality. This decisionism is radically insufficient, since one cannot be satisfied (as often occurs within deontological and expert approaches) to choose and pose principles of action and, from those, justify the action (that will in return justify the principles). Here we find ourselves in a circular argument that is clearly unfalsifiable, because the result is justified by the principle, and the principle is legitimised by the result. The ethical step cannot be limited to provide the elements to establish the justification of a decision, since ethics must assume a normative authority, free of any contextual constraints. By this we mean that it is not the norm that is free of any contextual constraints, but the relationship that the individual has with the norm. If it is not, it is simply subject to the rules of instrumental rationality, and becomes a justification for objectives that are not ethically founded but are, for example, economically, politically, or scientifically driven instead.

In general there is a fragmentation of practices due to these ad hoc sectorial responses to artificially isolated specific contexts. This fragmentation is generally dealt with by inflating ethical approaches and responses (for example, deontological codes, ethical guidelines and ethical charts). The mechanisms that give responses expected by these specific contexts are not questioned, since the justified context becomes itself the justification of the social function of ethics.

The classical deontological, consequentialist and utilitarian ethical theories are in trouble: they talk about values without having contextualised these values (suspending this contextualisation), with the context later appearing mysteriously (and conveniently). This lack of foundation of the context means that it escapes the rationalisation proposed by ethical theory.

Proceduralism

What about proceduralism (as a philosophical approach, and not as a political method)? The option of proceduralism is generally taken to be a solution to the problem of the contextual limits of the other ethical theories. Indeed, it appears as a sort of synthetic combination of requirements that are usually separate among the two competing traditions of moral philosophy, the deontological and the teleological, in the broad sense of the word. If one takes the example of Habermas' discourse ethics and, especially, the principles of discussion and of universalisation (Principle D and Principle U) (1981: 81–2; 145–46), it is undeniably an original procedural combination of requirements. It combines criteria that are both deontological (criteria of universality and obligation) and teleological (criteria of ends and consequences). The strength of this procedural approach in ethics is that it provides a dialogical grounding of the moral rules, as opposed to the supposedly monological one in Kant. It then links up the individual and the community's will, without supporting any particular, substantial statement on the content of the moral rule or the ethical end.

It has to be remembered that dialogism in practical philosophy, including ethics and politics, functions as a powerful influential framework and can be termed the 'dialogical revolution'. Thus, the dialogical paradigm gathers a great variety of authors and streams, forming an impressive set of dialogical sub-paradigms or models. This is obvious within the contemporary philosophy rooted in the 'Aristotelian' triptych (logic, dialectic and rhetoric). One can mention, for instance, the 'dialogical logic' of Lorenzen and Mackenzie,[21] the 'new rhetoric'

---

21   See for example: JD Mackenzie, 'The dialectics of logic', *Logique et Analyse n.s.*, vol 24, 1981, pp 159–77; or P Lorenzen and K Lorenz *Dialogische Logik*. 1978, Wissenschaftliche Buchgesellschaft: Darmstadt.

of Perelman[22] or the 'new dialectics' of Van Eemeren and Grootendorst.[23] This is also the case in contemporary philosophy and human sciences, including linguistics, psychology, sociology, or anthropology. One can mention the critical rationalism of Popper, the phenomenology of communication of Lanigan, the polyphony of Bakhtin, the anthropology of communication of Winkin, the dialogical sociology of technology in Callon and Latour.[24]

It happens that some streams and authors, which are critical towards the dialogical paradigm, such as the current researches of Axel Honneth elaborating on the notion of struggle for recognition, are refinements of the Habermassian model grounded on the activity of communication (1996). Some others like Pierre Bourdieu (1997), for instance, who is certainly, with Rorty, one of the most critical authors of Habermas, also advocates for an ideal of free social communication. This norm for dialogical situations, which is quite different in the method of discussion from the ideal speech situation, is, surprisingly enough, close to that of Habermas. Today, it is difficult to develop theoretical and conceptual investigations as well as practical and experimental implementations in ethics without referring to the dialogical paradigm and, further, to the procedural approach. Dialogical proceduralism has become to some extent the ideological grounding of the social exchange in general and of the discursive exchange in particular.

Yet, dialogical procedures, while being a means to avoid the pitfalls of both deontological and teleological theories, face, at least on some crucial points, the same kind of limit. The major limit is that of the context and, more precisely, the relationship between the rational justification of norms and the context of application of norms. Dialogical procedures cannot avoid criticism of the limited relevance of norms to the context, and so their supporters claim for the moral norms to be regarded as universal rules or, at least, universalisable ones. The problem is that the procedure of discussion within a community, even an informed community like a parliament, for instance, is not sufficient to warrant the relevance of the norm to a wider community, who is supposed to apply this norm within the proper life form. There must be something more than a mere procedural discussion to elaborate a rational justification of a norm and almost simultaneously adapt it to the specificity of a social and cultural context.[25]

---

22 See for example : Chaïm Perelman and L Olbrechts-Tyteca, *The new rhetoric: A treatise on argumentation*, University of Notre Dame Press, 1991.

23 See F Van Eemeren and R Grootendorst, *A systematic theory of argumentation, the pragma-dialectical approach*, Cambridge University Press, 2004.

24 M Callon and B Latour, 'Unscrewing the big Leviathan: How actors macro-structure reality and how sociologists help them to do so', in K Knorr-Cetina & AV Cicourel (eds), *Advances in social theory and methodology: Towards an integration of micro- and macro-sociologies*, Boston: Routledge & Kegan Paul, 1981, pp 277–303.

25 S Lavelle, contribution to the EGAIS deliverable — P Goujon, C Flick, 'Deliverable 2.1 grid-based questionnaire development'.

In order to address this contextual blindness, we will criticise the procedural reason behind it. An ethical norm needs to be conceived with a feedback mechanism that is induced by the anticipation of the norm's insertion into the coherence of an application field. It must be questioned, from its point of view, about its capacity to participate in the emergence of a way of life. If we take the conditions seriously, all such Kantian schema approaches are fundamentally insufficient because they are still dominated by a logic of subsumption (deductive relation). The context itself refers the norm to its effective possibilities, confronting it with all of the possibilities in which it can be accommodated within the world. This fundamental and radical epistemological reversal is contained by all normative elaborations and, by itself, allows the norm to be translated in effective power in a given context. This is because the normative elaboration process integrates the reflexivity on the conditions that ensure the effective expression of the norm from the beginning.

A governance process, therefore, is needed that integrates 'learning' through the process of development and broadens the view beyond the technical, opening questions and debate regarding values. To do this we need to address the question of the conditions required for the effective insertion of an ethical norm into a context, from the very beginning.

To be able to confront these practices with this question, we need to understand the limitations of actual ethical and theoretical trends, in particular the limitations of proceduralism.

Limitations of proceduralism

Proceduralism (particularly that of Habermas, 1981) holds that a normative statement's semantic content is not important, but the approval process by concerned parties is the validating step. Thus, free approval needs to be gained for a normative statement to be considered valid. In exploring the limitations of proceduralism we rely on the analysis of Lenoble and Maesschalck (2003) in their book *Toward a theory of governance: the action of norms*.

Proceduralism seems to solve the problems outlined above because it proposes an internalist, rather than externalist, point of view. The current Habermasian (1981) approach takes into account the reversibility of the operation of justification and the application of rules. This dimension of reversibility indicates a co-dependence between these two operations, that is, that the application of the rule act retrospectively upon its justification. Therefore, the choice of a relevant norm necessitates a reflexive return on its context of application. This, however, leads to a supposition that the elements which condition the application of the rule are subsumable under rules that guide formal discursivity of the mechanism which is mobilised by the formal calculation of the relevant rule. Procedural relations, however, can resolve situations where two conventions of relations to the norm come into conflict, which is not a situation that can be resolved with the conventional relation. Every judgement, and thus every norm, since a norm is the result of a judgement, can only be applied at the end of an operation. The operation is an exchange between the effects expected by the abstract norm and the possible effects raised by the coherences specific to the existing way of life; that is, we encounter the difficulty of actually expressing abstract norms in reality: that it does not necessarily invoke the effects expected, and the effects that are realised are quite specific to the context in which the expression takes place.

The norm can thus only be inscribed in reality, and can only make sense by being supported by particular perceptions of the way of life, and particularly those whom the rule affects. It fails to allow for any reflexive capacity for the actors to identify the various effective possibilities on which the operation of the selection of the relevant norm will be carried out. There is also an implicit limitation of this approach, which is that the discursive and rational construction of the definition of the relevant norm is capable by itself to take into account all possibilities of the social context to be regulated. Although there is a reflexive approach to rationality, there is no reflexivity relating to the context.

Although this does not invalidate the discursive procedure for reasoned elaboration of relevant rules, it does require that the arrangements for normative production be more complex. The choice of a relevant rule consists of a making a decision about a real-life solution that is supposed to optimise the ideal objective drawn out by the anticipation of an idealised way of life. The choice of this solution, however, rests precisely on an operation of the selection of possibilities that does not exhaust the possibilities of the context within which the idealised way of life would have been realised. Transforming the context with the view to incorporating an ideal norm within it will only result from operations of interpretation that cannot be formalised. Nothing within the formal mechanism can guarantee that the choice of possibilities taken into account in order to define the choice of a norm will ensure that the realisation of this ideal will correspond to any one of the diverse possibilities perceived by the actors concerned. It is therefore necessary for a rational procedure of the calculation of norms to be intersected by reflexive incentives. This would allow for the reconstruction of the problems that condition peoples' practical acceptance of the transformation of their way of life. Without this prior reflexive effort, the processes that would enable the effective expression of relevant norms would probably not be taken into account.

The operation of the selection of possibilities carried out by the production of each norm cannot therefore be restricted by the rules, because of the guarantee of the discursive operation of formal reason. The effective possibilities drawn out by the social application of a norm are a function of the conditionality that depends on the structure of the context. These can only be reconstructed within a concrete act of community reflexivity whose conditions are not anticipated by the formal rules of the discursive act.

Using the Louvainist contextual pragmatic theory (Lenoble & Maesschalck, 2003; 2006) we can underline the limitations of every theory that presupposes the conditions that makes the exercise of reason possible. This is true for all the theoretical ethical approaches that are characterised, according to Lenoble and Maesschalck's terms, as 'intentionalist, mentalist, and schematising' presuppositions.[26] Even if individuals are able to revise or adapt these conditions, they do not take into account the reversible or reflexive character that allows for these revisions or adaptations.

The criticism we level here emphasises the necessity of understanding the reference to the background as a speculative and transcendental logical constraint of the operation. This allows us to better understand the consequence of our approach to the reflexivity of judgement on the level of the construction of governance arrangements.

Theoretical consequences for an ethical governance

As Lenoble and Maesschalck point out in *Toward a theory of governance* (2003, 91–3) 'Every norm aims to institute a way of life that is judged to be rationally more acceptable'.[27] The formal rules that condition the rationality of this choice, such as calculation of optimisation, argumentative rules, or any formal mechanism, don't guarantee, by themselves, the transformation of existing ways of life. The realisation of the ideal way of life that is called for by the norm is conditioned by something other than the simple formal validity of the rule.

The norm can only be expressed in reality by establishing a reflexivity on the perceptions of the ways of life that are lived by and accepted by those to whom the norm is addressed. To suppose that the adaptation of the dominant

---

26　The 'intentionalist' presupposition is so named because the norm's effects are supposed to be deducible from the simple intention directing its adoption; and the 'mentalist' and 'schematising' presuppositions are so named because what enables the determination of the effect of a norm is supposed to be linked to the rules (or schemes, in Kant's language) located in every mind. In this presupposition, there is a function of mental capacities, therefore, that do not at all depend on a thinking subject's exterior context. The operation of application can be considered, in this case, as a simple formal operation of deduction on the basis of the rule itself.

27　The following background and justification for a reflexive approach is largely based on the large amount of work done on this subject by Lenoble and Maesschalck in the Centre de Philosophie du Droit (CPDR), at the University of Louvain.

perception and the corresponding actions in reality will happen automatically or can be directly linked to the simple implementation of a formal mechanism conditioning the social acceptability of the norm is misunderstanding this reflexivity.

The insufficiency of proceduralism, as explained previously, is evident in that the arrangements that are necessary for organising the reflexive capacity for the actors to identify the various effective possibilities on which the operation of the selection of the norm will be carried out are problematic. Whether a norm is effective in modifying a way of life in a rationally acceptable way presupposes an independence from the discursive procedures that are used to select what is rationally acceptable. That is, all the procedural mechanisms and rational approaches to the determination of a norm cannot by themselves assure the modification of the way of life.

If we were to increase the capacities for reflexivity with regard to the conditions for the production of the norm, the effectiveness of norm expression could be measured according to the incentives required to enable the reflexive reconstruction by the actors, driven by what motivates their institution of a new way of life.

Without the organisation of a common reflexive capacity, and the form of negotiation it involves between the norms to be constructed, the normative injunction risks remaining ineffective, even if the objective is judged relevant and legitimate. The operation of judging the conditions of the choice of the rationally acceptable idealised way of life, that is, the rational determination of the norm that is supposed to enable the realisation of this objective, and the effective transformation of this way of life by the application of the norm, is distinct and asymmetric. Asymmetry is the way in which the social meanings of a norm are conditioned by an operation that cannot be anticipated by formal variables of reasoning (variables that condition the norm's relevance). Therefore every reconstruction of the process that was enacted by the production of a norm itself mobilises two operations, which do not respond to the same conditions of production. The intersecting articulation of this asymmetry is the focus for governance arrangements.

In order to do this, it is necessary to organise the reflexive capacity of the actors by constructing the capacities of the reflexivity in such a way as to not presuppose it as already existing due to a formal method, such as argumentation, deliberation, debate or discussion. All of these formal methods presuppose their own required conditions and, as such, do not necessarily involve reflexivity. It is therefore important to make sure that every application of a norm presupposes not only a formal moment of choice of its acceptable normative constraints, but an operation of the selection of the possibilities according to the acceptable way of life within the community concerned.

Without a negotiated construction of the moment of reflexivity that is specific to the conditions for the application of the norm, however, there will be no control of the process of the expression of the norm, and it will be left to the dominant common culture to express. Thus, what is often presented as the only effective choice is always conditioned by an operation such as the above (including in the construction of the deontological codes). Criticism of this reconstruction of the reflexivity used in the construction of the social norm also affects the moral approaches to legitimacy. Economic theories often obliterate the operation of the choice of possibilities that already condition the effects of rational decisions,[28] but the deliberative or communicative approaches also miss the question of the conditions for an effective expression of the ethical objectives they intend to promote.

Institutional cooperative arrangements are necessary for the effectiveness of the expression of norms in concrete situations, as well as for the legitimisation process for the norm. These arrangements result from the contextual limitation as an inescapable part of the reflexive operator of modality.

According to our analysis, what is needed is a profound change in the 'modes of inquiry' (Stiglitz 2002, 244; Bohman 2004, 347) that respect the requirement of the fact that the context has to be constructed in the ethical analysis respecting its complexity and secondly respect the ethical normativity characteristic.

Conclusion

In this chapter, we have argued that the current methods for approaching ethics in technical and scientific projects are insufficient because they lack a thorough mechanism for implementing ethical change within these projects due to a separation between technology and ethics, that is, that they are highly formalist approaches.

To take into account these limitations and achieve second-order reflexivity, we need to escape the binds of formalism, which constrains ethics with its presuppositions, that is, that the determination of ethical issues provides a method for resolution of these ethical issues, and internal limitations. To more effectively incorporate ethical norms into contexts, it is necessary to construct the framing of the context in relation to the norm (ie not presuppose it), then

---

28   This blind point affects the rational choice theory (efficiency) framework.

open up this context so that we can have a reflexivity on the opening of this framing (that is, a feedback mechanism). In order to do this, there is a need to reconstruct, from a normative perspective, how research projects should reconstruct the two-way relationship between the norm and the context to overcome the fundamental limitations outlined above in order to achieve a second-order reflexivity, an issue that will be addressed by the EGAIS project and by the authors of this chapter.

References

Bauman, Z, 1995, *Life in fragments: essays in postmodern morality*, Blackwell Publishers, Oxford.

Barker, JR, 1993, *Tightening the iron cage: concertive control in self-managing teams*, Administrative Science.

Bohman, James, 'Realizing deliberative democracy as a mode of inquiry: Pragmatism, social facts, and normative theory, *The Journal of Speculative Philosophy*, vol 18, no 1, pp 23–43.

Bourdieu, P, 1997, *Méditations pascaliennes*, Seuil, Paris.

Callon, M & Latour, B, 1981, *Unscrewing the big Leviathan: how actors macro-structure reality and how sociologists help them to do so*, in K Knorr-Cetina & AV Cicourel (eds), *Advances in social theory and methodology: Towards an integration of micro- and macro-sociologies*, Boston: Routledge & Kegan Paul, pp 277–303.

Cetron Marvin J, Connor Lawrence W, 1972, 'A method for planning and assessing technology against relevant national goals in developing countries', in Marvin J Cetron, Bodo Bartocha, *The methodology of technology assessment*, Gordon and Breach Science Publishers, New York.

Coppens, P & Lenoble, J, 2000, *Démocratie et Procéduralisation du Droit*, Bruylant, Bruxelles.

Crutzen, Cecile KM, 2003, 'ICT-representations as transformative critical rooms', in Gabriele Kreutzner & Heidi Schelhowe (eds) *Agents of change: Virtuality, gender, and the challenge to the traditional university*, Leske Budrich: Opladen, pp 87–106.

Joseph F Coates, 1995, 'Technology assessment: Here today, gone tomorrow, in *Technological Forecasting & Social Change*, 49, pp 321–23.

Drenth, PJD , Honnefelder, L, Schroots, JJF & Sitter-Liver, B (eds), 2006, *In search of common values in the European research area*, ALLEA, Netherlands.

European Commission (EC) (2002), *Science and society: action plan*, European Commission, Brussels.

Friedman, Batya, Kahn, Jr, Peter H, & Borning, Alan, 2006, *Human–computer interaction and management information systems: Foundations*. ME Sharpe.

Gadamer, HG, 1976, *Philosophical hermeneutics*, ed and trans by DE Linge, University of California Press, Berkeley.

Gentil, E, 1997, *Penser la modernité, Essai sur Heidegger, Habermas et Eric Weil*, Presse Universitaire de Namur.

Habermas, J, 1981, *Theory of communicative action*.

———, 1992, *De l'éthique de la discussion*, Passages, Cerf.

———, 1997, *Droit et démocratie, entre faits et normes*, Galimard.

Honneth, A, 1996, *Struggle for recognition. The moral grammar of social conflicts*, Polity Press.

Isaac, H, & Kalika, M, 2001, 'Organisation, nouvelles technologie et vie privée', *Revue Française de gestion*, Juillet- Aout, pp 101–06.

Ladrière, J, 1993, 'Les incertitude's de la conscience historique', *Cahier de l'école des sciences philosophiques et religieuses*.

Ladrière, J & Van Parijs, Ph, 1984, *Fondements d'une théorie de la justice. Essais critiques sur la philosophie politique de John Rawls*, Éditions de l'Institut Supérieur de Philosophie, Louvain-la-Neuve.

Lenoble, J & Maesschalck, M, 2003, *Toward a theory of governance: the action of norms*, Kluwer Law International.

———, 2006, *Beyond neo-institutionalist and pragmatic approaches to governance*, REFGOV, FP6

Lorenzen, P and Lorenz, K, 1978, *Dialogische Logik*. Wissenschaftliche Buchgesellschaft: Darmstadt.

Lyon, D, 1993, 'An electronic panopticon? A sociological critique of surveillance theory', *Sociological review*, vol 41, no 4, pp 653–78.

Mackenzie, JD, 1981, 'The dialectics of logic' *Logique et Analyse n.s.*, vol 24, pp 159–77.

————, 2006, *Theorizing surveillance: the panopticon and beyond*, Willan Publishing.

Maesschalck, M, 1998, *Habitus et lien social*, authors' translation.

————, 2001, *Normes et Contextes*, OLMS.

Orlikowski, WJ, 1991, 'Integrated information environment or matrix of control? The contradictory implications of information technology', *Accounting Management and Information Technology*, vol 1, no 1, pp 9–11.

Perelman C & Olbrechts-Tyteca L, 1991, *The new rhetoric: a treatise on argumentation*, University of Notre Dame Press.

Punie, Yves, 2003, A *social and technological view of ambient intelligence in everyday life: what bends the trend?* IPTS EMTEL2 key deliverable Work Package 2 EU FP5 HPRN-CT-2000-00063, September, Technical Report EUR 20975 EN.

Ramberg, B & Gjesdal, K, 2005, 'Hermeneutics', *Stanford encyclopedia of philosophy*.

Rawls, J, 1971, *A theory of justice*, Harvard University Press, Cambridge (Mass).

Ricœur, P, 1990, *Soi-même comme un autre*, Le Seuil, Paris.

Schütz, A, 1951, 'Choosing among projects of action', Philosophy and Phenomenological Research, vol 12, pp 161– 84.

————, 1954, 'Concept and theory formation in the social sciences', *Journal of Philosophy*, 51, pp 257–72.

————, 1959, 'Type and eidos in Husserl's late philosophy', *Philosophy and Phenomenological Research*, vol 20, pp 147–65.

Stiglitz, Jospeh, 2002, *Globalization and its discontents*, London: Penguin Books.

Teubner, G, 1995, *Droit et réflexivité*, ed. juridiques Kleuver.

van den Hoven, J, 2007, 'ICT and value sensitive design', in Ph Goujon et al (eds), *The information society: innovation, legitimacy, ethics and democracy*, Sprinter, p 68.

Van Eemeren, F, Grootendorst R, 2004, *A systematic theory of argumentation, the pragma-dialectical approach*, Cambridge University Press.

www.ingramcontent.com/pod-product-compliance
Lightning Source LLC
LaVergne TN
LVHW071356070326
832902LV00028B/4627